WITHDRAWN FROM STOCK

From the Proceedings of
Will We Use The Oceans Wisely—
The Next Fifty Years in Oceanography
A Symposium on the Future of Oceanography,
September 29—October 2, 1980, at the
Woods Hole Oceanographic Institution,
Woods Hole, Massachusetts, USA, on the
occasion of the Fiftieth Anniversary of
the founding of the Institution

Oceanography
The Present and Future

Edited by
Peter G. Brewer

With 123 Figures

Springer-Verlag
New York   Heidelberg   Berlin

Peter G. Brewer
Woods Hole Oceanographic Institution
Woods Hole, Massachusetts 02543 USA

Library of Congress Cataloging in Publication Data
Main entry under title:
Oceanography, the present and future.
  Subtitle: From the proceedings of Will we use the
oceans wisely—the next fifty years in oceanography, a
symposium on the future of oceanography, held September
29-October 2, 1980, at the Woods Hole Oceanographic
Institution, Woods Hole, Massachusetts, USA, on the
occasion of the fiftieth anniversary of the founding of
the Institution.
     Includes bibliographies and index.
     1. Oceanography—Congresses. I. Brewer, Peter.
II. Woods Hole Oceanographic Institution.
GC2.028   1982   551.46   82-10480

© 1983 by Springer-Verlag New York Inc.
All rights reserved. No part of this book may be translated or reproduced
in any form without written permission from Springer-Verlag, 175 Fifth
Avenue, New York, New York 10010, U.S.A.
The use of general descriptive names, trade names, trademarks, etc., in
this publication, even if the former are not especially identified, is not
to be taken as a sign that such names, as understood by the Trade Marks
and Merchandise Marks Act, may accordingly be used freely by anyone.

Typeset by Ms Associates, Champaign, Illinois
Printed and bound by Halliday Lithograph, West Hanover, Massachusetts
Printed in the United States of America.

9 8 7 6 5 4 3 2 1

ISBN 0-387-90720-3   Springer-Verlag New York Heidelberg Berlin
ISBN 3-540-90720-3   Springer-Verlag Berlin Heidelberg New York

Preface

Oceanography: The Present and Future is the proceedings of a symposium held at the Woods Hole Oceanographic Institution, Woods Hole, Massachusetts, on September 29-October 2, 1980 on the occasion of the fiftieth anniversary of the founding of the Institution. The symposium was immediately preceded by the Third International Congress on the History of Oceanography, also held at Woods Hole, and the proceedings of that Congress, *Oceanography: The Past,* also published by Springer-Verlag, forms a companion volume to this book.

The editorial responsibilities were handled by Ms. Kate Eldred, who worked extraordinarily hard on this volume, while the scientific editing was performed by Dr. Peter G. Brewer. The organizing committee of scientists charged with responsibility for the symposium was: Dr. Peter G. Brewer, chemistry; Dr. Arthur E. Maxwell, geology and geophysics; Dr. Robert W. Morse, marine policy; Dr. David A. Ross, marine policy and marine geology; Dr. Peter B. Rhines, physical oceanography; Dr. John A. Teal, marine biology; and Dr. Robert Spindel, ocean engineering. They were faced at the outset with the problem that science proceeds with intense effort and competition within a disciplinary peer group but that, particularly in ocean science, the results of this work often have completely unforseen and important consequences in a totally unrelated area.

Who could have foreseen, for instance, that the theory of plate tectonics and the quest for finding its thermal signature could have led to the discovery of radically new biological fauna living at high temperature and pressure in the ocean abyss? Yet this has indeed happened, and it is only a hint of oceanic processes yet to be discovered. Talks, therefore, were scheduled not along narrow disciplinary lines, but as groups by scale of oceanic processes: the small or molecular scale; the medium scale, covering events and distributions out to oceanic basin range; the global or planetary scale of climatic interest; and the human scale of engineering and the human use of oceanic resources. Each day's session concluded with a talk and panel discussion on a marine policy issue to examine how the institutions of man utilize or affect the activities of ocean scientists. The papers in this volume appear in this format, much as they were presented.

Contributors to the symposium were invited to address the current status and future trends in their area of ocean science. It is, of course, impossible to cover the complexity of the oceans completely in one symposium. Nor is it possible to look very far into the future. What then is covered, and what do we see? J. Stewart Turner examines the smallest scales of ocean dynamics, where molecular differences in transport processes affect such diverse events as the mixing of Mediterranean water with the Atlantic Ocean and the debouching of 350°C brine on the floor of the Pacific. H.D. Livingston and W.J. Jenkins report on the fate of radioactive waste put into the oceans by man since the dawn of the atomic age, and attempt to chart its future course. Biologists G.R. Harbison and J.J. Childress, in separate papers, point out that our knowledge of the community of deep-sea biological species is being revolutionized by the simple ability now to carry out visual observations from submersibles or with scuba equipment. There are omissions too, and the editor feels keenly the loss of geological and geophysical contributions due to illness and unforeseen events affecting invitees. The volume is not a complete text, but a view of ocean science and policy as it stands today with a view of the future from key figures in the field.

The meetings were held in the Lillie Auditorium of the Marine Biological Laboratory, Woods Hole and were characterized by such entertaining presentations and stimulating discussions that a true anniversary spirit prevailed. Particular thanks go to Mr. Charles S. Innis, the hero of organization, and Mrs. Florence Mellor for attention to a thousand details. The scientific editor is particularly grateful to Ms. Kate Eldred, who bore the brunt of editing, of tracking down lost figures, and bringing order to diverse presentations. All members of the organizing committee reviewed papers in their area. Dr. Philip Manor and Ms. Ronnie Frankel of Springer Verlag were patient and professional in all their dealings with this group.

Support for the symposium was provided by the Office of Naval Research; the United States Geological Survey; the National Sea Grant Program; and the Woods Hole Oceanographic Institution.

Peter G. Brewer

Contents

Part I. Small- and Local-Scale Oceanography

1. Molecular Processes in the Marine Environment
   John M. Wood 3

2. The Structure of Planktonic Communities
   G. Richard Harbison 17

3. Oceanic Fine- and Microstructure
   J. Stewart Turner 35

4. Experiments with Free-Swimming Fish
   Frank G. Carey 57

5. Coastal Dynamics, Mixing, and Fronts
   Christopher Garrett 69

6. Shoreline Research
   Orrin H. Pilkey 87

7. The Ocean Nearby: Environmental Problems and Public Policy in the Next Fifty Years
   Evelyn Murphy 101

Part II. Regional-Scale Oceanography

8. Acoustics and Ocean Dynamics
   J. Walter Munk 109

9	Oceanic Biology: Lost in Space? *James J. Childress*	127
10	Eddies and the General Circulation *H. Thomas Rossby*	137
11	Radioactive Tracers in the Sea *Hugh D. Livingston and William J. Jenkins*	163
12	Fisheries and Productivity Studies *Peter A. Larkin*	193
13	The Impact of Oceanography on the Military and Security Uses of the Ocean *Alan Berman*	205

Part III. Global-Scale Oceanography

14	Large Scale Geochemistry *Heinrich D. Holland*	219
15	General Circulation of the Oceans *Pearn P. Niiler*	231
16	Remote Sensing of the Oceans from Space *John A. Whitehead, Jr.*	255
17	United States Distant-Water Oceanography in the New Ocean Regime *Edward J. Miles*	283

Part IV. The Human Scale

18	Changing Global Biogeochemistry *Bert Bolin*	305
19	Innovative Ocean Energy Systems: Prospects and Problems *Abrahim Lavi*	327
20	Aquaculture: Potential Development *Hillel Gordin*	347

21	Technology and Communications: New Devices and Concepts for Ocean Measurements *D. James Baker, Jr.*	363
22	Institutional and Educational Challenges *John H. Steele*	377
Index		381

Contributors and Their Affiliations

D. James Baker, Jr. University of Washington, School of Oceanography, University of Washington, Seattle, Washington 98195, USA

Alan Berman Director of Research, U.S. Naval Research Laboratory, Washington, D.C. 20390, USA

Bert Bolin University of Stockholm, Department of Meteorology, Arrhenius Laboratory, S-10691 Stockholm, Sweden

Frank G. Carey Woods Hole Oceanographic Institution, Woods Hole, Massachusetts 02543, USA

James J. Childress Department of Biology, University of California, Santa Barbara, California 93106, USA

Christopher Garrett Department of Oceanography, Dalhousie University, Halifax, Nova Scotia, Canada

Hillel Gordin Marine Acquaculture Research Station, Israel Oceanographic and Limnological Research, Ltd., Elat, Israel

G. Richard Harbison Woods Hole Oceanographic Institution, Woods Hole, Massachusetts 02543 USA

Heinrich D. Holland Department of Geology, Harvard University, Cambridge, Massachusetts 02139, USA

William J. Jenkins Woods Hole Oceanographic Institution, Woods Hole, Massachusetts 02543, USA

Peter A. Larkin Institute of Animal Resource Ecology, University of British Columbia, Vancouver, B.C., Canada

Abrahim Lavi Carnegie-Mellon University, Pittsburgh, Pennsylvania 15213, USA

Hugh D. Livingston Woods Hole Oceanographic Institution, Woods Hole, Massachusetts 02543, USA

Edward J. Miles School of Oceanography, University of Washington, Seattle, Washington 98195, USA

Walter J. Munk Scripps Institution of Oceanography, La Jolla, California 92039, USA

Evelyn Murphy Massachusetts Institute of Technology, Department of Urban Studies, Cambridge, Massachusetts 02139, USA

Pearn P. Niiler School of Oceanography, Oregon State University, Corvallis, Oregon 97331, USA

Orrin H. Pilkey Department of Geology and Marine Laboratory, Duke University, Durham, North Carolina 27706, USA

H. Thomas Rossby Graduate School of Oceanography, University of Rhode Island, Kingston, Rhode Island 02881, USA

John H. Steele Woods Hole Oceanographic Institution, Woods Hole, Massachusetts 02543, USA

J. Stewart Turner Australian National University, Research School of Earth Sciences, P.O. Box 4, Canberra 2600, Australia

John A. Whitehead, Jr. Woods Hole Oceanographic Institution, Woods Hole, Massachusetts 02543, USA

John M. Wood Gray Freshwater Biological Institute, P.O. Box 100, Navarre, Minnesota 55392, USA

Part I
Small- and Local-Scale Oceanography

Molecular Processes in the Marine Environment

John M. Wood

1 Introduction

In the past decade the possible fate of pollutants in the aquatic environment has received much attention. The pathways for heavy metals, chlorinated organic compounds, radioactive wastes, and atmospheric pollutants have been studied in both terrestrial and marine environments.[1] Ironically, much of this research, although analytically significant, has been performed without a full appreciation of the metabolic capabilities of organisms which live in aquatic systems. In order to understand the routes taken by man-made chemicals and pollutants we need some basic knowledge of the metabolic capabilities of aquatic biota. We need to answer some crucial questions of the marine environment such as: What are the biosynthetic pathways for halogenated natural products in the marine environment? How are halogenated natural products degraded? Do marine organisms have metabolic capabilities which are different from terrestrial organisms? What are the natural biological cycles for trace elements in the sea? What are the rate-limiting steps for metabolic processes in the sea? How important is a kinetic, rather than a thermodynamic, approach in studying marine ecosystems? Answers are crucial if we hope to understand the fates of pollutants in saltwater systems. I shall attempt to give a few hints on how some of these questions may be answered by adopting a classical biochemical approach. Although our knowledge is sketchy, there are sufficient examples to show that marine biota have evolved to deal with metabolism in a halide ion-rich environment.

2 Halogenation By Marine Biota

As early as 1940, Clutterbuck et al. showed that certain fungi were capable of using halide ions in the biosynthesis of halo-organic compounds as secondary metabolites. Later Shaw and Hager (1960) isolated an enzyme from the fungus

---

[1] For a general introduction see Fates of Pollutants, Commission on Natural Resources, National Academy of Sciences, Washington, D.C. (1977) and the Nature of Seawater, Physical & Chemical Sciences Research Report No. 1, Dahlem Konferenzen.

Caldariomyces fumago which catalyzed the oxidation of Cl^-, Br^-, and I^- to give an enzyme-bound electrophilic halogenating agent. This halogenating reagent was shown to react with a variety of nucleophiles to give halogenated reaction products. Since the enzyme has a heme prosthetic group, and since it catalyzes a peroxidase reaction, it was given the name chloroperoxidase (Hager et al., 1975). Recently, Edwards et al. (1980) have shown that the filamentous bluegreen alga *Scytonema hoffmanii* synthesizes a chlorinated cytotoxin (cyanbacterin I) which is lethal to at least 12 different species of bluegreen algae. Filamentous cyanophytes are now known to contain chloroperoxidase. This is the limit of the work in terrestrial systems, where Cl^- availability is crucial for the biosynthesis of the above secondary metabolites.

However, in the marine environment significant evolutionary pressures have led to the utilization of halide ions. Hager et al. (1980) have conducted a survey of over 900 marine animal and plant species for the presence of halogenated organic compounds. Approximately 25% of species tested had lipid extracts which contained greater than 10 μg of organic halogen per gram wet weight of tissue. The *Rhodophyta* (red algae) were found to be particularly rich in halogenated organic compounds. Also, most of these halocarbons were shown to be cytotoxic to microorganisms (Table 1). In all cases antimicrobial activity was found to be a function of the halogen content of these lipid extracts (Fig. 1). Therefore, it is likely that organisms in the marine environment have evolved mechanisms to detoxify these halogenated organic compounds through special metabolic pathways. Studies by Suida et al. (1975) with extracts of the red alga *Bonnemaisonia hamifera* have demonstrated the presence of a number of halogenated heptanones which arise by direct halogenation with a bromoperoxidase. By analogy with chloroperoxidase the reaction sequence in Scheme I explains the biosynthesis of brominated heptanones from 3-oxo-octanoic acid.

Hewson and Hager (1980) surveyed 72 different species of marine algae for

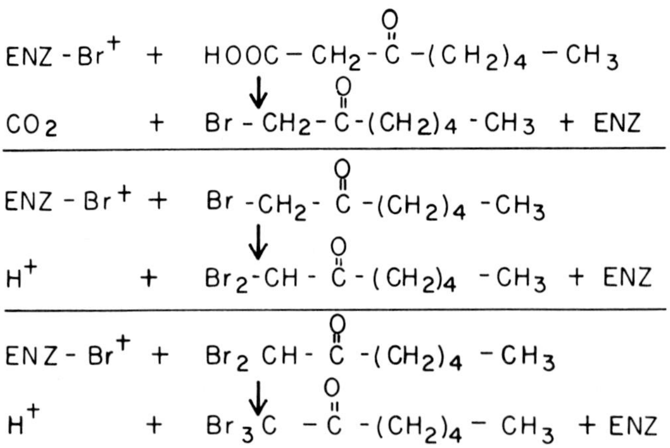

$$ENZ-Br^+ + HOOC-CH_2-\overset{O}{\underset{\shortparallel}{C}}-(CH_2)_4-CH_3$$
$$\downarrow$$
$$CO_2 + Br-CH_2-\overset{O}{\underset{\shortparallel}{C}}-(CH_2)_4-CH_3 + ENZ$$

$$ENZ-Br^+ + Br-CH_2-\overset{O}{\underset{\shortparallel}{C}}-(CH_2)_4-CH_3$$
$$\downarrow$$
$$H^+ + Br_2-CH-\overset{O}{\underset{\shortparallel}{C}}-(CH_2)_4-CH_3 + ENZ$$

$$ENZ-Br^+ + Br_2CH-\overset{O}{\underset{\shortparallel}{C}}-(CH_2)_4-CH_3$$
$$\downarrow$$
$$H^+ + Br_3C-\overset{O}{\underset{\shortparallel}{C}}-(CH_2)_4-CH_3 + ENZ$$

Scheme I. Biosynthetic route for the synthesis of brominated heptanones.

Table 1. Organic halogen content and antimicrobial activity in marine animal and plant lipids

Phylum	Number of species examined	Average organic halogen content*			
		Total halogen by direct assay (μmol/g)	Chlorine (μg/g)	Bromine (μg/g)	Chlorine plus bromine (μg/g)
Animal					
Porifera	71	0.55	14.1	19.4	33.5
Cnidaria	72	0.26	8.8	1.6	10.4
Clenophora	3	0.06	2.0	0.0	2.0
Platyhelminthes	4	0.14	5.0	0.3	5.3
Nemertina	4	0.22	8.5	0.0	8.5
Annelida	37	0.49	13.2	0.5	13.7
Mollusca	199	0.34	10.1	7.0	17.1
Arthropoda	97	0.18	6.9	0.4	7.3
Sipuncalida	4	0.11	4.0	0.0	4.0
Entoprocta	1	0.46	7.6	20.0	27.6
Ectoprocta	13	0.21	6.0	0.4	6.4
Chaetognatha	1	0.52	18.5	0.0	18.5
Echinodermata	83	0.30	9.4	4.0	13.4
Chordata	81	0.21	7.7	0.4	8.1
Animal summary	670	0.31	9.6	5.1	14.7
Plant					
Chlorophyta	31	0.11	3.7	0.3	4.0
Phactophyta	46	0.16	4.2	0.3	4.5
Rhodophyta	104	1.03	30.7	32.4	63.1
Cyanophyta	2	0.10	3.5	0.0	3.5
Tracheophyta	4	0.14	2.0	0.0	2.0
Angiosperms	2	0.13	2.3	5.0	7.3
Plant summary	189	0.64	19.8	19.1	38.9
All species summary	859	0.38	11.9	8.3	20.2

*Averages for chlorine and bromine are based on approximately 75% of the number of species collected.

the presence of bromoperoxidase, and 55 species were found to have high levels of this enzyme. The *Rhodophyta* were found to be better brominators, having the highest levels of bromoperoxidase and the greatest lipid halogen content. The *Phaeophyta* were the poorest halogenators. Besides containing a great variety of halogenated aliphatic compounds, marine organisms synthesize a multitude of halogenated aromatic compounds. For example, red algae of the

Table 1. (continued)

	Antimicrobial activity (% active species)				
Phylum	E. coli	B. subtilis	S. cerevisiae	P. atrovenetum	Active against at least one organism
Animal					
Porifera	18	32	13	17	37
Cnidaria	6	15	6	3	21
Clenophora	0	0	0	0	0
Platyhelminthes	0	0	0	25	25
Nemertina	0	0	0	0	0
Annelida	3	16	3	5	16
Mollusca	4	14	5	9	16
Arthropoda	0	1	0	0	1
Sipuncalida	0	0	0	0	0
Entoprocta	0	0	0	0	0
Ectoprocta	8	23	0	8	23
Chaetognatha	0	0	0	0	0
Echinodermata	0	17	27	16	43
Chordata	0	6	1	1	6
Animal summary	4	13	7	7	18
Plant					
Chlorophyta	0	10	0	0	10
Phactophyta	2	28	11	7	28
Rhodophyta	1	14	4	4	14
Cyanophyta	0	0	0	0	0
Tracheophyta	0	25	0	0	25
Angiosperms	50	50	0	50	100
Plant summary	2	17	5	4	18
All species summary	3	14	7	7	18

genus *Laurencia* synthesize a great variety of brominated and iodinated aromatic compounds (Izac and Sims, 1979) (Scheme II).

Several of these halogenated natural products are cytotoxic, and some of them resemble synthetic medicaments, insecticides, or pesticides. Since these halogenated natural products are synthesized by marine organisms, it follows that such organisms must have dehalogenation mechanisms to degrade them. Most halogenated synthetic compounds of industrial origin are regarded as

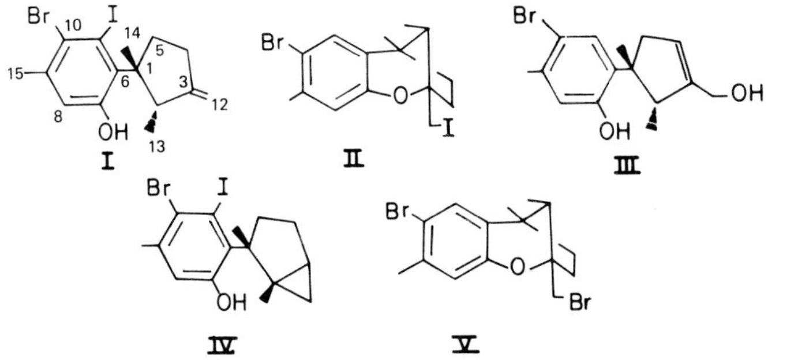

Scheme II. Some halogenated aromatic compounds from the genus *Laurencia*.

"persistent" or even "recalcitrant" in the terrestrial environment. Can the same be assumed when compounds of industrial origin contaminate the marine environment?

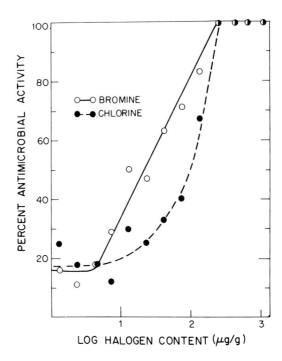

Figure 1. Antimicrobial activity as a function of organic halogen content of marine organism extracts. A plot of percentage of the species yielding extracts that show antimicrobial activity as a function of organic halogen content of extracts.

3 Dehalogenation Mechanisms

Dehalogenation reactions can occur either as a result of volatilization of low molecular weight halocarbons or by biological elimination of halide ions from organohalogen metabolic intermediates. Probably the most surprising discovery is the volatilization of dibromomethane, bromoform, and pentylbromide by the red alga *Bonnemaisonia hamifera*. Tribromoheptanone is a likely substrate for the biosynthesis of bromoform (Theiler et al., 1978) (Scheme III). Pentyl bro-

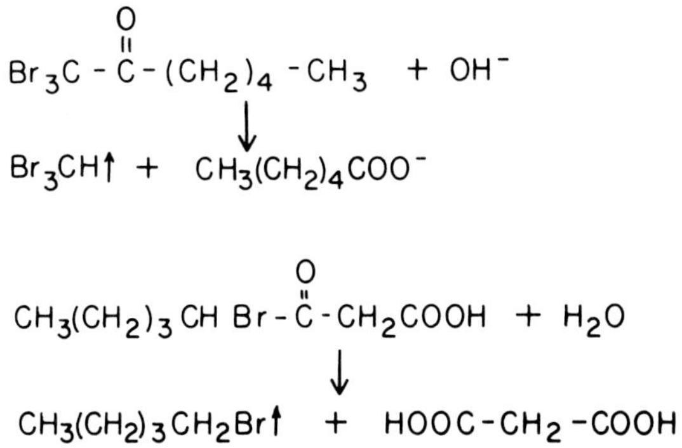

$$Br_3C-\overset{O}{\overset{\|}{C}}-(CH_2)_4-CH_3 + OH^-$$
$$\downarrow$$
$$Br_3CH\uparrow + CH_3(CH_2)_4COO^-$$

$$CH_3(CH_2)_3CHBr-\overset{O}{\overset{\|}{C}}-CH_2COOH + H_2O$$
$$\downarrow$$
$$CH_3(CH_2)_3CH_2Br\uparrow + HOOC-CH_2-COOH$$

Scheme III. The biosynthesis of bromoform and tribromoheptanone by the red alga *Bonnemaisonia hamifera*.

mide probably arises by the bromination of oxo-octanoic acid and by bromoperoxidase followed by hydrolysis of the brominated intermediate (Scheme IV).

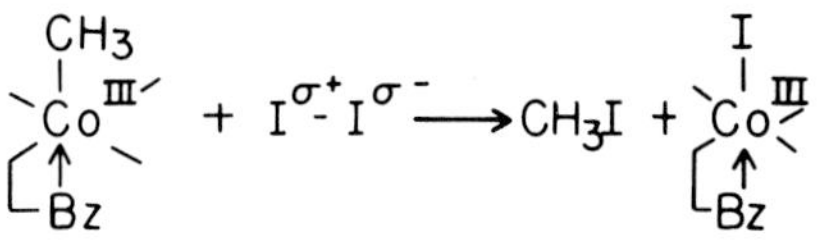

$$\overset{CH_3}{\underset{Bz}{\overset{|}{Co}^{III}}} + I^{\sigma+}I^{\sigma-} \longrightarrow CH_3I + \overset{I}{\underset{Bz}{\overset{|}{Co}^{III}}}$$

Scheme IV. A mechanism for the biosynthesis of methyl iodide.

The research I have described above is largely the product of almost 20 years of work in Lowell Hager's laboratory at the University of Illinois. Peroxidative halogenation provides us with one mechanism for both the biosynthesis and the biodegradation of halogenated organic compounds. It should be emphasized here that the enzymes responsible contain heme, are classical peroxidases, and rely on the availability of the transition metal iron for their synthesis. Cobalt has also been implicated in the biosynthesis of simple alkyl halides. Five years ago we demonstrated the biosynthesis of methyl iodide as a consequence of electrophilic attack by molecular iodine on the Co–C bond of methyl-B_{12} (Wood et al., 1975). Methyl iodide is now implicated in the synthesis of methyl mercury and tetramethyl lead in the aquatic environment (Wood et al., 1975; Ahmad et al., 1980) (Scheme V). Lovelock et al. (1973) were the first to show that methyl iodide was present in surface layers in the open ocean.

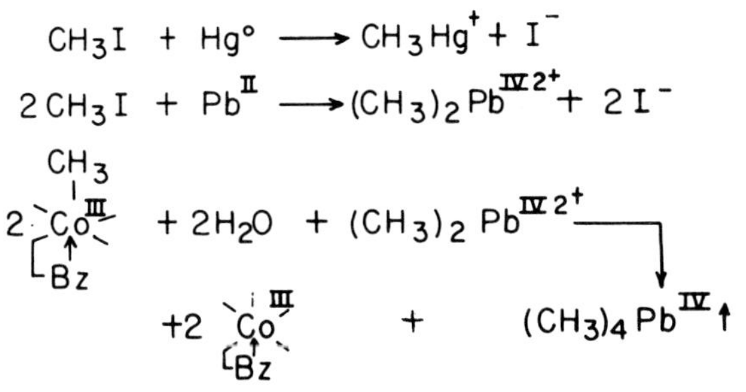

$$CH_3I + Hg^\circ \longrightarrow CH_3Hg^+ + I^-$$

$$2\,CH_3I + Pb^{II} \longrightarrow (CH_3)_2Pb^{IV\,2+} + 2I^-$$

$$2\begin{bmatrix}CH_3\\Co^{III}\\Bz\end{bmatrix} + 2H_2O + (CH_3)_2Pb^{IV\,2+}$$

$$+2\begin{bmatrix}Co^{III}\\Bz\end{bmatrix} + (CH_3)_4Pb^{IV}\uparrow$$

Scheme V. The formation of methyl mercury and tetramethyl lead from methyl iodide.

The metabolism of halogenated aromatic compounds has been widely studied, especially since the environmental aspects, first with DDT and later with PCBs, have come to our attention. A great deal is known about the general metabolism of aromatic compounds. In the aerobic world the oxygenases play a central note in the direct oxidation of aromatic compounds to aliphatic products (Dagley, 1975, 1976). Although the function of oxygenases is to catalyze the degradation of natural products, these enzymes often are nonselective, oxidizing industrial chemicals, carcinogens, pollutants, drugs, and so on. Microorganisms often convert lipid-soluble aromatic pollutants to more water-soluble metabolites. For example, the oxidation of the simplest aromatic compound, benzene, to catechol yields a water-soluble product which becomes diluted and is metabolized by aerobes, especially *Bacillus* and *Pseudomonas*. In some cases monohydroxylated aromatic compounds can be substrates for direct ring cleavage with molecular oxygen. For example, 5-chlorosalicylic acid is cleaved by a dioxygenase to give a ring fission product which eliminates the chlorine atom as chloride ion through lactonization and hydrolysis (Crawford et al., 1979; Wood, 1980) (Scheme VI). Even stable, cytotoxic chlorinated aromatic compounds of industrial origin can be efficiently converted to aliphatic products which enter central metabolic pathways.

Que et al. (1975) have shown that persistence to degradation for aromatic compounds is most likely determined by the electron-donating or electron-withdrawing nature of substituents on the benzene ring (i.e., the Hammett $\sigma\rho$ relationship). The presence of more than one strong electron (withdrawing) group, such as $-Cl$, makes it virtually impossible for dioxygen to add to the substrate, an important prerequisite for ring cleavage. Likely the elimination of halide ions from aromatic compounds proceeds through ring cleavage, lactonization, and hydrolysis. (Note that the rate-limiting step for these reactions is always oxidative cleavage of the aromatic ring.) Fortunately, the haloaromatic

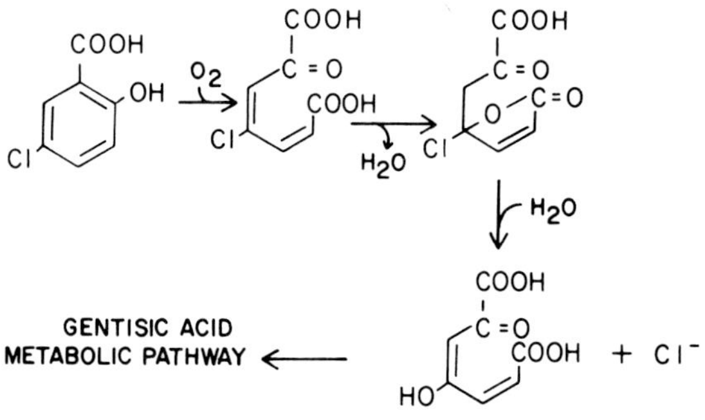

Scheme VI. Degradation of 5-chlorosalicylic acid by *Bacillus brevis*.

compounds of natural origin in the sea only contain one halogen atom per benzene ring; this is due to the restrictions placed on halogenation of nucleophilic centers by the halogenating peroxidases.

4 Sulfate Reduction

Some of the anaerobic reactions which occur in estuaries, coastal waters, and salt marshes are worth examining here. Erosion and sedimentation in the coastal areas add to sanitation practices to produce an ever-increasing aerobic zone. A survey of the flow rate of the major rivers demonstrates the scale of terrestrial silting. The reduction of sulfate to sulfide in coastal waters is important since it influences so many other metabolic processes. The production of sulfide ions has a profound effect on the availability of trace metals and directly influences their uptake by marine biota. The reduction of sulfate to hydrogen sulfide and the reduction of carbon dioxide to methane relies on the transfer of electrons from fermentation products (i.e., fatty acids and alcohols) up a potential gradient to molecular hydrogen. For example, the production of hydrogen from lactic acid by *Desulfovibrio* species requires the transfer of electrons up a potential gradient from −185 millivolts to −410 millivolts (Scheme VII).

A thermodynamic standpoint provides a compelling argument against the production and utilization of molecular hydrogen. However, kinetic effects are overriding and molecular hydrogen is produced and utilized as the electron source for methanogenic bacteria. An interesting molecular symbiotic relationship occurs between the sulfate-reducing bacteria *Desulfovibrio* and the methanogens, in that sulfate is used as the electron acceptor until the concentration is low. At this point electrons flow to produce molecular hydrogen, which is kept at a low partial pressure by the methanogenic bacteria in their reduction of CO_2 to CH_4. The regulation of electron flow is mediated by the low molecular weight multiheme cytochrome c_3. Cytochrome c_3 contains four hemes which operate at four different redox potentials. This unique cytochrome regulates

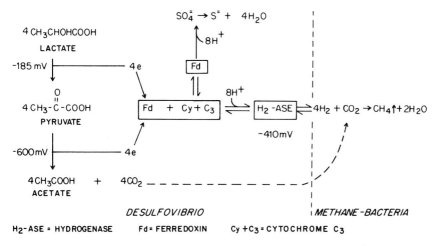

Scheme VII. The connection between sulfate reducers and methanogens.

electron flow in a multifunctional mode with pathways to sulfite reduction and H_2 formation. These critical processes in the anaerobic environment are regulated by kinetic rather than thermodynamic considerations, providing us with a classic example of why a study of molecular processes is crucial to our understanding of reactions in sediments.

5 Biological Cycles for Trace Metals

In order to understand the interactions of trace metals in the marine environment it is necessary to have information on the biochemical mechanisms used to achieve uptake selectivity and the physiochemical basis for interactions of trace metals with biological molecules or structures. With this information (for both aerobic and anaerobic systems) we can understand why certain heavy metals bioaccumulate. In addition to obtaining information on biochemical mechanisms for transport, we must understand the abundance and availability of heavy metals. Physical, chemical and biological parameters determine the uptake of trace metals and toxic heavy metals.

Physical parameters of concern are: natural occurrence; volatility; adsorption and desorption on particulates; and diffusion into and through biological membranes. Chemical parameters include: formation and stability of coordination complexes and organometallic complexes; redox properties as they relate to precipitation and/or solubilization; sedimentation processes (i.e., precipitation as sulfides, "humic" complexes, etc.); and stability and persistence of coordination complexes and organometallic complexes in the marine environment.

Biological parameters of concern are: toxicity of coordination complexes and organometallics to marine biota; rates of bioconcentration in marine food webs;

Table 2. Classification of hard and soft acids and bases (from Pearson, 1968 and Förstner and Wittmann, 1979)

Hard acceptor	Intermediate	Soft acceptor
$H^+, Na^+, K^+, Be^{2+}, Mg^{2+}$	$Fe^{2+}, Co^{2+}, Ni^{2+}$	Cu^+, Ag^+, Au^+, Tl^+
$Ca^{2+}, Mn^{2+}, Al^{3+}, Cr^{3+}$	$Cu^{2+}, Zn^{2+}, Pb^{2+}$	Hg^{2+}, CH_3Hg^+
$Co^{3+}, Fe^{3+}, As^{3+}$		
Hard donor	Intermediate	Soft donor
H_2O, OH^-, F^-, Cl^-	Br^-, NO_2^-, SO_3^{2-}	SH^-, S^{2-}, RS^-
$PO_4^{3-}, SO_4^{2-}, CO_3^{2-}, O^{2-}$		$CN^-, SCN^-, CO,$
		R_2S, RSH, RS^-

retention times in different marine organisms; and microbial metabolism (e.g., oxidation-reduction reactions, biomethylation and demethylation, volatilization and precipitation).

The properties of trace metals can be used to predict interactions in a complex environment. Pearson (1968) has summarized the preference of metal cations in his chemical theory of hard and soft acids and bases. Table 2 outlines ligand preferences for a number of trace elements. It should be noted here that many of the toxic heavy metals are "soft" acids and therefore complexation with sulfide is preferred. For example, the stability of $Cu^{(II)}$ follows

$$F^- < Cl^- < Br^- < I^- < SO_4 < NH_3 < PO_4^{2-} < OH^- < CO_3^- < CN^- < S^{2-}$$

The important role played by sulfide in the biological cycle for mercury is presented in Scheme VIII. Hydrogen sulfide appears to play a central position in

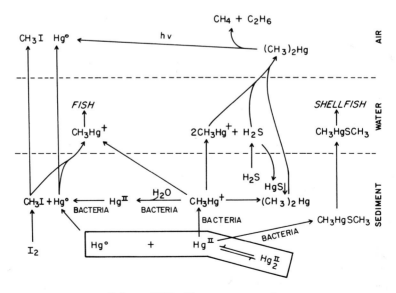

Scheme VIII. The mercury cycle.

the volatilization and precipitation of mercury through disproportionation. The same is true for the volatilization and precipitation of organo-lead compounds (Scheme IX). Clearly, these important processes aid in the mobilization of toxic

$$2 CH_3Hg^+ + H_2S \longrightarrow (CH_3)_2Hg\uparrow + HgS\downarrow$$

$$2(CH_3)_3Pb^+ + H_2S \longrightarrow (CH_3)_4Pb\uparrow + (CH_3)_2PbS\downarrow$$

Scheme IX. Disproportionation reactions of methyl mercury and trimethyl lead by H_2S.

elements from the aqueous environment to the atmosphere. However, it should be pointed out that such reactions occur only in polluted (anaerobic) lakes, rivers, coastal zones, estuaries, and salt marshes. In the open ocean the principal mercury species is $HgCl_3^-$ (66%).

6 General Conclusions

Although of necessity this chapter is brief, I hope that it serves a purpose in showing that there is a tendency for marine organisms to detoxify their environments of halo-organic compounds and heavy metals by volatilization, mineralization, and precipitation processes. The recognition that chloroperoxidases and bromoperoxidase have evolved, and that these enzymes play a role in both the biosynthesis and degradation of natural halogenated organic compounds in the sea, serves to emphasize the evolutionary pressures faced by marine biota in the salt-rich systems. Similarly, the aqueous ion chemistry for trace metals in marine environments is complicated by the presence of sulfate, its reduction to sulfide, and the subsequent volatilization of organo-mercury and -lead compounds. The small number of examples of processes which I chose to discuss certainly could be extended to include metals such as tin, and the volatilization of methyl tin compounds (Ridley et al., 1977), as well as the biosynthesis of organo-arsenolipids and the volatilization of methyl arsenic compounds, and the complex interactions between selenium and mercury with volatilization of methyl selenium compounds (Wood et al., 1975).

All this shows us that a number of metabolic processes are unique to marine biota. The rules for biodegradation, bioaccumulation, and persistence of natural and man-made substances appear to be different when taken from terrestrial systems and applied to marine systems. Clearly we need much more basic information on the metabolic capabilities of marine organisms. If we simply try to keep an inventory of marine pollutants and ignore the different biochemical processes in the sea, then we shall continue to oversimplify biological cycles. While oversimplification has helped in the development of mathematical models for cycling of elements in the biosphere, the impacts of individual chemical species in cycles are often overlooked. In the next 50 years of the development of Woods Hole Oceanographic Institution I expect to see the application of more sophisticated chemical and biochemical techniques in the elucidation of metabolic pathways and catalytic processes which are unique to the marine world.

Acknowledgements

I am indebted to Lowell P. Hager for making some of his unpublished work available to me. Also, I wish to thank Antonio Xavier for lively discussions on sulfate reduction and methane formation. Some of the work reported here was supported by a grant from the National Institutes of Health, AM 18101.

References

Ahmad, I., Y. K. Chau, P. T. S. Wong, A. J. Cartey, and L. Taylor. 1980. Chemical alkylation of lead (II) salts to tetraalkylead. Nature (London) In press.

Clutterbuck, P. W., S. C. Mukhopadhyay, A. E. Oxford, and H. Raistrick. 1940. Microbial halogenation. Biochem. J. 34, 664.

Crawford, R. L., T. Frick, and P. Olson. 1979. Catabolism of 5-chlorosalicylate by a bacillus. J. Bacteriol. 38, 379–384.

Dagley, S. 1975. Degradation of aromatic compounds in the biosphere. Am. Scientist 63, 681.

Dagley, S. 1976. Metabolism of aromatic compounds by micro-organisms. In: Essays in Biochemistry. (P. N. Campbell, Ed.). Biochemical Society, London, England, G.B. 81–137.

Edwards, K., F. K. Gleason, C. Mason, and J. M. Wood. 1980. Cytotoxins from filamentous blue green algae. (Unpublished data.)

Förstner, U., and G. T. W. Wittman. 1979. Metal Pollution in the Aquatic Environment. Springer-Verlag, New York, Berlin.

Hager, L. P., P. F. Hollenberg, T. Rand-Meir, R. Chiang, and D. Doubek. 1975. Mechanism of action of chloroperoxidase. Ann. New York Acad. Sci. 244, 80–93.

Hager, L. P., R. H. White, D. L. Doubek, P. F. Hollenberg, P. D. Shaw, K. L. Rinehart, Jr., R. D. Johnson, R. C. Brusca, G. E. Krejcarek, J. F. Suida, R. Guerrero, W. O. McClure, G. R. Van Blaricom, J. J. Sims, W. O. Fenical, and J. Rude. 1980. Halogenated natural products and antimicrobial activity in marine organisms. J. Phycol. In press.

Hewson, W. D., and L. P. Hager. 1980. Bromoperoxidases and halogenated lipids in marine algae. J. Phycol. 16:3, 340–345.

Izac, R. R., and J. J. Sims. 1979. Iodinated sesquiterperes from the genus *Laurencia*. J. Am. Chem. Soc. 101(20), 6136–6137.

Lovelock, J. E., R. J. Maggs, and R. J. Wade. 1973. The biosynthesis of methyliodide in the open ocean. Nature 241, 194–196.

Pearson, R. 1968. Hard and soft acids and bases. J. Chem. Educ. 45, 643–648.

Que, L., J. D. Lipscomb, E. Münck, and J. M. Wood. 1975. Kinetics and mechanism of protocatechuate 3,4 oxygenase. Biochim. Biophys. Acta 485, 60–71.

Ridley, W. P., L. J. Dizikes, and J. M. Wood. 1977. Biomethylation of toxic elements in the environment. Science 197, 329–332.

Shaw, P. D., and L. P. Hager. 1960. The peroxidation of halogens. J. Biol. Chem. 236, 1626–1630.

Suida, J. F., G. R. Van Blaricom, P. D. Shaw, R. D. Johnson, R. H. White, L. P. Hager, and K. L. Rinehart, Jr. 1975. The isolation of brominated natural products from marine organisms. J. Am. Chem. Soc. 97, 937.

Suida, J. F., R. Guerrero, W. O. McLure, G. R. Van Blanicum, J. J. Sims W. O. Fenical, and J. Rude. 1980. Brominated compounds from marine organisms. J. Phycol. In press.

Theiler, R., J. C. Cook, L. P. Hager, and J. F. Suida. 1978. Halohydrocarbon synthesis by bromoperoxidase. Science 202, 1094–1097.

Wood, J. M., H. J. Segall, W. P. Ridley, A. Cheh, W. Chudyk, and J. S. Thayer. 1975. Metabolic cycles for toxic elements in the environment. In: Proceedings of an International Conference on Metals in the Environment. (T. C. Hutchinson, Ed.). Plenum Press, New York, pp. 49–67.

Wood, J. M. 1980. Recent progress on the mechanism of action of dioxygenases in metal ion activation of dioxygen. In: Metal Ion Activation of Dioxygen. (T. G. Spiro, Ed.). John Wiley and Sons Inc., New York, pp. 163–180.

The Structure of Planktonic Communities

G. Richard Harbison

1 Introduction

Over the past 50 years, the paths of terrestrial and plankton ecology have diverged markedly. *In situ* observations, directed sampling, and field experiments have become important tools in the study of the structure of terrestrial and near-shore benthic marine communities. However, the primary observational tool of most zooplankton ecologists remains the towed, remotely operated net. These nets are easily used from ships, and they collect a relatively large number of organisms. They are most effective in the study of the distribution of planktonic organisms over large scales, but at smaller scales their effectiveness diminishes. Further, remotely operated sampling devices cannot be used to make field observations and experiments. Thus, the average plankton ecologist has fewer available tools to study life in the ocean than the terrestrial ecologist has to study life on land. Over the past few years, an increasing number of people have begun to study oceanic organisms *in situ*, with self-contained underwater breathing apparatus (SCUBA) and submersibles (Hamner, 1977). These methods work well for the study of the behavior and microscale distribution of plankton but lose their effectiveness at larger scales.

As we look ahead to the next 50 years of plankton ecology, we need to assess our various methods for the study of planktonic community structure with regard to effective scale and the sort of information each method provides. I feel that many present opinions about the nature of life in the open ocean are narrow and distorted. If we fail to use all of the methods available with efficiency and resolution, I fear we will continue to have an unrealistic view of the nature of the open ocean ecosystem.

All of the methods presently available for the study of planktonic organisms may be grouped into three categories: laboratory studies, *in situ* studies, and remote sampling studies. I will not deal with laboratory studies, as they have been widely used in conjunction with both *in situ* and remote-sampling methodology. Laboratory studies provide information about what an organism is capable of doing, which may be very different from what it actually does in nature. Everyone is aware of the power of laboratory work, provided it·is checked against the "ground truth" of observations of the real world. I will discuss mainly the way in which this "ground truth" is obtained—either directly, through *in situ* observations or indirectly, through remote sampling.

2 The Haeckel-Hensen Controversy—A Question of Scale

Since the inception of quantitative sampling methodology, the proponents of remote-sampling techniques have often engaged in vitriolic controversies with the proponents of *in situ* techniques. The first of these controversies was between Victor Hensen, the father of quantitative plankton sampling, and Ernst Haeckel, a proponent of the nineteenth century equivalent of *in situ* methods. Haeckel, who studied plankton from rowboats, using dipnets to collect animals for study, was most interested in the systematics, life history, and behavior of planktonic organisms. Hensen, who was the first to use the towed net as a quantitative tool, was most interested in large-scale patterns of distribution of plankton. Haeckel (1890) initiated the controversy with a bitter attack on Hensen, asserting that Hensen's methods were doomed to failure because planktonic organisms were not regularly distributed, but occurred in patches of all different sizes. He dismissed Hensen's quantitative comparisons between different organisms and between different regions of the sea as meaningless. Haeckel believed strongly in Darwin's theory of natural selection. To him, quantitative comparisons between different species were absurd, since such comparisons denied the existence of qualitative differences. In essence, denying these qualitative differences meant denying that evolution had any relevance in the open ocean. Such a denial was no problem to Hensen, who was opposed to Darwin's theory anyway (Hensen, 1891). Further, in his response to Haeckel's attack, he dismissed patchiness as unimportant in the open sea and showed, through the results of the Plankton Expedition, that Haeckel held erroneous views about the distribution of living organisms in the ocean.

The results of the Plankton Expedition were so impressive that Hensen's victory over Haeckel in the early part of this century was almost total. Not until several decades had passed was it shown that Haeckel had been correct about the patchy distribution of plankton (Hardy, 1936). In summarizing their controversy, Hardy said,

> It was a pity that Haeckel's attack was so bitter and that he did not confine his criticism to the question of irregularity in distribution; as is well known,

he disputed also one of the most important conclusions of the work of Hensen's Plankton Expedition: that the plankton is on the whole more abundant in the colder and temperate regions than in the tropical seas. Since Haeckel's views on this latter question were shown to be in error, and it was clear that he had misunderstood some of Hensen's other conceptions, his remarks on the uneven distribution of the oceanic plankton lost much of their force.

Hardy's paper showed that Haeckel had indeed been right about patchiness: it continues as a problem in the interpretation of remote sampling data and will remain a problem in the foreseeable future. However, more important than who was right in the controversy are the reasons that Hensen and Haeckel were each partially right and partially wrong. Where did both of them fall into error?

The key to understanding this lies in scale. Haeckel's microscale methods enabled him to see quite easily that planktonic organisms were not evenly distributed. As he went out day after day, he could see that sometimes large numbers of a certain animal were present, and on other days the same animals were entirely absent. Hensen took his plankton samples over a much larger scale, so the patchiness that Haeckel could easily see was nothing but minor noise to him. Hensen could see from his samples that he caught a greater mass of plankton in temperate than in tropical waters. Haeckel, with his *in situ* methods, could not make quantitatively correct conclusions about the scale Hensen worked on. Indeed, Hardy could only confirm that Haeckel was correct about the uneven distribution of oceanic plankton by developing a remote-sampling device (the Hardy Plankton Recorder) that worked on a scale as fine as the one Haeckel studied. Thus, both Haeckel and Hensen were correct when they talked about the scales their techniques permitted them to easily study, and both were incorrect when they talked about scales that were inappropriate to their methods. Phenomena that were obvious to Haeckel were but dimly perceived, if at all, by Hensen, and vice versa, and each held the erroneous belief that his method was appropriate to the study of life in the ocean on all scales.

3 No Single Method Works Equally Well at Every Scale

To the terrestrial ecologist, the methods of the biogeographer obviously must differ from the methods of the behaviorist, and each has a great deal to contribute to understanding how communities are structured. Apparently this is less obvious to plankton ecologists, since a great deal of meaningless controversy is still going on between the remote-sampling and the *in situ* schools of plankton research. Members of each group imply that their methods are appropriate at every scale, ranging from the very large to the very small. What is true in terrestrial ecology must also be true in plankton ecology—it is absurd to suggest that the methods of the biogeographer and the behaviorist must always be the same. As we look ahead to the next half-century of biological oceanography, first we must evaluate our methods and determine the scales over which they can be most successfully applied. We waste time and creative energy attempting to study phenomena with inappropriate methods if alternative, superior methods are available. There already is ample evidence that *in situ* methods are most appropriate for the study of microscale phenomena, and that remote-sampling methods are most appropriate over larger scales. To illustrate this, I will give two examples from my own experience; many other examples could be given.

I would like to discuss a group of planktonic crustaceans called hyperiid amphipods, most species of which are found in the open ocean. However, the first hyperiid amphipod described was a nearshore species, *Hyperia medusarum* (Strøm, 1762). As its name suggests, it was found in association with a jellyfish. Throughout the nineteenth century, other hyperiid amphipods were found living on a variety of gelatinous organisms (including jellyfish, ctenophores, siphonophores, colonial radiolarians, and salps). Although these observations were not common, the bulk of evidence, in conjunction with studies of mouthpart morphology, led Pirlot (1932) to conclude that most species of hyperiid amphipods were probably parasitic on gelatinous organisms. However, Pirlot's viewpoint was not widely shared by biological oceanographers (the majority regarding hyperiid amphipods as free-living, randomly drifting plankton), since the quantitative sampling revolution had occurred. Using Hensen's methods, the study of these associations, which had been relatively easy with earlier techniques, had become extremely difficult because the scale over which these associations occur is too fine to be effectively studied with remote-sampling devices (Fig. 1).

When I began to use SCUBA diving to collect salps for experimental work in 1972, the small crustaceans on them and on other gelatinous organisms were obvious. I was amazed that so little was known about these associations and began to study the nature and specificity of the relationship (Harbison, 1976; Harbison and Madin, 1976; Madin and Harbison, 1977; Harbison et al., 1977; Harbison et al., 1978). These associations are often highly specific, stereotyped, and central to understanding the role of hyperiid amphipod species in planktonic communities (Harbison et al., 1977; Laval, 1980). These associations show the importance of behavior in the formation and maintenance of microscale "patchiness," and they violate at least two common generalizations of biological oceanographers: animals that live in the open ocean should be food generalists, since

---

Figure 1. Examples of microscale "patchiness of hyperiid amphipods." (a) Juvenile *Oxycephalus* sp. on the ctenophore *Ocyropsis cystallina*. Scale line = 1.0 cm. (b) Juvenile *Eupronoe* sp. on the siphonophore *Forskalia edwardsii*. The juvenile amphipods appear as small dots on the nectophores (swimming bells) of the siphonophore. One is indicated with an arrow. Scale line = 1.0 cm. (c) Closeup of a juvenile *Eupronoe* sp. in a *F. edwardsii* nectophore. The amphipod is embedded in the mesogloea of the nectophore, and the ventral part of its pereion is so enlarged that the animal can only make feeble movements when it is removed from the nectophore. Scale line = 1.0 mm. "Patchiness" occurring at scales as fine as in these examples cannot be effectively studied with remote sampling.

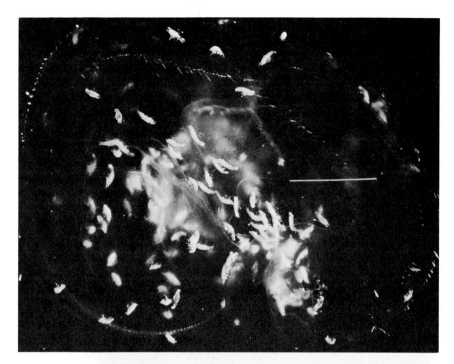

(c)

the food there is so dilute; and larger organisms prey on smaller ones in the open ocean, in contrast to the terrestrial environments. These generalizations cannot be very meaningful if they are violated by the third or fourth most abundant crustacean zooplankton group in the open sea!

If one peruses the literature on microscale "patchiness," one will largely look in vain for a reference to these amphipods. Today, "patchiness" seems to be defined solely in terms of the statistical manipulation of data obtained from remote-sampling devices. Thus, "patchiness" of the kind seen in Figure 1 lies outside its definition—it's too real! I believe that it will be extremely difficult to invent a remote-sampling device that can collect a single siphonophore or ctenophore with its associated hyperiid amphipods. Why bother inventing such a device anyway? With SCUBA, we can collect individual siphonophores, identify them to species, precisely count the hyperiid amphipods that are on them, and identify the amphipods to species. We can study the behavior of the amphipods on the siphonophore and see how the behavior of the siphonophore is altered by the presence of the amphipods. For the study of the associations of hyperiid amphipods with gelatinous zooplankton, *in situ* methods are clearly superior to any present or imaginable remote-sampling device, simply because these associations occur on a scale that can be easily studied with SCUBA or submersibles.

To show how *in situ* techniques break down as the scale becomes larger, I will discuss the phylum Ctenophora, or "comb jellies." Nineteenth century zoologists described a number of species of ctenophores, which became lost to biological oceanography after the advent of quantitative sampling, because net collection and subsequent preservation reduces them to unrecognizable bits of goo. Our much gentler collection techniques showed us that ctenophores did indeed exist in the open sea. Further, we were able to study their feeding behavior, discover some of their symbionts and predators, and study their microscale distribution patterns with ease (Harbison et al., 1978) (Fig. 2). However, we also wanted to express, in some way, their patterns of distribution across the

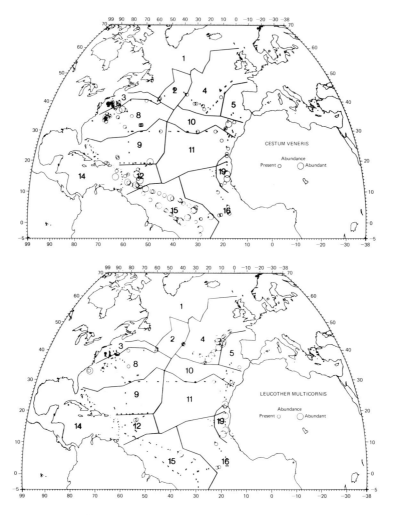

Figure 2. Distribution of two species of oceanic ctenophores. The crosses represent negative stations. Although these maps (which were the results of 441 SCUBA diving stations) indicate that *Cestum veneris* is more ubiquitous in its distribution than *Leucothea multicornis,* their quantitative value is nil, because of the highly subjective criteria for determining abundance (Harbison et al., 1978). *In situ* methods are as inappropriate for studying biogeography as remote sampling methods are for studying microscale "patchiness" and animal behavior.

entire North Atlantic, patterns too large to be studied with SCUBA. We could not possibly collect all of the ctenophores we saw, and those that we did not collect could not be identified to species. We chose to identify ctenophores to species, which resulted in the generation of quantitatively useless distribution maps. We could have chosen to count all of the ctenophores we saw, in which case our distribution maps would also have been worthless, because we could not then have identified ctenophores to species. Here, then is a fine example of a scale that is far too large for *in situ* techniques. The only justification for the publication of these quantitatively worthless distribution patterns is that they indicate a sampling problem for ctenophores. These and several other organisms that we regularly encounter represent a challenge to the remote-sampling community, for their methods are the only way to gain meaningful data on vertical and horizontal distribution and abundance over large scales. I expect that some sort of towed camera system will be used to solve some of these sampling problems.

My criterion, then, for establishing whether or not a particular method works is simple—can one obtain, with relative ease, both statistically meaningful counts and precise species identifications? If one must be given up in order to obtain the other, the method is inappropriate for the scale that is being studied.

4 The Importance of Species Identification

In the remote-sampling literature of today, I see an increasing tendency to discount the importance of precise species identifications to biological oceanography. "Alternative taxonomies" are proposed when it is discovered that both species identifications and meaningful counts are unobtainable at the microscale with remote-sampling devices. These "alternative taxonomies" are usually incredibly crude—they are usually simple, easily measured parameters such as "chlorophyll," "size," "displacement volume," and so forth. The thought of proposing such parameters as these as alternatives to species identifications would be ludicrous were it not becoming so common.

No biologist can afford to ignore species. Identification to species constitutes the most precise characterization of an organism for the plankton ecologist. All experience with the ecology, behavior, physiology, or molecular biology of terrestrial organisms indicates that identification to species is often of overriding importance. I am dismayed that so many appear to be rejecting the species concept. By doing this, they are implicitly denying that animal diversity has any relevance to plankton ecology.

An analogy perhaps best illustrates the importance of the species concept to plankton ecology. The Fissipeda are a suborder of the Carnivora, just as the Hyperiidea are a suborder of the Amphipoda. Dogs, bears, raccoons, pandas, weasels, skunks, otters, civet cats, hyenas, domestic cats, and leopards belong to the Fissipeda (Table 1). What sort of generalizations can we make about fissipeds? Even though they belong to the order Carnivora, they are not all carnivores, since pandas eat bamboo. Although otters are aquatic, many fissipeds don't like getting wet (domestic cats, for example). Some fissipeds are nocturnal, others are diurnal. Some are gregarious, and some are solitary. Some hunt prey by running it down, some stalk their prey, some lie in wait, and some eat carrion left by other fissipeds. There are very few statements that can be made about fissipeds in general, since so much is known about them. We must conclude that the ecology of fissipeds cannot be studied in general, since there is no such thing as a general fissiped.

Why should we expect hyperiid amphipods, which are more diverse morphologically than fissipeds, (see Figure 3), to be any less diverse and complex in their behavior patterns and life histories than fissipeds? In fact, all evidence indicates that the life styles of hyperiids are at the very least just as diverse and complex as the life styles of fissipeds. [For example, several species belonging to the genus *Lycaea* live as parasites of salps (Madin and Harbison, 1977). One species, *Lycaea nasuta,* is found only on a single species of salp, *Cyclosalpa affinis;* another species, *Lycaea vincentii,* lives primarily on a different salp, *Salpa cylindrica;* and a third species, *Lycaea pulex,* lives on a wide variety of salps, including those that harbor the first two species (Madin and Harbison, 1977).] As we learn more and more about hyperiids, I am confident that we will find that they are every bit as complex and interesting as fissipeds.

If one were to try to understand the community structure of a terrestrial ecosystem by classifying fissipeds by displacement volume, rather than by species, not much would be learned. While domestic cats usually eat prey smaller than themselves, wolves, which hunt in packs, can bring down animals that are considerably larger than they are. Suppose that terrestrial ecologists said, "Maybe size is not a very good alternative taxonomy, but identification to genus will be good enough, since species identifications are so difficult and time consuming." For genera with only one or two species, such a procedure might work pretty well, but for more complex genera, such as *Felis* (including, among others, domestic cats, cheetahs, lions, and leopards), such a course wouldn't work very well. While the list of generalizations one can make about species in the genus *Felis* is longer (that is what hierarchical classifications are all about, after all) than the list of generalizations about species in the suborder Fissipeda, a great deal of important information relevant to their ecology is lost. For some questions, then, species identifications may not be necessary, but we cannot decide, *a priori,* in which cases they are irrelevant. We can make rational decisions only after species identifications have been made, never before.

The only reason that people can contemplate discarding the species concept for hyperiid amphipods, or copepods, or euphausiids, or lanternfishes, or any other pelagic organisms, is because so little is known about them. Since nineteenth century naturalists did not have aqualungs and submersibles at their disposal, the whole body of careful observations made in forests and tidepools was not duplicated in the open ocean. Those of us interested in plankton ecology must become like those nineteenth century naturalists, accumulating information on the diversity of form, function, and life history of planktonic

Table 1. A comparison of the Fissipeda, a suborder of the Carnivora, with the Hyperiidea, a suborder of the Amphipoda*

Order Carnivora
Suborder Fissipeda

Superfamily Miacoidea
Family Miacidae (Miacis, Vulpavus)

Superfamily Canoidea
Family Canidae (Canis, Vulpes)
Family Ursidae (Ursus)
Family Procyonidae (Procyon, Ailurus, Ailuropoda)
Family Mustelidae (Mustela, Meles, Taxidea, Mephitis, Lutra, Martes)

Superfamily Feloidea
Family Viverridae (Viverra, Herpestes)
Family Hyaenidae (Hyaena)
Family Felidae (Felis, Hoplophoneus, Smilodon)

Order Amphipoda
Suborder Hyperiidea
Infraorder Physosomata

Superfamily Lanceoloidea
Family Lanceolidae (Prolanceola, Metalanceola, Paralanceola, Megalanceola, Scypholanceola, Lanceola)
Family Chuneolidae (Chuneola)
Family Microphasmidae (Mimonecteola, Microphasma, Microphasmoides)

Superfamily Scinoidea
Family Archaeoscinidae (Archaeoscina)
Family Scinidae (Scina, Ctenoscina, Acanthoscina, Spinoscina)
Family Mimonectidae (Mimonectes, Pseudomimonectes)
Family Proscinidae (Proscina, Mimoscina)

Infraorder Physocephalata
Superfamily Vibilioidea
Family Cystisomatidae (Cystisoma)
Family Vibiliidae (Vibilia, Cyllopus)
Family Paraphronimidae (Paraphronima)

Superfamily Phronimoidea
Family Hyperiidae (Pegohyperia, Iulopis, Hyperoche, Parathemisto, Hyperiella, Hyperia, Bougisia, Lestrigonus, Hyperionyx, Themistella, Phronimopsis, Hyperioides, Hyperietta)

Family Dairellidae (*Dairella*)
Family Phrosinidae (*Primno, Phrosina, Anchylomera*)
Family Phronimidae (*Phronimella, Phronima*)
Superfamily Lycaeopsoidea
  Family Lycaeopsidae (*Lycaeopsis*)
Superfamily Platysceloidea
  Family Pronoidae (*Pronoe, Paralycaea, Eupronoe, Parapronoe, Sympronoe*)
  Family Anapronoidae (*Anapronoe*)
  Family Lycaeidae (*Lycaea, Tryphana, Pseudolycaea, Thamneus, Brachyscelus*)
  Family Oxycephalidae (*Rhabdosoma, Calamorhynchus, Leptocotis, Glossocephalus, Simorhynchotis, Cranocephalus, Streetsia, Tullbergella, Oxycephalus*)
  Family Platyscelidae (*Amphithyrus, Tetrathyrus, Paratyphis, Platyscelus, Hemityphis*)
  Family Parascelus (*Thyropus, Schizoscelus, Euscelus*)

*Genera are in parentheses. Whereas the Fissipeda contains 3 superfamilies, 8 families, and 20 genera (1 superfamily, 1 family, and 4 genera are extinct), the Hyperidea contains 6 superfamilies, 21 families, and 71 genera. The greater number of taxa reflects a greater diversity in morphology as compared with fissipeds (see Fig. 3). Classification from Young (1962) and Bowman and Gruner (1973).

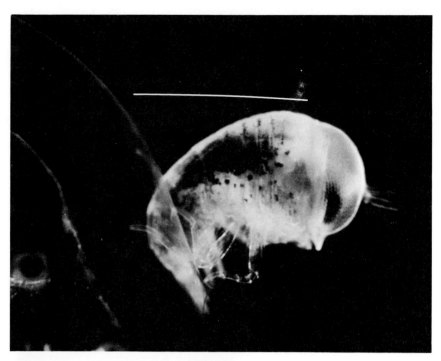

(a)

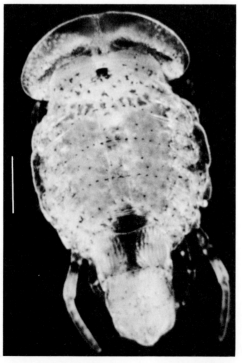

(b)

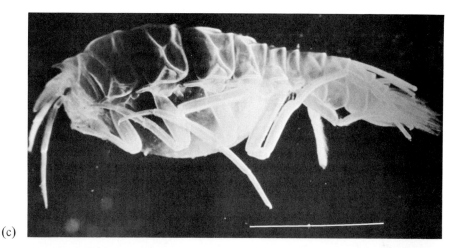

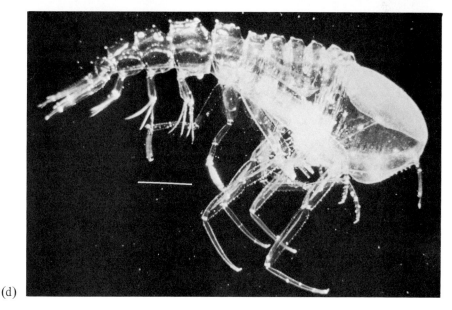

Figure 3. Examples of the great differences in size and morphology between hyperiid amphipods. (a) *Lestrigonus schizogeneois* living on the medusa *Proboscidactyl ornata* is less than 3 mm long when adult. Scale line = 1.0 mm. (b) *Lycaea nasuta* is specifically associated with the salp *Cyclosalpa affinis*. Scale line = 1.0 mm. (c) *Lanceola sayana* is found on the scyphomedusa *Pelagia noctiluca*. Scale line = 1.0 cm. (d) *Cystisoma* sp., one of the larger hyperiids, is over 30 times larger than *L. schizogeneios*. Its host is unknown. Scale line = 1.0 cm. Even the most skeptical must admit that these amphipods differ from one another in morphology at least as much as dogs, pandas, otters, and tigers differ. Why should it be expected that the behavior of hyperiids should be any less diverse than the behavior of fissipeds?

organisms. This is an exciting prospect, for there are many discoveries yet to be made; we are indeed fortunate to be working in plankton ecology now.

I think that plankton ecologists are extremely fortunate in another way—we have two independent criteria for evaluating the efficacy of our methods, and determining that we are obtaining "ground truth." For our data to have biological meaning and relevance, both precise identification to species and meaningful quantitative data are necessary. When it is difficult to obtain both at once, we know that we are working at scales inappropriate to our methods. Presently, the "quantitative planktologists" are trying to convince us that the species concept has no relevance to oceanic organisms; but I have little doubt that some *"in situ* planktologists," trying to work on inappropriately large scales, will soon be attempting to convince us of the same thing. Rather than defend the dubious proposition that species don't matter for plankton ecology, they should use appropriate methods for the scale they wish to study.

5 Can the Composition of Planktonic Communities Be Known?

Our first priority should be the objective evaluation of our methods. We must determine the scales over which our methods are most powerful, and the amount of overlap between them. We have so few ways of seeing into the ocean that we must not dismiss any of them out of hand. All methods that provide internally consistent, replicable data should be regarded as potential tools for the study of planktonic organisms.

If one uses a plankton net to study the fauna of the southern Sargasso Sea, the composition of the catch can be predicted in a qualitative way. Repeated tows will catch many of the same organisms; this internal consistency tempts one to conclude that a representative sample of organisms has been obtained. However, if one jumps off the ship and collects animals by SCUBA diving, another set of internally consistent data that has little overlap with the plankton net collection data set is obtained. The same is true with water bottle samples, long-line collections, or purse-seine collections. The problem is, how do these sets of data interrelate? Each is giving us a picture of life in the open sea, yet each is incomplete and biased. Can we put them all together and come up with a complete, or mostly complete, view of life in the open ocean?

The fundamental question that we should ask is, can we know what organisms are present in the open sea? If we cannot answer this question then all of our subsequent research must change; the most we can ever hope to do is to define our questions in terms of specific organisms, always realizing that we may be missing some of the most important information relevant to their lives. To answer this question, we need to bring together all biologically valid sampling methods in one place, at one time.

We need to study the composition of the plankton with towed nets (ranging in size from tiny plankton nets to giant midwater trawls), purse seines, water bottles, acoustics, diving and submersible observations and collections, towed camera systems and other remote-sensing devices, and any other method that could give us information about pelagic organisms.

Obviously it is impossible for one individual, or even several, to embark on such a task. Practically every portion of the biological oceanographic community is needed. Much of the equipment is still undeveloped. For example, collection of midwater animals from submersibles is still difficult, there are no ships in the University National Oceanographic Laboratory System (UNOLS) fleet capable of towing giant midwater trawls, and remotely operated towed cameras are still in the early stages of their development.

All of us must admit that no single method gives a complete view of the plankton. However, let us perfect each method to its ultimate resolution, using as our criterion the ability to obtain precise species identifications *together with* the ability to obtain precise quantitative information. Then let us bring all these methods together, perhaps 10 or 20 years hence, in one place at one time, in order to see the way they interrelate. The chosen place should be as simple and constant as possible physically (perhaps the southern Sargasso or the North Pacific Central Gyre) so that most of what we will be studying will be biology.

By using all of these methods simultaneously, we should be able to begin interrelating our perceptions of life in the open sea. The many scales at which we study organisms in the oceanic environment should provide intercomparisons, and the simultaneous use of all these techniques should indicate the extant major sampling problems. These sampling problems will challenge the ingenuity of future plankton ecologists, but they must be solved before we can move on to further questions about planktonic community structure. From *in situ* observations, I know that many organisms, including colonial radiolarians, siphonophores, and ctenophores, are not adequately sampled with any existing method; squids, many fishes, and many nanoplankters are also very poorly known.

This "intercalibration of perceptions of life in the open sea" would, at the very least, provide biological oceanographers with choices as to which method is most appropriate for answering specific questions about pelagic organisms and studying how these organisms interact. After such comparisons, people wouldn't use towed nets to answer questions about the behavior and microdistribution of zooplankton, nor would they apply *in situ* methods to the study of zoogeography. This, in itself, would constitute significant progress in plankton ecology.

6 Toward the Development of an Undersea Research Vessel

People most concerned with microscale phenomena (the interactions of planktonic organisms) have the farthest to go in developing new methods. At present, most *in situ* research on plankton is done with SCUBA, which can only be used effectively to a depth of 30 meters or so meaning only the smallest fraction of the life present in the open sea can be observed directly. Within the next 50 years, I expect that we will observe planktonic organisms *in situ* without disturbing them, collect them, and perform field experiments anywhere throughout the water column. Several research submersibles have been used in midwater, but submersibles lack the maneuverability and precision of a diver. One-atmosphere diving suits (such as are presently used by the petroleum industry) may provide an answer to some of the logistic problems, and present technology permits divers (using mixed-gas apparatus) to work in ambient pressures equivalent to depths of several hundred meters. However, with all of the present systems, the plankton ecologist is still nothing more than an infrequent visitor to the midwater environment.

The ultimate goal of *in situ* exploration of the midwater regions of the open sea should be to lengthen the duration of our stays as well as to increase the depth of our excursions. The first step toward lengthening our stays should be with shallow (10- to 20-meter deep) "planktonic habitats," which would permit scientists to observe and collect planktonic organisms on a round-the-clock basis. Even at such shallow depths, we can study the behavior of vertically migrating animals that spend the day much deeper but come to the surface at night. In order to work at ambient light levels, image intensifying devices will be necessary, but they should be relatively easy to develop, using existing technology.

The first planktonic habitat, as I envision it, would be moored at a depth of 10 meters in the open sea, attached to a surface tender, such as the Woods Hole Oceanographic Institution's (WHOI) R/V *Lulu* or the Scripps Institute of Oceanography's (SIO) R/V *FLIP*. This first habitat would be small, with laboratory and observational facilities, and with easy means to enter and exit. Planktonic organisms could be studied *in situ* almost continuously, and field and laboratory experiments could be performed at depth. All cooking and sleeping would be done on the surface tender, since the divers could freely move between the habitat and the surface. This first habitat would be used to develop techniques for the collection, observation, and experimental manipulation of planktonic organisms under truly ambient conditions. All of the techniques to be used in the first habitat could be easily modified for use at greater depths.

Subsequent habitats would range to greater depths, and should be entirely free of surface support. Ultimately, and in far less time than 50 years, should come the development of an undersea research vessel. Such a vessel would resemble the *Oceanlab* concept proposed a few years ago. This research vessel would allow scientists to work *in situ* anywhere between the surface and 600 meters.

An undersea research vessel such as *Oceanlab* would allow biologists to study the behavior of organisms *in situ*, to conduct physiological experiments at ambient pressures, and to study the microscale distribution of zooplankton in a way that cannot now be done and will never be done with remote sampling. I am certain that direct observations will reveal the existence of a varied fauna which remote sampling methods cannot detect. The gains to biological oceanography would be very great, but an undersea research vessel such as *Oceanlab* would also contribute a good deal to chemical and physical oceanography. With such a vessel, microscale phenomena could be directly measured and studied, rather than merely inferred.

I know that the scientific potential of such a vessel is very great, for my few dives in *Alvin* have shown me that no one knows very much about the possible forms life can take in the open sea. This largest of all environments on Earth is yet to be explored, and all previous work there represents but the first few hesitant steps into it. The time has come to try to live and work in the open ocean. I hope that the means to do so are developed well before the next 50 years elapse, so that I will be able to have some part in it.

Acknowledgements

I thank all those who commented on this paper. Figures 1a, 1b, 3b, 3c, and 3d by L. P. Madin. Figures 1c and 3a by N. R. Swanberg.

References

Bowman, T. E., and H. E. Gruner. 1973. The families and genera of Hyperiidea (Crustacea: Amphipoda). Smithsonian Contrib. Zool., No. 146.
Haeckel, E. 1890. Planktonic studies: a comparative investigation of the importance and constitution of the pelagic fauna and flora. Jena Z. Naturissw. 25(1-2).
Hamner, W. M. 1977. Observations at sea of live, tropical zooplankton. Proceedings of the Symposium on Warm Water Zooplankton. Special Publication, Natl. Inst. Oceanogr., Goa pp. 284-296.
Harbison, G. R. 1976. The development of *Lycaea pulex* Marion, 1874 and *Lycaea vincentii* Stebbing, 1888 (Amphipoda, Hyperiidea). Bull. Mar. Sci. 26(2), 152-164.
Harbison, G. R., and L. P. Madin. 1976. Description of the female *Lycaea nasuta* Claus, 1879 with an illustrated key to the species of *Lycaea* Dana, 1852 (Amphipoda, Hyperiidea). Bull. Mar. Sci. 26(2), 165-171.
Harbison, G. R., D. C. Biggs, and L. P. Madin. 1977. The associations of Amphipoda Hyperiidea with gelatinous zooplankton. II. Associations with Cnidaria, Ctenophora and Radiolaria. Deep-Sea Res. 24, 465-488.
Harbison, G. R., L. P. Madin, and N. R. Swanberg. 1978. On the natural history and distribution of oceanic ctenophores. Deep-Sea Res. 25, 233-256.
Hardy, A. C. 1936. Observations on the uneven distribution of oceanic plankton. "Discovery" Rep. 11, 511-538.
Hensen, V. 1891. Die Plankton-Expedition und Haeckel's Darwinismus. Lipsius & Tischer, Kiel and Leipzig, 87 pp.
Laval, P. 1980. Hyperiid amphipods as crustacean parasitoids associated with gelatinous zooplankton. Oceanogr. Mar. Biol. Ann. Rev. 18, 11-56.
Madin, L. P., and G. R. Harbison. 1977. The associations of Amphipoda Hyperiidea with gelatinous zooplankton. I. Associations with Salpidae. Deep-Sea Res. 24, 449-463.
Pirlot, J. M. 1932. Introduction à l'étude des Amphipodes Hypérides. Ann. Inst. Oceanogr., Paris, N.S. 12, 1-36.
Strøm, H. 1762. Physisk og Oeconomisk Beskrivelse over Fogderiet Sondmor, beliggende i Bergens Stift, i Norge. Sorøe, 2 vols., 572 pp., 4 plates.
Young, J. Z. 1962. The Life of Vertebrates. Oxford University Press, New York, Oxford. 820 pp.

Oceanic Fine- and Microstructure

J. Stewart Turner

1 Introduction

As recently as 20 years ago, we assumed that such oceanic parameters as temperature (T), salinity (S), and other chemical properties varied smoothly with depth. Earlier observational data came from widely spaced water bottle samples and reversing thermometers, and curves drawn through the discrete points obtained in this way were taken to represent the actual state of the ocean. Newly developed instruments have shown, however, that the vertical distributions of properties are often very far from smooth and typically consist of a series of quasihomogeneous, nearly horizontal layers, separated by regions in which the gradients are much larger. These variations, with layer scales ranging from about a meter to several hundred meters, are now called the oceanic finestructure; they are most prominent in the vicinity of fronts, across which there are large horizontal variations of T and S. Fluctuations of temperature, salinity, and velocity representing variations on a scale of about 10 centimeters and smaller have also been measured using rapidly responding sensors, and these constitute the turbulent microstructure.

Other common assumptions which have had to be revised in the light of new observations are: the ocean, because of its large scale, is everywhere turbulent; and molecular diffusion must therefore be entirely negligible. Instead, the ocean is characteristically so stably stratified that overturning motions are inhibited. In the interior of the deep ocean turbulent microstructure occurs only intermittently and in patches, while the level of fluctuations through most of the volume is extremely small for most of the time. The turbulent patches tend to be thin and elongated horizontally and they are often associated with internal wave activity.

Microstructure measurements should always be interpreted in relation to the finestructure, and motions on the two scales can interact in significant ways. Molecular processes can be important, and even dominant, at density interfaces as well as in thin surface boundary layers. Particularly striking phenomena are observed when there are two (or more) separate components, with different molecular diffusivities and opposing effects on the density, and these "double-diffusive" processes will receive special attention in this review.

All the ideas introduced above have important implications for the understanding and prediction of oceanic mixing, but as they have become gradually accepted, the individual smaller scale mixing processes have been mainly studied in isolation. No overall theoretical framework exists, and certainly no consensus about the relative importance of the various small-scale mixing processes. Before we look to the future, we will examine why this should be so, by tracing the development of a few of the ideas underlying this new field. I will not attempt to be comprehensive; topics of special interest to me will be used here as examples of more general tendencies. For more thorough recent reviews of the subject and its literature, the reader can refer to Sherman et al. (1978), Gregg and Briscoe (1979), Garrett (1979), and Turner (1980, 1982).

A vital factor in the rapid advance of the subject has been the invention and refinement of suitable instruments to resolve the smallest scales of motion, and of temperature and conductivity fluctuations. Probably the most influential of these have been the two generations of conductivity-temperature-depth (CTD) instruments developed by Neil Brown. The lack of such instruments was a strong constraint on earlier workers and led to the parameterization of the unresolved scales in terms of eddy mixing coefficients which shed no light on the physical mechanisms actually responsible for the mixing. The availability of detailed measurements has produced a corresponding increase in activity by theoretical and laboratory modelers, who have interacted closely with the observers to the benefit of both groups. Many of the most exciting developments have been based on identifying and comparing well-defined events in the ocean with those studied in the laboratory, using experiments designed to test specific physical ideas in a carefully chosen oceanic environment. I believe that such individual initiatives, and cooperative testing of hypotheses, will remain the most important factors for future progress in this field, though it is time to give more conscious attention to the wider implications of the results.

2 Mixing Due to Shear or Waves

The first type of process is that of instability due to shearing motions in the ocean, particularly across the sharper gradient regions of the finestructure. These are "mechanical" mixing events, driven by the kinetic energy of the mean motion. We must also consider the sources of this shear and of the interfaces on which it can act.

The mechanism of instability is now well understood at a relatively sharp transition region in a fluid with substantial density and velocity differences across it. If the velocity and density profiles have comparable vertical length scales, and the shear (du/dz) is gradually increased, a parallel stratified flow can become unstable. A series of Kelvin-Helmholtz billows forms when the minimum gradient Richardson number

$$\text{Ri} = N^2 / \left(\frac{du}{dz}\right)^2 \tag{1}$$

falls below $\frac{1}{4}$; here N is the buoyancy frequency, defined by

$$N^2 = -(g/\rho_0)(d\rho/dz),$$

where g is the acceleration due to gravity, ρ is the density, z is the vertical coordinate and u is the horizontal velocity. Laboratory experiments (Scotti and Corcos, 1969) have confirmed the linear stability theory, and the wavelength and growth rate it predicts.

The growth beyond the initial instability has also been studied in a continuing series of experiments (Sherman et al., 1978). For larger (constant) values of the shear, the billows roll up and merge, as shown in Figure 1; subsequently the interfacial fluid is mixed both by gravitational instability and by horizontal interleaving to produce a region of nearly constant density gradient. There is a theoretical upper limit to the possible thickening of the interface, such that the change in potential energy just equals the kinetic energy available locally to produce it. The process is not perfectly efficient, because energy is lost to viscosity, and the final interface thickness δ is only a fraction of the limiting value; it is given approximately by

$$\delta = 0.3\rho_0 U^2/g\Delta\rho \tag{2}$$

in terms of the velocity difference U and density difference $\Delta\rho$ across the interface. The above result implies that this kind of instability is self-limiting, and that mixing due to the Kelvin-Helmholtz mechanism alone is small; some other process is needed to transport the mixed fluid away from the interfaces.

At the base of a surface mixed layer (see Section 3) the shear produced by a wind stress may be large enough to make the flow unstable, but these processes become important for the deep ocean only when they are considered in conjunction with internal waves. The shear needed to reduce Ri to unstable values is often produced by the passage of an interfacial wave. A notable step in the development of this subject was taken by Woods (1968), who made direct visual observations of billows in the ocean generated in this way. These concentrated

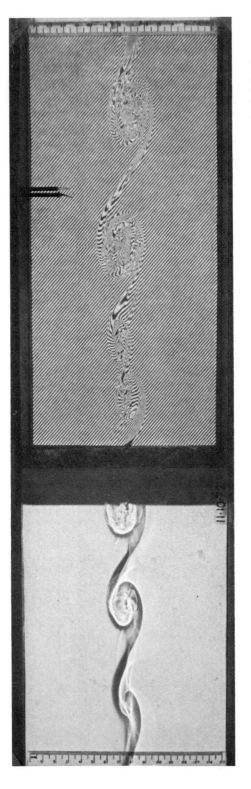

Figure 1. Photographs of a Kelvin-Helmholtz instability in a stratified shear flow, from Thorpe (1973). The flow was to the left in the upper layer and to the right in the lower. At the left is a shadowgraph, and at the right a "moiré" pattern formed by viewing the small-scale density gradients against a background of diagonal parallel lines.

attention on the need to study individual mixing events and also demonstrated the relevance of simple experiments in the laboratory and the ocean. This work has been followed by more sophisticated laboratory experiments (Thorpe, 1978a), and measurements in the ocean (Eriksen, 1978; Dillon and Caldwell, 1979) which show that overturning regions are indeed associated with large shears on high-gradient parts of the finestructure. It has become clear, however, that overturning, followed by collapse and spreading of the mixed regions, cannot explain the thicker layers in the oceanic finestructure since deep overturned regions are energetically impossible.

The processes so far considered need a preexisting sharp interface to become effective, and hence they cannot be used to explain the initial formation of layered structures in a smoothly stratified fluid. As pointed out by Turner (1973), the deep ocean is so stable (the mean Richardson number is very large) that there is not enough kinetic energy available in the mean motion to produce mixing. Thus extra energy must be propagated into the region from the boundaries in the form of inertial or internal gravity waves if mixing is to be sustained.

The study of internal waves *per se* is in a far better state than that of the associated mixing. Good measurements of current velocities and temperature fluctuations have been obtained from moored arrays, and from towed sensors and profiling instruments. The most important step in the collation and interpretation of the data has been the introduction of a model to describe the wave number-frequency spectra; a recent review article by the authors of this model (Garrett and Munk, 1979) describes its development and current form. They point out that many theoretical problems are still to be solved, but the study of mixing has not even reached the stage where there is an adequate framework for the description of all the diverse physical processes involved.

Some of the fluctuations on vertical profiles, detected using moored instruments, are in fact produced by internal wave straining of a smooth profile. This effect is reversible, and it will not be included in our definition of finestructure, which implies an inherent quasisteady nonuniform structure. Various mechanisms have been identified, however, which produce an instability of a field of waves in a stratified fluid, thus leading to local overturning and mixing, and persistent nonuniformities in the gradient. Nearly always, the source of wave energy at a point in the ocean is not identifiable but is the result of superposition of waves from many sources (Wunsch, 1976).

Strong interactions between an arbitrary pair of waves of large amplitude can feed energy rapidly into small-scale forced waves that overturn locally. Resonant interactions are more selective and require a special relation between the wave numbers and frequencies. In this field too, laboratory experiments have played an important role (Sherman et al., 1978), and those by McEwan (1971, 1973) and McEwan and Robinson (1975) are particularly noteworthy. In the earlier experiments McEwan found that irreversible intensification of small-scale density gradients (which he called "traumata") formed under conditions such that no resonant interaction had been predicted (Fig. 2). In the 1975 paper, a theory was developed and supported by an elegant experiment, which explained this case in terms of a "parametric" instability. This is in fact another resonant mechanism, not previously considered, which depends on the modulation by long waves of the effective component of gravity. When the forcing frequency is twice that of a disturbance, energy is fed into a range of much shorter wavelengths than that of the forcing wave, and these can grow against internal viscous dissipation when the forcing amplitude is large enough.

The detailed consequences of these models for mixing in the ocean have yet to be tested, though McEwan and Robinson (1975) extended Garrett and Munk's universal wave spectrum to compute the mean square slopes of the isopycnals, which they deduced is large enough to excite the parametric instability. The mechanism does seem to be capable of transferring wave energy to much smaller scales, and thus ultimately creating patches of mixing in an initially smoothly stratified ocean. McEwan (personal communication) is currently extending this work to measure directly the partition of energy between dissipation and buoyancy diffusion in a saturated internal wavefield [cf. Equation (2)].

Many observations have been made of the microstructure in the turbulent patches which result from the processes discussed above. Most measurements are of temperature fluctuations, interpreted in terms of spectra, but there are also some records of conductivity and velocity fluctuations. A continuing debate relates to theoretical and observational ways to distinguish between "active" turbulence, where there are strong velocity as well as temperature and salinity fluctuations, and "fossil" turbulence, consisting of T-S microstructure persisting after the turbulent velocities have decayed. Recent theoretical work suggests, however, that even in this latter case persistent motions will result from the release of potential energy stored in the perturbation density field, which play a crucial role in maintaining the temperature and salinity fluctuations. The question will only be resolved by further instrumental developments, enabling the very small velocity gradients to be measured simultaneously with the T-S gradients. Strong differences of opinion are held about the relative merits of towed instruments, which produce longer records but are subject to vibration, and vertical profilers, which give shorter records through the thin patches of microstructure.

Various arguments have been proposed to relate the effective vertical transport coefficients to the rate of energy dissipation ϵ or the rate of dissipation of temperature variance. The method of Osborn and Cox (1972) (now widely used) implies that there is a balance between the mechanical production and molecular destruction of variance and leads to a relation for the vertical eddy diffusivity K_z

$$K_z = \kappa \overline{(\partial T'/\partial z)^2} (\partial \overline{T}/\partial z)^{-2} = C\kappa \qquad (3)$$

where $\overline{T}$ is the mean temperature, T' is the temperature fluctuation, and κ is

Figure 2. Shadowgraph photograph due to McEwan (1971), showing the breakdown of a large-scale internal wave with the formation of "traumata," or irreversible intensifications of the density gradient.

the molecular diffusivity. The "Cox number" C varies from approximately 10^3 in actively turbulent regions, to 10 or less in quiet regions, and though its magnitude gives some indication of turbulent intensity, the concept (like that of eddy diffusivity itself) must be used with caution. It is appropriate only when the motion is steady and horizontally homogeneous and it is possible to think of a gradient diffusion process based on temperature alone. In patches of "fossil" turbulence the microstructure activity is not a proper indicator of turbulent transport of heat at all (Gibson, 1980), and Gargett (1978) has pointed to similar difficulties when there are horizontal intrusions with coupled T and S anomalies.

3   Surface Mixing Processes

We will discuss more briefly the fine- and microstructure produced by inputs of energy across the sea surface. Similar ideas can be applied near the bottom boundary of the ocean, though fewer measurements are so far available there, and rapid developments can be expected in this field in the next few years.

Because of the overall static stability, the direct effect of any form of energy input through the surface is to produce a nearly homogeneous surface layer with a region of increased density gradient below it. Though this gradient region is called the diurnal or seasonal thermocline, there can also be significant variations of salinity and velocity which should be regarded as part of the near-surface finestructure. The currently most successful theoretical models of the mixing through this region are one-dimensional integral formulations (Niiler and Kraus, 1977) though a great deal of work has been done recently on turbulent closure models (Launder, 1976) which will be worth watching closely.

Laboratory experiments have played a significant role in this field too. The effect of the wind on the surface has been simulated either by imposing a stress at a solid, moving boundary (which produces a shear flow) or by stirring with grids (and so creating smaller scale turbulence which decays with depth, but without a mean shear). In both cases, the ratio of the deepening rate, or entrainment velocity u_e, to a reference velocity u_1 (which may be the mean velocity or a turbulent velocity scale) has been found to be a function of an overall Richardson number Ri_0. This parameter can be defined [cf. Equation (1)] by

$$\mathrm{Ri}_0 = \frac{g \Delta \rho}{\rho} \frac{\ell_1}{u_1^2} \qquad (4)$$

where $\Delta \rho$ is the density difference across the interface, and ℓ_1 is either the depth of the upper layer or the length scale of the turbulent motions. These models take no account of the likely interaction between wind-driven currents and surface waves, though recent work suggests that such an interaction can be the mechanism of production of Langmuir cells. Since these systematic motions extend through the depth of the surface layer, they can have an important effect on the mixing. This process has been included explicitly in the model proposed recently by Leibovich and Paolucci (1980).

Energy for mixing can in principle also be provided by an unstable buoyancy flux through the surface; that is, by surface cooling or evaporation. Both this stage, when the thermocline is deepening, and the preceding heating period, when the thermocline is building up and becoming shallower, have been included in time-dependent models based on the one-dimensional heat and mechanical energy conservation equations. But in fact Gill and Turner (1976) have shown that observations made during the cooling period are better described in terms of a "nonpenetrative" model, that is, by assuming that the convection contributes very little directly to the entrainment. It is very important, nevertheless, in decreasing the density difference $\Delta\rho$ at the bottom of the surface layer and in keeping it well mixed.

Other models based on energy arguments (e.g., Pollard et al., 1973; Thompson, 1976) emphasize the kinetic energy produced by the shear across the thermocline instead of by surface processes. Some of the apparent conflict between these alternative formulations is resolved by recognizing that different processes may dominate at different stages of the mixing. On a time scale of half a pendulum day, the inertial currents build up by the wind certainly dominate the mixing, as shown by the measurements of Price et al. (1978). On longer time scales, on the order of weeks or a season, the one-dimensional models based on surface fluxes give good predictions of the variation of sea-surface temperature (Denman and Miyaki, 1973).

The most unsatisfactory feature of the mixed layer theories to date is the paramaterization of the dissipation. It is assumed that the energy used for entrainment is some fixed fraction of that produced in each source region, and laboratory or field data are used to evaluate the constants. However, the decay of turbulent energy with depth below the surface is not adequately described in this way. Indications are that the mixing at an interface can be substantially reduced by the presence of a gradient region below, due to the generation of internal waves which carry energy downwards. This process can produce "breaking" and further mixing in the thermocline below the surface layer, and neglecting it leads to poor predictions of the layer depths. Although most of the existing observations of turbulent microstructure in the ocean (particularly those by Russian oceanographers) have been made in or close to the surface layer, these have not yet been related closely enough to the finestructure parameters to shed much light on these questions.

The above discussion suggests that the theoretical models have got somewhat ahead of observations which could be used to distinguish between them. More detailed measurements are urgently required of intermediate-scale structures and motions in the surface layer, and of the small-scale responses to them. An impressive example has been set by Thorpe (1978b) and his co-workers, who have made a comprehensive set of measurements in Loch Ness. These have shown that many of the mixing events at the bottom of the surface layer are consistent with a Kelvin-Helmholtz type of instability (see Fig. 1); Dillon and Caldwell (1979) have reported preliminary results from the Mixed Layer Experiment (MILE) in the North Pacific which suggests the same kind of structures. Further results from this and the Joint Air Sea Interaction (JASIN) 1978 Experiment in the North Atlantic should soon become available.

A strong warning, however, is contained in the results reported by Gregg (1976). He followed the development of the layer depth and the temperature microstructure during the passage of a storm. In spite of strong surface-driven vertical mixing, the change in heat content of the upper 200 m was dominated by lateral effects, with intrusions carrying a large fraction of the net heat flux into the area studied. Thus truly one-dimensional models of mixed layer evolution are really inadequate, and models must be developed which take account of horizontal variability (see also Section 5). Detailed, repeated observations of T and S variations along horizontal sections will be required, and it is not yet even clear what horizontal scales will be relevant. A promising new technique which should be valuable for such studies has been demonstrated by Haury et al. (1979), who made simultaneous observations of an overturning internal wave, using a CTD and a high-frequency acoustic echo sounder towed at 4 m/sec.

4 Double-Diffusive Convection

The convective processes discussed above in the context of surface mixing have an obvious energy source—namely, cooling from the atmosphere above. We return now to interior mixing mechanisms and consider processes which are also driven by convection, but in a more subtle way. When T and S both increase or both decrease with depth, one of the properties is "unstably" distributed (in the hydrostatic sense). The basic fact about double-diffusive convection is that the difference in molecular diffusivities of the two properties affecting the density allows potential energy to be released from the component that is heavy at the top. This is still true when the mean density distribution is "hydrostatically stable"; and in fact, since the driving component (the one which is providing the potential energy) is transported faster than the driven, the vertical density difference is increased following this kind of mixing, instead of decreased, as it is during a mixing event driven by the release of kinetic energy. As Garrett (1979) has pointed out, this net downward mass flux implies that vertical double-diffusive fluxes alone cannot solve the problem of maintaining the main oceanic pycnocline.

In recent years, the literature documenting the significance for the ocean of molecular processes, particularly double-diffusive effects, has multiplied rapidly. Their relevance is now so widely accepted that a review starting from first principles is no longer justified here, and we can just refer to Turner (1973, 1979,

1982) for an early and more up-to-date discussions of the principles and progress. Again much of the detailed understanding has come from theory and laboratory experiments, though the recent oceanic applications and observations are emphasized here. We start by discussing results which focus on the one-dimensional coupled transports of two or more components across interfaces separating convecting layers.

In the case where both salinity and temperature increase with depth, interfaces of the "diffusive" type readily form. They are most prominent in the ocean when the density ratio $R_\rho = \beta \Delta S / \alpha \Delta T$ is small (between 1.5 and 5), where $\beta \Delta S$ and $\alpha \Delta T$ are the contributions to the fractional density difference across an interface due to salinity and temperature, respectively. Undoubtedly the regular series of layers observed under the Arctic ice, above the Red Sea brine pools, and in various salt-stratified lakes are formed by double-diffusive convection for which heat is the destabilizing property. Foster and Carmack (1976) have observed layers and diffusive interfaces in the Weddell Sea (see Fig. 3), but at low temperatures the nonlinearity of the equation of state also allows a "cabbeling" instability, and they interpreted the thicker layers (those observed at greater depths) in these terms. McDougall (1981) has recently conducted a careful series of laboratory experiments which incorporate the nonlinear effects and show a continuous transition between the two types of behavior—it is not

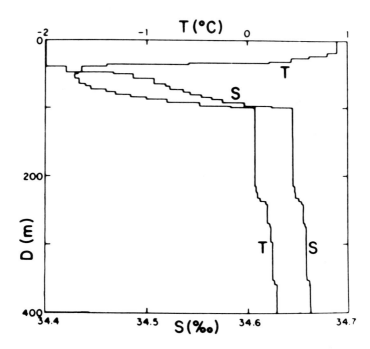

Figure 3. Temperature and salinity profiles observed in the upper 400 m of the Weddell Sea, near the turning point of the current gyre (from Foster and Carmack, 1976).

possible to explain the observations by neglecting double diffusion and considering the conventional cabbeling instability alone. He has also shown that interface migration is an essential element of this process and has discussed the kind of observational program which will be needed to test his ideas.

In this and other regions, more attention should be paid in the future to the study of the time history of a series of layers. In certain special situations, such as a salt-stratified lake, laboratory measurements of coupled interfacial fluxes appear to give a quantitatively accurate description of the steady state (Huppert and Turner, 1972). However laboratory observations and associated theoretical studies suggest that unsteady situations may be important. An example of an evolving state is a linear salinity gradient heated from below; new layers form sequentially at the top of a growing convection region while lower down adjacent layers merge (Huppert and Linden, 1979). Linden (1976) studied the case of a destabilizing "temperature" gradient already present in the interior and documented two kinds of interfacial breakdown. The interface can remain stationary, while the density difference across it decreases to zero, or it can migrate vertically to the adjacent interface (see also McDougall, 1981). When the first process is acting, Huppert (1971) showed theoretically that an interface should be stable to small perturbations if R_ρ is greater than about 2, and unstable otherwise. Observations of stable convecting layers in the ocean are generally consistent with this criterion, but the influence of unstable layers on the vertical fluxes, as they intermittently break down and reform, has not yet been properly assessed.

Clearly the "transport coefficients" K_i, defined as the vertical fluxes divided by the corresponding mean gradients, must be different for heat and salt when the transports are produced by double-diffusive processes, and eddy diffusivity concepts are not appropriate in that case. It is not so widely appreciated that the individual fluxes of several dissolved species can depend strongly on molecular diffusivity. Griffiths (1979) has predicted that K_1/K_2 for two components, both of which are driven aross an interface by convective heating, should be proportional to $\tau^{1/2} = (\kappa_1/\kappa_2)^{1/2}$ (where κ_1 and κ_2 are the molecular diffusivities) at low total solute-heat density ratios R_ρ, and to τ at higher R_ρ. Experiments at interfacial density ratios R_ρ between 2 and 4 are consistent with $K_1/K_2 = \tau$, but as shown in Figure 4, there is an even greater (and so far unexplained) separation of components at higher R_ρ. These results are potentially of great importance for the interpretation of geochemical data, as will be discussed further in the final section.

A newly discovered small-scale process of interest to geochemists may be relevant here: the direct observation from R/V *Alvin* of hot plumes emerging from the sea floor. Turner and Gustafson (1978) have suggested that double-diffusive and nonlinear mixing effects could be important in determining the composition of the outflow and the distribution of precipitation in such plumes, and an interdisciplinary study of the phenomenon could be very productive.

A second kind of double-diffusive process is active when warmer, salty water overlies cooler fresher water. Deep convecting layers are still observed in this

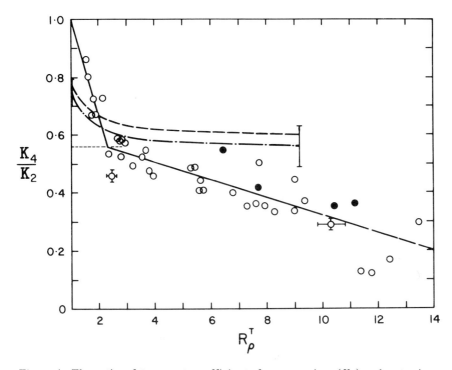

Figure 4. The ratio of transport coefficients for magnesium (K_4) and potassium (K_2) measured by Griffiths (1979) across a diffusive interface. The ratio of molecular coefficients is 0.60, and the upper curves represent theoretical predictions.

case, but the transports across the interfaces between them are effected by "salt fingers," long narrow convection cells which form because of the different rates at which salt and heat diffuse horizontally between them. The first observations of step structures were made under the Mediterranean outflow, but there are now many other measurements which are demonstrably the result of salt fingering (Fedorov, 1978). The detection of salt fingers in the interfaces, using an optical method (A. J. Williams, 1975) and conductivity probes (Magnell, 1976) has more recently provided a most convincing direct confirmation of this interpretation.

Regular layered structures are only observed when the density ratio $R_\rho^* = \alpha \Delta T/\beta \Delta S$ (the inverse of the R_ρ^* introduced previously) is less than about 1.7. Theory predicts that salt fingers can readily form up to values of R_ρ^* of order 10^2, so they should be extremely common below any region where warm salty water is formed at the surface by solar heating and evaporation. Indirect evidence also shows that they are often associated with intrusions (see the following section), though they have not been observed directly in the main thermocline. Their presence implies that locally the vertical "transport coefficient" for salt is greater than that for heat. In many cases, however, they are not vigorous enough to produce layering, nor can they remain steady in the presence of other mixing mechanisms. They could interact in an important way with the internal wave field. Schmitt and Evans (1978) have shown that when the temperature and salinity gradients are nearly equal, salt fingers grow rapidly enough to become active on the high-gradient regions produced by internal wave straining and can thereby produce irreversible finestructure with alternating high and low gradients. They predict that this salt fingering should be intermittent but nevertheless calculate mean salt fluxes (using laboratory data and measured S and T profiles) which are comparable to the surface input of salt due to evaporation (that is, they deduce that fingers can account for all the vertical flux in the ocean). Lambert and Sturges (1977) reached similar conclusions about the importance of the finger fluxes through a system of layers and interfaces observed below the core of a warm saline intrusion in the Caribbean Sea.

5 Intrusions, Fronts, and Eddies

The previous section has emphasized one-dimensional, vertical transport processes, through double-diffusive interfaces which already exist. But it is now clear that the initial formation of the finestructure is dominated by intrusive motions and that these are often associated with horizontal variations of S and T. Here, too, laboratory experiments have shed light on the mechanisms involved, and several recent examples will be mentioned before turning to the oceanic observations.

Sidewall heating of a salinity gradient has long been known to produce a series of nearly horizontal convecting layers, growing in from the heated wall. Huppert and Turner (1980) have extended these results by showing that a vertical ice surface, melting into a salinity gradient dS/dz, produces layers of the same scale as heating alone:

$$\ell = 0.65 \frac{\alpha \Delta T}{\beta dS/dz} \tag{5}$$

where $\alpha \Delta T$ is the horizontal buoyancy difference evaluated at the mean salinity. In both cases, this scale is determined by the vertical density gradient due to salinity and the horizontal buoyancy anomaly due to temperature, and it is little affected by the melting. The meltwater, however, spreads out into these layers rather than rising to the surface, and this behavior will clearly influence the way icebergs can affect the water structure in the Antarctic Ocean.

The intrusion of one fluid into a gradient of another has been studied explicitly by Turner (1978), using the sugar/salt laboratory analog of the salt/heat system. When the source fluid has a different molecular diffusivity but the same density as that of the gradient into which it is released, there is strong vertical convection near the source, followed by spreading at several levels above and below the source (Fig. 5). "Warm salty" source fluid produces a series of noses

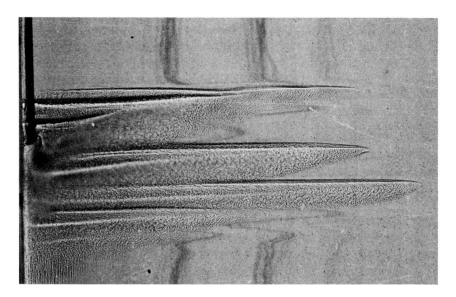

Figure 5. The flow produced by releasing sugar solution at its own density level into a salinity gradient. Strong vertical convection occurs, followed by intrusion at several levels, with a slight upward inclination across isopycnals.

with diffusive interfaces above and fingers below, which tilt upwards across isopycnals as they extend, and cold fresh intrusions tilt downwards—a very different behavior from the thin, horizontally spreading intrusion of, say, salt solution released into a salinity gradient. The motion across isopycnals is the direct result of the quasivertical double-diffusive fluxes across the interfaces, since the net buoyancy flux through the fingers is greater than that across the diffusive interface.

A related but much more controlled experiment, using a geometry of direct relevance to the ocean, has been reported by Ruddick and Turner (1979). They set up identical vertical density distributions on two sides of a barrier, one produced with sugar solution and the other with salt, and observed the development of a series of regular, interleaving layers when the barrier was withdrawn. The depth and speed of advance are both proportional to the horizontal property differences, and again there is a tilt indicating cross-isopycnal motion. The vertical scale is again given by Equation (5) (with a different multiplying constant), where $\alpha\Delta T$ is now the horizontal property anomaly across the front. All the experiments described support the same general conclusion: the formation and propagation of interleaving double-diffusive layers are self-driven processes, sustained by the local density anomalies set up by vertical double-diffusive fluxes, which act much more rapidly than the advective time scale.

There are now many observations in the ocean in which the influence of the processes described above can be identified. The earlier measurements have been well summarized by Fedorov (1978), who emphasizes that the strongest layering is found near boundaries between water masses of different origin. Layers are most prominent when there are large horizontal contrasts in T and S but a small net density difference; in these circumstances, motions and associated temperature inversions readily form. A few recent examples will be discussed here.

Various measurements at the Antarctic polar front (Gordon et al., 1977; Joyce et al., 1978) have revealed inversions that decrease in strength with increasing distance away from the front. The latter paper documents a definite slope of the intrusions relative to density surfaces and favors an explanation in terms of double-diffusive processes. A remarkable implication is that molecular effects, by driving intrusive motions across the frontal zone, could thus play a significant role in the large-scale balance of salt in the Southern Ocean. Voorhis et al. (1976) have shown that coastal fronts between colder, fresher water on a continental shelf and warmer, salty water offshore also exhibit strong double-diffusive interleaving. They measured directly, using sensors mounted on neutrally buoyant floats, the salt and heat transports into a cooler fresher layer from a warmer saltier layer above; these transports were unequal (in density terms) and consistent with salt fingering.

A very detailed three-dimensional mapping of a shallow thermohaline intrusion, just below the base of the seasonal thermocline, has recently been reported by Gregg (1980). A cold, fresher tongue about 6 km long, 2 km wide and 7 to 15 m thick was shown to slope downwards by about $1°$ relative to isopycnal surfaces, away from its origin at a front, though the feature was almost undetectable in the density profiles. The sense of this slope and the scale of the intrusion are in good agreement with the predictions made by Ruddick and Turner (1979) and support double-diffusion as the most probable driving mechanism.

Various observers (Gregg, 1975; G. O. Williams, 1976; Gargett, 1976) have demonstrated the importance of measuring microstructure with a full knowledge of the finestructure. The regions of most intense activity are the upper and lower boundaries of intrusions and are associated with finestructure inversions. The undersides of the salt-stabilized temperature inversion layers are found to have the highest level of temperature fluctuations, and it is plausible to associate this feature with salt fingers, though there is as yet no direct evidence for them in those regions. We note again the point made by Gargett (1978) that quantitative deductions made from Equation (3), using temperature fluctuations alone, will not be valid when the mixing is associated with double-diffusive intrusions.

This review is written from the point of view of one who is already convinced of the effectiveness of double-diffusion as a driving mechanism for interleaving across fronts. In the volume edited by Woods (1982), however, another mechanism has been proposed: that the process of frontogenesis, in a region having a net horizontal density gradient, could itself lead to a folding of isotherms to produce local inversions. Double-diffusive microstructure would then form above and below these "kinematic" intrusions, but as a passive response to the distortion, rather than the active driving mechanism. This controversy has pointed to the need to distinguish carefully between fronts which are strongly "thermoclinic" but have a small net horizontal density difference, and those that are "baroclinic," and to devise observations to assess the relative importance of these two mechanisms for the formation and subsequent extension of intrusions.

Frontal regions which would repay more attention in the future are those bounding persistent eddies, such as the Gulf Stream rings. Laboratory studies of double-diffusive fronts with this geometry in a rotating system show that axisymmetric interleaving motions can persist when there is a shear across a front, but the stability of these fronts, and the mixing rate which results from coupling double-diffusive with baroclinic mechanisms, remain to be investigated thoroughly. The interaction of eddies with internal waves and finestructure at their edges could provide a mechanism for the decay of these mesoscale motions.

Several groups have begun to use such eddies as a traceable water mass, in which the interaction of physical, chemical, and biological processes can be followed. For example Tranter et al. (1980) have made repeated surveys of a warm-core eddy shed from the East Australian Current. Biological production remained low as long as deep convection and mixing were present, and during the winter the phytoplankton concentration was lower inside the eddy than outside it. However, as soon as a persistent thermocline had formed on top of the eddy, and as the surface warmed, the nutrients previously mixed upwards from below produced a rapid plankton bloom which was confined to the upper layers where the light level was high. A systematic study of the relation between biological production and other features of the finestructure, particularly possible interleaving layers across the frontal boundary of an eddy, would be well worth while.

6 Looking Ahead

What can we learn from the very short history of this field which will help us evaluate a likely or desirable course it might follow in the future? It is certain that the development of new instruments will continue to play an important role. Acoustic methods, and arrays of sensors measuring different properties simultaneously, have not yet been fully exploited, for example, but it would be unwise to impose a development program too firmly in advance. Innovative individuals, exploiting new technology in consultation with the users, should be free to pursue their own ideas. For similar reasons, it also seems vital not to concentrate entirely on large cooperative measurement programs, but to continue to couple these closely with laboratory and theoretical work.

We still need the identification and detailed study of individual physical processes in particular regions; for example, the direct detection of salt fingers in the main thermocline would be a significant step. To this end, we should systematically search for distinctive "signatures" of various processes, so that they can be recognized from fluctuation records, or in spectra calculated from these. This will allow one to distinguish between different mixing mechanisms and to assess their relative importance under various conditions. Researchers generally agree that finestructure and microstructure activity in the ocean is greatest near the external boundaries (the surface or the bottom), or across internal boundaries (fronts), and that processes on these two scales should be investigated together. Not enough is known, however, about the temporal and spatial scales of the intrusions which are a vital element in the mixing, and their variability in each region.

The need to distinguish between different physical processes is not just an academic question. Deductions about vertical transports made on the basis of the Cox number are invalid if the turbulence has a double-diffusive rather than "mechanical" origin, or if it has decayed to a fossil stage rather than being active. The overall consequences of different mechanisms may be quite different and perhaps profound. Schmitt (1981) has suggested, for instance, that the $T:S$ relationships of the central water masses of the oceans are better fitted by a curve of constant R_ρ^* (as defined in Section 4) than by a straight line. This is consistent with the dominance of mixing by double-diffusive processes, and inconsistent with mechanical mixing (which implies equal eddy diffusivities for heat and salt). To give another example, if we could establish that active double-diffusive driving, rather than passive advection, is responsible for producing interleaving layers at fronts, then we should be able to relate the horizontal transports of heat and salt to the horizontal gradients of T and S and the quasi-vertical fluxes of these properties.

As yet no satisfactory theoretical synthesis of fine- and microstructure observations has been made. The task is an important one, but it may well be more difficult than it was for internal waves, since so many different physical mechanisms are involved, and we are not even certain that we have discovered all of them. In particular regions, however, one process or another may be found to dominate the mixing, and then a simpler picture could emerge. Until that stage is reached, individual ideas and initiatives should continue to be encouraged.

The physical oceanographers' increased understanding of fine- and microstructure has not so far had much effect on other fields of marine science. In the previous section reference was made to the possible importance for biological production studies, through its effect on the distribution of nutrients and on mixing. We also referred earlier to the potentially far-reaching implications for geochemical studies. When a tracer is used to mark a water mass, it is assumed that its changing concentration is a measure of the mixing rate for the water mass as a whole. But if "diffusive" interfaces are important anywhere between the source and the sampling point, the transport of a tracer having a different molecular diffusivity is not necessarily a good indicator of the flux of a major component, much less of heat. The prevalence of double-diffusive phenomena in finestructure observations suggests at least that a single eddy diffusivity for all properties must be used with great caution.

Acknowledgments

I am grateful to H. E. Huppert, T. J. McDougall, and B. R. Ruddick for helpful comments on an earlier draft of this paper.

References

Denman, K. L., and M. Miyaki. 1973. Upper layer modification at Ocean Station Papa: observations and simulation. J. Phys. Oceanogr. 3, 185-196.
Dillon, T. M., and D. R. Caldwell. 1979. Catastrophic events in a surface mixed layer. Nature (London) 276, 601-602.
Eriksen, C. C. 1978. Measurements and models of fine structure, internal gravity waves and wave breaking in the deep ocean. J. Geophys. Res. 83, 2989-3009.
Fedorov, K. N. 1978. The thermohaline finestructure of the ocean (English translation). Pergamon, Oxford, 170 pp.
Foster, T. D., and E. C. Carmack. 1976. Temperature and salinity structure in the Weddell Sea. J. Phys. Oceanogr. 6, 36-44.
Gargett, A. E. 1976. An investigation of the occurrence of oceanic turbulence with respect to finestructure. J. Phys. Oceanogr. 6, 139-156.
Gargett, A. E. 1978. Microstructure and finestructure in an upper ocean frontal regime. J. Geophys. Res. 83, 5123-5134.
Garrett, C. J. R. 1979. Mixing in the ocean interior. Dynam. Atmos. Oceans 3, 239-265.
Garrett, C. J. R., and W. H. Munk. 1979. Internal waves in the ocean. Ann. Rev. Fluid Mech. 11, 339-369.
Gibson, C. H. 1980. Fossil temperature, salinity and vorticity turbulence in the ocean. In: Marine Turbulence (J. Nihoul, Ed.). XI Ocean Hydrodynamics Colloquium Liège, May 1979. Elsevier, Amsterdam, pp. 221-257.
Gill, A. E., and J. S. Turner. 1976. A comparison of seasonal thermocline models with observation. Deep-Sea Res. 23, 391-401.
Gordon, A. L., D. T. Georgi, and H. W. Taylor. 1977. Antarctic polar front zone in the western Scotia Sea—summer 1975. J. Phys. Oceanogr. 7, 309-328.
Gregg, M. C. 1975. Microstructure and intrusions in the California current. J. Phys. Oceanogr. 5, 253-278.
Gregg, M. C. 1976. Temperature and salinity microstructure in the Pacific Equatorial Undercurrent. J. Geophys. Res. 81, 1180-1196.
Gregg, M. C. 1980. The three-dimensional mapping of a small thermohaline intrusion. J. Phys. Oceanogr. 10, 1468-1492.
Gregg, M. C., and M. G. Briscoe. 1979. Internal waves, finestructure, microstructure and mixing in the ocean. Rev. Geophys. Space Phys. 17, 1524-1548.
Griffiths, R. W. 1979. The transport of multiple components through thermohaline diffusive interfaces. Deep-Sea Res. 26A, 383-397.
Haury, L. R., M. G. Briscoe, and M. H. Orr. 1979. Tidally-generated internal wave packets in Massachusetts Bay. Nature (London) 278, 312-317.
Huppert, H. E. 1971. On the stability of a series of double-diffusive layers. Deep-Sea Res. 18, 1005-1021.
Huppert, H. E., and P. F. Linden. 1979. On heating a stable salinity gradient from below. J. Fluid Mech. 95, 431-464.
Huppert, H. E., and J. S. Turner. 1972. Double-diffusive convection and its implications for the temperature and salinity structure of the ocean and Lake Vanda. J. Phys. Oceanogr. 2, 456-461.

Huppert, H. E., and J. S. Turner. 1980. Ice blocks melting into a salinity gradient. J. Fluid Mech. 100, 367-384.

Joyce, T. M., W. Zenk, and J. M. Toole. 1978. The anatomy of the Antarctic polar front zone in the Drake Passage. J. Geophys. Res. 83, 6093-6113.

Lambert, R. B., and W. Sturges. 1977. A thermohaline staircase and vertical mixing in the thermocline. Deep-Sea Res. 24, 211-222.

Launder, B. E. 1976. Heat and mass transport. In: Turbulence (P. Bradshaw, Ed.). Springer-Verlag, Berlin, pp. 231-287.

Leibovich, S., and S. Paolucci. 1980. The Langmuir circulation instability as a mixing mechanism in the upper ocean. J. Phys. Oceanogr. 10, 186-207.

Linden, P. F. 1976. The formation and destruction of fine-structure by double-diffusive processes. Deep-Sea Res. 23, 895-908.

Magnell, B. 1976. Salt fingers observed in the Mediterranean outflow region ($34°N$, $11°W$) using a towed sensor. J. Phys. Oceanogr. 6, 511-523.

McDougall, T. J. 1981. Double-diffusive convection with a non-linear equation of state. II. Laboratory experiments and their interpretation. Prog. Oceanogr. 10, 91-121.

McEwan, A. D. 1971. Degeneration of resonantly-excited standing internal gravity waves. J. Fluid Mech. 50, 431-448.

McEwan, A. D. 1973. Interactions between internal gravity waves and their traumatic effect on a continuous stratification. Boundary Layer Meteorol. 5, 159-175.

McEwan, A. D., and R. M. Robinson. 1975. Parametric instability of internal gravity waves. J. Fluid Mech. 67, 667-687.

Niiler, P. P., and E. B. Kraus. 1977. One-dimensional models of the upper ocean. In: Modelling and prediction of the upper layers of the ocean (E. B. Kraus, Ed.). Pergamon, Oxford, pp. 143-172.

Osborn, T. R., and C. S. Cox. 1972. Oceanic fine structure. Geophys. Fluid Dynam. 3, 321-345.

Pollard, R. T., P. B. Rhines, and R. O. R. Y. Thompson. 1973. The deepening of the wind mixed layer. Geophys. Fluid Dynam. 3, 381-404.

Price, J. F., C. N. K. Mooers, and J. C. van Leer. 1978. Observation and simulation of storm-induced mixed layer deepening. J. Phys. Oceanogr. 8, 582-599.

Ruddick, B. R., and J. S. Turner. 1979. The vertical length scale of double-diffusive intrusions. Deep-Sea Res. 26, 903-913.

Schmitt, R. W. 1981. Form of the temperature-salinity relationship in the Central water: evidence for double-diffusive mixing. J. Phys. Oceanogr. 11, 1015-1026.

Schmitt, R. W., and D. L. Evans. 1978. An estimate of the vertical mixing due to salt fingers based on observations in the North Atlantic Central water. J. Geophys. Res. 83, 2913-2920.

Scotti, R. S., and G. M. Corcos. 1969. Measurements on the growth of small disturbances in a stratified shear layer. Radio Sci. 4, 1309-1313.

Sherman, F. S., J. Imberger, and G. M. Corcos. 1978. Turbulence and mixing in stably stratified waters. Ann. Rev. Fluid Mech. 10, 267-288.

Thompson, R. O. R. Y. 1976. Climatological numerical models of the surface mixed layer of the ocean. J. Phys. Oceanogr. 6, 496-503.

Thorpe, S. A. 1973. Experiments on instability and turbulence in a stratified shear flow. J. Fluid Mech. 61, 731-751.

Thorpe, S. A. 1978a. On the slope and breaking of finite amplitude internal gravity waves in a shear form. J. Fluid Mech. 85, 7-31.

Thorpe, S. A. 1978b. The near-surface mixing layer in stable heating conditions. J. Geophys. Res. 83, 2875-2885.

Tranter, D. J., R. R. Parker, and G. R. Cresswell. 1980. Are warm-core eddies unproductive? Nature (London) 284, 540–542.

Turner, J. S. 1973. Buoyancy Effects in Fluids. Cambridge University Press, Cambridge, 367 pp.

Turner, J. S. 1978. Double-diffusive intrusions into a density gradient. J. Geophys. Res. 83, 2887–2901.

Turner, J. S. 1979. Laboratory models of double-diffusive processes in the ocean. Proc. 12th Symposium on Naval Hydrodynamics. National Academy of Science, Washington, D.C., pp. 596–606.

Turner, J. S. 1980. Small-scale mixing processes. In: Evolution of Physical Oceanography (B. Warren and C. Wunsch, Ed.). MIT Press, Cambridge, Mass., pp. 236–262.

Turner, J. S. 1982. The influence of molecular processes on turbulence and mixing in the ocean. In: Turbulence in the Ocean (J. D. Woods, Ed.). Springer-Verlag, Berlin.

Turner, J. S., and L. B. Gustafson. 1978. The flow of hot saline solutions from vents in the sea floor: some implications for exhalative sulfide and other ore deposits. Econ. Geol. 73, 1082–1100.

Voorhis, A. D., D. C. Webb, and R. C. Millard. 1976. Current structure and mixing in the shelf/slope water front south of New England. J. Geophys. Res. 81, 3695–3708.

Williams, A. J. 1975. Images of ocean microstructure. Deep-Sea Res. 22, 811–829.

Williams, G. O. 1976. Repeated profiling of microstructure lenses with a midwater float. J. Phys. Oceanogr. 6, 281–292.

Woods, J. D. 1968. Wave-induced shear instability in the summer thermocline. J. Fluid Mec. 32, 791–800.

Woods, J. D. (Ed.). 1982. Turbulence in the Ocean. Springer-Verlag, Berlin.

Wunsch, C. 1976. Geographical variability of the internal wave field: a search for sources and sinks. J. Phys. Oceanogr. 6, 471–485.

Experiments with Free-Swimming Fish

Frank G. Carey

1 Introduction

Biological oceanography has been strongly oriented toward ecology; its concerns have been energy flow, productivity, and communities of organisms. Probably in the future the simplifications which were necessary to obtain an overall view will give way to more detailed studies of what is actually going on. We need to know something about the characteristic requirements and reactions of individuals in order to understand the assemblages of organisms which live in the sea. Some of this knowledge can be gained from laboratory experiments which allow precise control and repetition; we can also observe and do experiments at sea, where the animal can be studied in its own environment. Harbison (this volume) has told us some of the insights gained by direct examination of zooplankton *in situ*. This examination of organisms in the sea will become an increasingly important aspect of biological oceanography in the future (Kanwisher, 1978).

We have been successful in conducting experiments with free-swimming pelagic fish which we follow using acoustic telemetry techniques. Many of these fish, the tunas, the lamnid sharks, and some of the billfish, are warm and an interest in their body temperature and thermal relationships stimulated us to begin these experiments. While we learned much about temperature effects, it soon became apparent that we were also going to learn about the natural history of these fish and this has become an interesting and rewarding aspect of the work. While I lack the breadth of knowledge, the vision, and even the interest which would allow me to make reasonable predictions of the course of biological oceanography during the next 50 years, I do think that our current research on these fish is a reasonable example of the kinds of *in situ* studies which will be done. By describing some of our results during the past 5 years, I may be giving the flavor of what can be achieved in the future.

2 The Choice of Telemetry

Although large pelagic fish are fascinating and beautiful, they make difficult experimental subjects; but with proper facilities it is possible to maintain them in an aquarium. For some time several species of small tunas have been maintained in a research facility at the NOAA-NMFS Kewalo Basin Laboratory in Honolulu (Nakamura, 1972), and several groups of Japanese researchers have successfully reared tunas. The large public aquaria are interested in displaying tuna and pelagic sharks and will probably be successful in capturing, transporting, and holding a variety of such fish during the next decade. Such efforts will always require a large capital investment, however, and assurance of use by a continuing program to make it worthwhile.

We elected to study these fish in nature and invested our time and effort in developing a telemetry system for three principal reasons: the practical reason was that we could start our experiments without first promoting elaborate and costly facilities which seemed beyond the reach of individual research grants; also, many of the fish that we were most interested in were too big, too fragile, and sometimes too ferocious, to bring ashore and handle with present techniques; and the best reason: that animals show different reactions and behavior in nature than they do in stressful captive or laboratory situations. Kanwisher (1978) has emphasized that one can learn quite different things from free-ranging animals than from captive ones. As the activities and behavior of these fish became more interesting to us, the telemetry methods became more attractive as a method of learning what they do in the sea. Kanwisher's enthusiasm for telemetry from free-ranging animals and his skill with instrumentation were largely responsible for the development of our telemetry experiments.

As a way of studying animals the telemetry technique has some severe limitations. The transmitters do not tell us much about what is going on. Only one, two or a few channels of data are practical in present systems and the transmitters give us a very narrow range of information. The view of these fish which we obtain is not at all comparable to what has been possible for some terrestrial ethologists, who can live with a group of animals and observe every nuance of behavior. We must combine the telemetry data with every scrap of supporting information that we can obtain on board and fill in the gaps by inference. Our telemetry system is technically primitive and there is great room for improvement, but it has given us an exciting new way to study open-ocean animals and has revealed a number of unsuspected phenomena.

A limit to telemetry is size: the animal must be big enough to carry the transmitter. Considering the noise level in the sea and the power available from any conceivable battery, it will never be a technique for plankton. We are fortunate that most of the animals we are interested in are large, and it is relatively easy to devise powerful transmitters which are less than 1% of the body weight.

3 Telemetry Experiments

Our approach has been to plan an experiment for a specific purpose, usually concerned with some temperature relationship, and to see what we can learn about the fishes' activities and reactions in the process. Animals are capable of a bewildering variety of behavior, but we have been fortunate on a number of occasions in finding repetitive, stereotyped activities which we have been able to analyze. This approach, the unexpected behavior which we found, and our attempts to analyze it are illustrated in the following accounts of four recent experiments.

The first set of experiments was designed to examine the normal range of water temperature encountered by swordfish and are described in Carey and Robison (1981). We harpooned transmitters into swordfish as they swam slowly on the surface in the Pacific near the tip of Baja California. Two of the fish which had been found on an inshore bank moved between the 50-fathom curve on the bank during the day and on surface waters offshore at night. Figure 1 shows the depth record for these fish. A clear pattern of activity emerges, with the fish on the surface at night over deep water and near or on the bottom of the bank during the day. Light was an important factor in these movements: the fish left the surface at first light of dawn, about an hour before sunrise, and were back up on the surface about an hour after sunset. We think that this may be a feeding pattern, with the swordfish pursuing demersal fish on the bank during the day and feeding on squid on the surface at night, but we have no direct information on what they were eating.

We also followed several larger swordfish which were tagged in an area further offshore in the Pacific. These fish swam slowly west and at sunset were near the San Jose Canyon, a major submarine canyon at the top of Baja California. At sunset they turned and swam south along the axis of the canyon. As usual, the fish were on the surface at night. Their turning and swimming over the axis of the canyon 500 m below may only have been a coincidence. However, fishermen look for canyons and rugged bottom terrain as good places to catch swordfish, and some phenomena may have attracted the swordfish to the surface waters over the canyon. Marshall Orr at the Woods Hole Oceanographic Institution has shown that when coastal currents flow over submarine canyons, flow separation and turbulence may occur on a large scale. These flow patterns result in aggregations of biological sound scattering particles over the canyon which can be seen in high-resolution sonograms. Such aggregations might attract swordfish to the surface waters over the canyons far below. It will be interesting to include equipment for detecting acoustic scattering layers in future tracking experiments.

An oxygen minimum layer is present in the sea south of Baja California and

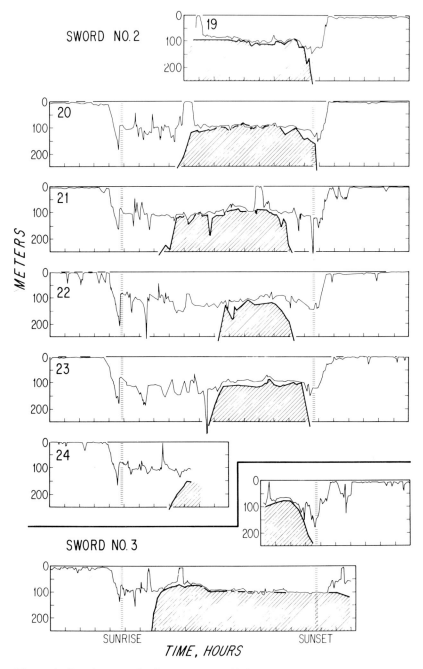

Figure 1. Depth records for two swordfish near the tip of Baja California, Mexico. The cross-hatched area indicates a bank. In this region the water is depleted in oxygen at depths below 70 to 100 m. The fish show a clear pattern of swimming on the surface at night and deep during the day. A dive just before sunrise each day carried the fish below their usual daytime depth.

Experiments with Free-Swimming Fish

the usual daytime depth of the swordfish was in the region where oxygen concentrations were only about 10% of those on the surface. We believe that "the morning dive" phenomenon, which can be seen in Figure 1, may be explained in terms of the oxygen minimum. (At dawn every day the swordfish swam down to a depth somewhat greater than their usual daytime depth, then came back up a short distance.) We think that in responding to increasing light by leaving the surface, the fish descended into a region where the oxygen level was uncomfortably low, then returned to a depth where the oxygen level was more tolerable. The poorly oxygenated water may also account for the fishes' coming to the surface during the day. They may have been basking in the well-aerated surface waters to recover from an oxygen debt incurred at depth. The fish which we followed offshore into the Pacific where the oxygen minimum layer is well developed came to the surface more frequently than the ones followed inshore near the Gulf of California where it is less well developed. A rough correlation exists between time spent below the thermocline in poorly oxygenated water and the succeeding time spent on the surface, as would be expected if this were a recovery from anoxia. Swordfish bask on the surface in other areas where there is no oxygen minimum, so there are clearly other reasons for this behavior; we will see if factors such as thermal stress from cold bottom waters prompt a return to the surface to rewarm.

We ran our second experiment in the Atlantic near Cape Hatteras with a fish taken on long-line fishing gear. The depth record for this experiment (Fig. 2) shows a clear response to light, particularly on the third day when the U-shaped record suggests that the fish was following an isolume. The most rapid vertical movements were at dawn and dusk, when natural illumination changes by six orders of magnitude in 2 hours. The greatest depth, 617 m, was reached at midday, the time of greatest surface illumination. This was an area of warm, clear Sargasso Sea and Gulf Stream water. If one assumes a low light attenuation factor of 0.028 for this water (Clarke and Kelley, 1964), the observed depth changes would have kept the swordfish at a light level which varied by less than a factor of two during the day. It is likely that the fish was adjusting its depth to an isolume.

The previous two days of the depth record in Figure 2 are not symmetrical but with the following considerations these records can still be taken as evidence that the swordfish was following constant light level in its vertical movements. On the second day the depth record is skewed, with the greatest depth occurring late in the day, rather than at noon. At this time, however, the deepening isotherms show that the fish was moving from Slope Water into the Gulf Stream, where greater water clarity may have required a greater depth to maintain the same light level. On the first day the swordfish was late in descending and came up in the middle of the day. We have no explanation for the late descent, except that the fish may have been late in going to work because of the effects of capture the night before. However, the rise toward the surface at midday can be interpreted as maintaining a constant light level, because it occurred at a time when the swordfish swam under the cold layer of the dark green shelf water which shows as a thermal inversion in Figure 2. To maintain the same light level

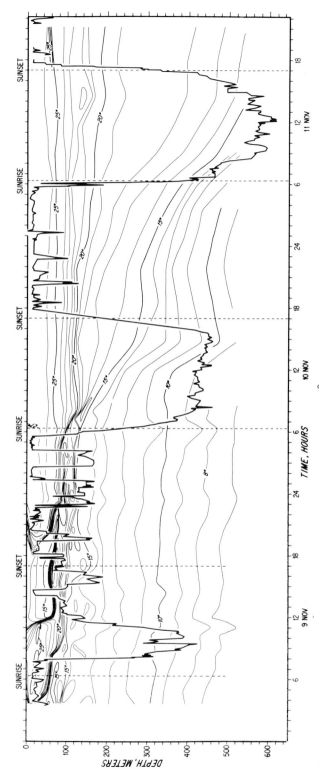

Figure 2. Depth record for a swordfish superimposed on a pattern of 1°C isotherms. The experiment took place in the Atlantic, northeast of Cape Hatteras. This fish also swam near the surface at night and deep during the day. The deepening isotherms indicate movement of the swordfish from the complex shelf and Slope Water into clear Gulf Stream water. In the clear water movements of the fish indicate that it was probably changing depth to remain at a constant level of illumination.

when it swam into this shadow, the fish would have to rise toward the surface. Future experiments with light and vertical migration in the swordfish are suggested. The study of vertical migration here offers special advantages over other systems in that it is possible to follow one positively identified animal over a 24-hour cycle and obtain a detailed minute by minute record of its depth. The swordfish is also large enough to carry a lightmeter and telemeter back to us the level of illumination that it is experiencing.

The original object of the swordfish experiments was to find their daily temperature range. In most of the experiments the fish readily passed through the thermocline and experienced large temperature changes in their daily vertical migrations. On the second day in Figure 2 the swordfish moved from surface water at $27°$ to $8°$ at 400 m, a $19°$ change in a few hours. This is a large temperature excursion for any fish to undergo and still remain active.

The thermocline can be an important factor in influencing fish movements, and other species (and perhaps swordfish in other situations) show a distinct preference for locating themselves on the thermocline. This was particularly apparent for some large bluefin tuna which were traveling out over the Continental Shelf from a bay on the Nova Scotia coast where they had been tagged for our third experiment. These fish spent most of the time swimming along the top of the steep temperature gradient which occurs at a depth of 10 to 30 m in those waters in the summertime (Carey and Lawson, 1973).

Similar behavior was seen in a large white shark which we followed for several days on the Continental Shelf south of Long Island, New York (Carey et al., 1982; Fig. 3). This fish swam in the steepest part of the thermal gradient during most of the several days that we followed it. While the thermocline might seem a good position for behavioral thermoregulation, the shark was not controlling its temperature. It changed both its thermal environment and its muscle temperature, which was $5°C$ warmer than the water. It appeared to be seeking the steepest part of the thermal gradient, rather than any absolute temperature. The thermocline is a major feature in what to us is a rather featureless environment and it would be surprising if fish did not react to it under some circumstances. Both the white shark and most of the bluefin tuna which we followed on the Continental Shelf moved at a relatively constant course and speed for a period of days as if they were heading to some goal. Possibly the thermocline made a useful navigational feature when traveling. The fish were not bound to the thermocline, however, for feeding bluefin in the same area ignored it and passed rapidly through it in frequent excursions between surface and bottom.

We have seen another interesting pattern of vertical movements in several shark experiments. These were designed to monitor the effect of changing water temperature on the body temperature of a warm shark, the mako (Carey et al., 1981), and to compare this with what happened in a cold fish, the blue shark. The animals cooperated in a most satisfactory fashion. The mako shark, a large female we followed for several days in the area north of the Bahamas, swam up and down between the surface and a depth of 500 m. These excursions were made on a regular 2- to 3-hour period, which took the shark through the thermo-

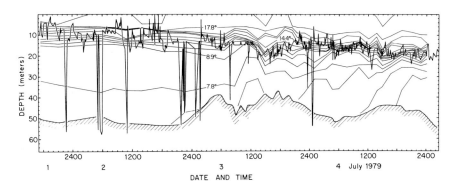

Figure 3. Depth pattern for a 1-ton white shark on the Continental Shelf south of Long Island, New York. During the first 2 days the shark remained in one area, feeding on a dead whale, then swam off on a southwest course for the remainder of the experiment. Horizontal lines are 1.1°C (2°F) isotherms which indicate a well-developed thermocline. The shark located itself in this temperature gradient throughout the experiment. A 10% error in the echosounder or in the depth telemetry incorrectly shows the shark at greater than bottom depth on a number of occasions.

cline. In 1979 we found similar behavior in two blue sharks which we followed in the Slope Water off the northeast coast of the United States. The blue sharks also made regular vertical movements with a 2- to 3-hour period, swimming through the thermocline between the surface and 250 m. This behavior, which brought the sharks through frequent large changes in water temperature, was ideal for our temperature experiments. The deep body temperature of the mako remained quite constant during changes in water temperature with a period of several hours (Fig. 4). This was expected, as the tissues are thermally isolated by a set of vascular heat exchangers which greatly reduce heat transfer by the circulation of the blood. The body temperature of the blue shark, which lacks these heat exchangers, followed the rapid changes in water temperature quite closely (Fig. 4). In this fish, convection by the blood is an important mode of heat transfer. This set of experiments provides a nice demonstration of the importance of the heat exchange system in isolating the animal from ambient temperature change.

The pattern of vertical movement which we saw in these sharks was quite unexpected, but we have since thought of two reasons why the sharks might do this. First, the vertical movements might be a hunting pattern which would be suitable for an animal which used olfaction to detect prey. With the increasing sensitivity and resolution of oceanographic instruments we have learned that the ocean is full of finestructure. (Dr. Turner, in the previous chapter, has discussed how some of these temperature and salinity gradients may be generated.) Current flow will be along these density interfaces and odor or taste emanating from an organism will tend to spread horizontally rather than diffuse spherically. The

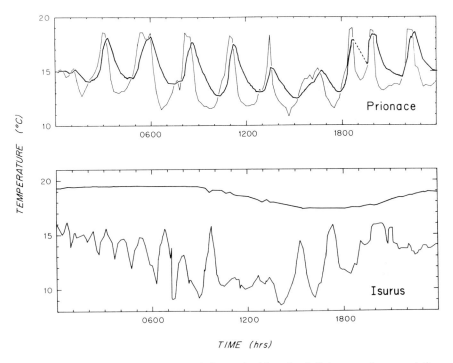

Figure 4. Body temperature records from the blue shark *Prionace glauca* and the mako *Isurus oxyrinchus*. Both of these sharks swam up and down through a range of several hundred meters in regular excursions and passed through the thermocline without hesitation. Temperature of the blue shark, which is a normal, poikilothermic fish, showed a rapid response to changes in water temperature. The warm-bodied mako was thermally isolated from rapid temperature change and followed only the slow changes in average water temperature.

odor trace will thus have a much larger cross-section in the horizontal plane than in the vertical. The sharks would encounter such traces more readily by swimming up and down than if they swam at one depth. We do not know if the mako shark was feeding, but the blue sharks probably were. Other blue sharks captured in the same time and place as the tagged ones had food in their stomachs. This usually consisted of squid and, more interestingly, of the deep water octopod, *Alloposis mollis*, which is usually found below 250 m. It seems that the vertical motions we observed were, at least in part, a hunting pattern.

Second, the vertical movements may be an energy-conserving mode of progression. Daniel Weihs, an Israeli aeronautical engineer with an interest in fish hydrodynamics, proposed that fish which were denser than water could save energy if, rather than swimming continuously at one level, they progressed by alternately swimming up and gliding down (Weihs, 1973). The effect is similar to that achieved by passerine birds which flutter up and glide down as they fly,

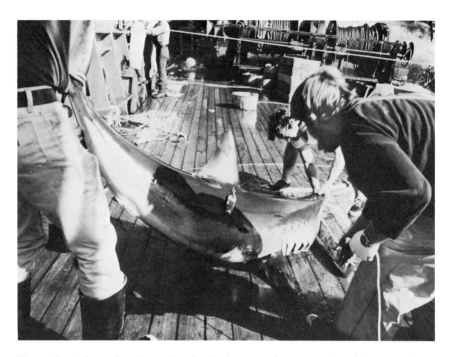

Figure 5. A large female mako shark about to be returned to the water in an acoustic telemetry experiment. The shark was lifted aboard with a rope around its tail and a seawater hose placed in its mouth to irrigate the gills and keep the fish quiet. A temperature transmitter followed by five mackerel were pushed into the stomach for an experiment in which we hoped to determine the effect of food on stomach temperature. A depth transmitter can be seen attached just below the dorsal fin. This shark was followed for a 5-day period during which it swam from near Cape Canaveral to a position northeast of the Bahamas. It frequently moved between the surface and 500 m depth and its stomach temperature was regularly 8°C warmer than the water. (Photo by Nancy Kohler, NOAA/NMFS, Narrangansett, R.I.)

rather than flap continuously. Apparently the drag while flapping or swimming is several times greater than the drag while gliding. The glide angles assumed by the blue sharks varied from 4:1 to 8:1 and are in an appropriate range for efficient progression by Weihs' model.

We are able to provide evidence that the blue sharks were swimming up and gliding down from data in Figure 4. Notice that the response of muscle temperature to changing water temperature is more rapid when the shark is swimming up into warm water than when gliding down into the cold. Heat transfer between the shark and its environment is almost entirely through convection by the blood. Using standard values for thermal conductivity of animal tissues we calculated that heat transfer by conduction accounts for less than one-twentieth of the heat flow at the site of measurement. Since convection by the circulatory system seems to be the important mechanism for changing body temperature, we can turn our argument around and use the rate of temperature change in the tissue to calculate blood flow. Doing this we get an average of 68 ml min^{-1}kg^{-1} when the shark is swimming up and 28 ml min^{-1}kg^{-1} when it is gliding down. These are reasonable values for blood flow in fish muscle and are different enough to indicate that the fish was working considerably harder going up than coming down. Weihs' prediction is probably correct for these sharks although we would like to check this with a device that telemetered tail beat, providing direct information on when the shark was swimming or gliding.

The vertical movements we observed in the two species of sharks are a good example of the unexpected phenomena so frequently found in these telemetry experiments and for which we can sometimes provide an interpretation or rationale.

4 Conclusion

The telemetry technique makes it possible to learn something about difficult animals in nature where we can study behavior and reactions which would not be available in a captive situation. It can be applied to suitable animals with relatively simple equipment and small initial investment (Fig. 5). Telemetry will benefit from the current rapid improvements in electronics technology. The next few decades will proably see an increase in the use of radio and acoustic techniques to study the physiology and behavior of marine animals. Beyond that, a great deal is possible, but it remains to be seen how interesting and worthwhile such research will be perceived to be by society. For those of us involved, the reward of learning more about the animals is sufficient justification. It seems likely that continued progress in this line of research will depend on the chance that a phenomenon of broad scientific or economic interest will be revealed during our pursuit of more specialized knowledge.

References

Carey, F. G., and K. D. Lawson. 1973. Temperature regulation in free-swimming bluefin tuna. Comp. Biochem. Physiol. 44A, 375-392.
Carey, F. G., and B. H. Robison. 1981. Daily patterns of activities of swordfish, *Xiphias gladius,* observed by acoustic telemetry. Fishery Bull. 9, 277-292.
Carey, F. G., J. M. Teal, and J. W. Kanwisher. 1981. Visceral temperatures of mackerel sharks. Physiol. Zool. 54, 334-344.
Carey, F. G., J. W. Kanwisher, O. Brazier, G. Gabrielson, J. G. Casey, and H. L. Pratt. 1982. Thermal ecology of the white shark, *Carcharodon Carcharias.* Copeia 1982 (2) 254-260.
Clark, G. L., and M. G. Kelley. 1964. Variations in transparency and in bio-

luminescence on longitudinal transects in the Western Indian Ocean. Bull. Inst. Oceanogr. Monaco 64, 1319.

Kanwisher, J. W. 1978. Monitoring free-ranging animals. Technol. Rev. June/July, 32–39.

Nakamura, E. L. 1972. Development and uses of facilities for studying tuna behavior. In: Behavior of Marine Animals, Vol. 2 (W. E. Winn and V. L. Olla, Eds.). Plenum Press, New York, pp. 245–277.

Weihs, D. 1973. Mechanically efficient swimming techniques for fish with negative buoyancy. J. Mar. Res. 31, 194–209.

Coastal Dynamics, Mixing, and Fronts

Christopher Garrett

> I have to say that in earlier years it was difficult to get physical oceanographers involved in shelf edge and coastal problems. Some were descriptive and trivial from a purely physical oceanographic standpoint, the sort that biologists could and should handle. Others were so intractable that studies of the open sea were more rewarding. However, that has now changed.
> —Gordon A. Riley (1979)

1 Introduction

Gordon Riley's comment, taken from his address at a 1979 workshop on the oceanography of the Gulf of Maine and adjacent seas, draws attention to the important trend toward more comprehensive and quantitative studies of the physical oceanography of shelf seas. Process-oriented physical oceanographers are finding that even the "descriptive and trivial" problems can be a source of inspiration and motivation, and that the "intractable" problems are beginning to yield better defined questions, if not answers. Also, perhaps, deep-sea dynamical oceanography has grown somewhat more complex over the years, so that shelf oceanography no longer seems so complicated by comparison.

Oceanographic research on the continental shelves is, of course, strongly motivated by practical problems of resource exploitation and management, navigation, and pollution. In many instances even a very rough answer from the physical oceanographer is adequate, or at any rate better than those obtainable on the biological or economic aspects of a problem, but in other cases the answers to some problem of environmental impact are severely limited by our ignorance of some physical process. I shall thus take it for granted that one would like to understand, and parameterize as well as possible, the factors affecting water properties, circulation, and mixing on the continental shelves.

Space does not permit a discussion of all of the consequences that can arise from the combined effects of river discharge, atmospheric and tidal forcing, bottom topography, and forcing by the circulation off the continental shelf. Instead, to provide some focus for a discussion of motivation, recent progress and future possibilities in some areas of shelf studies, I shall base my discussion on some of the questions that a physical oceanographer might ask (or be asked) about Georges Bank, an important area of the continental shelf close to the site of this symposium.

2 Georges Bank

At a Woods Hole Oceanographic Institution symposium on "The Next Fifty Years in Oceanography" it seems fitting to introduce my theme region with the classic picture of the summertime circulation of the Gulf of Maine (Fig. 1), published just over 50 years ago by Henry Bigelow, the first Director of Woods

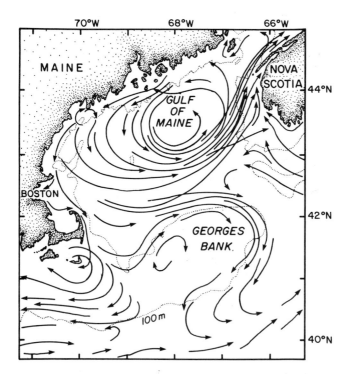

Figure 1. Schematic of the mean surface circulation in the Gulf of Maine in July and August (redrawn from Bigelow, 1927).

Hole Oceanographic Institution. It is only in the last few years that this schematic, which was based mainly on hydrographic data and observations of surface drift, has been fleshed out and that quantitative dynamical theories have been developed for such features as the clockwise gyre around shallow Georges Bank.

The present spate of scientific work on Georges Bank seems likely to continue. The bank is the site of an extremely productive commercial fishery (and, unfortunately, a minor international fish war) with the landed value of the catch approaching $200 million annually (Cohen et al., 1979) and is estimated to have oil and gas reserves worth about $600 million (Finn, 1980). Pressures to exploit all these resources in a compatible way make Georges Bank typical of shelf seas everywhere.

This area also provides an example of the complexity of circulation and mixing on continental shelves, as partly illustrated in Figure 2, an infrared image of the Gulf of Maine in late summer. The complicated patterns of surface temperature result from the effects of solar radiation, air-sea interaction, vertical and horizontal mixing, and eddying motions induced either locally or by the remote influence of the Shelf/Slope Water front and Gulf Stream to the south. Georges

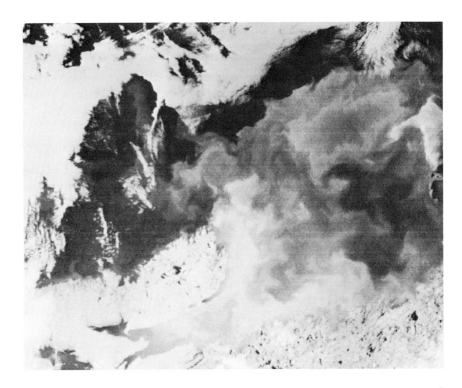

Figure 2. Infrared image of the Gulf of Maine on September 5, 1978, as observed from NOAA-5. The very light areas are land or cloud, the dark areas show warm water and the light grey areas show colder water. (Photograph courtesy of Atmospheric Environment Service, Canada.)

Bank itself is delineated in Figure 2 by the light ring of cold water surrounding a darker, warmer central patch.

For details of the oceanography of Georges Bank the reader is referred to the treatise by Bigelow (1927) and reviews by two other grand old men of Georges Bank, Dean Bumpus and Gordon Riley (Bumpus, 1976; Riley, 1980). There is also a spate of recent papers on different aspects of the bank; some will be referred to later in this paper.

Reasons for the high biological productivity of the bank are discussed in detail by Riley (1980) and Cohen et al. (1979). The basic hypothesis is that in the vertically well-mixed region over the top of the bank regenerated nutrients are made available to the phytoplankton in the euphotic zone. Retention of plankton, fish eggs, and larvae in the gyre over the bank and input of nutrients from deeper waters in the Gulf of Maine and offshore are also alleged to be important (though the compatibility of these two requirements is not clear).

There are clearly a number of aspects of the physical oceanography of the bank which are relevant to an improved understanding of the biology, or to a prediction of the environmental impact of events such as oil spills. The questions which I have in mind as a motivation for a more general discussion of shelf processes are:

1. Why is the water over the top of the bank vertically well mixed all year, and what are the vertical mixing rates in the well-mixed and stratified regions?
2. What happens at the front between well-mixed and stratified regions?
3. What drives the clockwise circulation round the bank?
4. What are the horizontal mixing rates on the bank and the exchange processes with water off the bank?

3 Vertical Mixing

Shallow areas of shelf seas generally become vertically well mixed in winter due to convective cooling, but in spring they tend to become stratified due to surface warming, except in areas where the wind or tidal currents are strong enough to maintain vertically mixed conditions. This generally leads to cool surface water, as shown by the light areas on Nantucket Shoals, around Georges Bank, off southwestern Nova Scotia, and in the Bay of Fundy in Figure 2, though in some locations (such as the center of Georges Bank and the upper reaches of the Bay of Fundy) the water is shallow enough that it becomes quite warm by September even though well mixed vertically.

Bigelow (1927) recognized the extent of well-mixed areas of the Gulf of Maine and attributed their existence to the effects of tidal stirring, as did Dietrich (1951) for similar areas in the English Channel. However, it is only recently that a simple quantitative criterion for the separation of well-mixed and stratified regions has been proposed. On the basis of an energy argument Simpson and Hunter (1974) suggested that the transition should occur at a critical value of

BH/U^3, where B is the surface buoyancy flux due to solar heating, H is the ocean water depth, and U the amplitude of the tidal current. In fact the importance of this parameter can be demonstrated purely as a consequence of dimensional analysis (Garrett et al., 1978); any deviation depends on the neglect of such effects as freshwater input and advection.

For a given shelf sea the heat flux, and hence B, is reasonably uniform in space, so fronts between well-mixed and stratified regions should occur at a critical value of H/U^3. Simpson and his colleagues (Simpson et al., 1977) and Pingree and Griffiths (1978) have applied this idea with great success around the British Isles and used it as a starting point for a number of studies.

This H/U^3 idea has since been applied in other parts of the world (including some where freshwater input or advection renders it inapplicable!) In the Bay of Fundy, Garrett et al. (1978) found that a summertime transition from well-mixed to stratified conditions occurs for log (depth/tidal dissipation) $\cong 1.9$ (Fig. 3). The equivalent value of H/U^3 is 70 m^{-2}s^3, corresponding very closely to that found around the British Isles.

Application to the Gulf of Maine of this value from the Bay of Fundy (Fig. 4) is consistent with the results of Bigelow (1927) and later authors, although the resolution of the tidal model used to produce Figure 4 is not adequate for a very detailed comparison. However, we can at least say that the well-mixed conditions over Georges Bank are consistent with the Simpson and Hunter H/U^3 criterion.

The success of this criterion shows the value of simple energetic or dimensional arguments, which can be very useful for some organizational or predictive purposes. However, the critical value of BH/U^3 (which corresponds to a conversion from kinetic to potential energy at an efficiency of less than 0.3%) is at present an empirically determined number. To explain its value, and for many other purposes, a more detailed understanding of vertical mixing is required.

Most studies of vertical mixing in shallow seas have been conducted in estuaries. Vertical fluxes of momentum and heat are usually modeled in terms of an eddy viscosity and eddy diffusivity, the values of which are again determined by matching observations to predictions obtained using formulas based on dimensional arguments, rather than on a full understanding of the physics. There is a great multiplicity of formulas, and notation, in the literature. A typical formula for the vertical eddy diffusivity K_v of heat or salt, averaged over a tidal cycle, is that of Kent and Pritchard (1959):

$$K_v = 5.5 \times 10^{-3} UH[z^2(H-z)^2 H^{-4}](1 + 0.276\text{Ri})^{-2} \qquad (1)$$

in which it is assumed that the mixing coefficient is basically proportional to the amplitude U of the tidal current and depth H of the water, but with a parabolic profile between the bottom at $z = 0$ and surface at $z = H$. Other terms can be added to represent mixing due to wind or waves, and some authors choose formulas to use at any phase of the tide, rather than for conditions averaged over a tidal cycle.

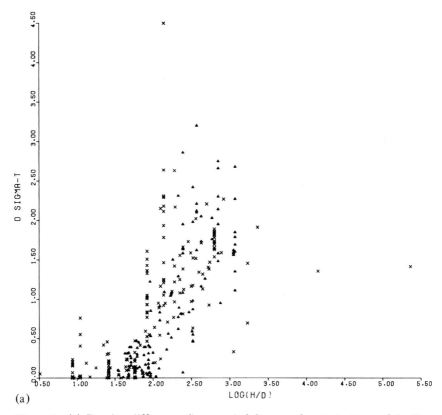

Figure 3. (a) Density difference (in σ_t units) from surface to bottom of the Bay of Fundy in July and August plotted against log (H/D), where H is the water depth in meters and D the tidal dissipation in W m^{-2} (from Garrett et al., 1978). (b) Location of stations for data displayed in Figure 3(a) (from Garrett et al., 1978).

The reduction of vertical mixing by stability of the water column is allowed for in Equation (1) by the dependence of K_v on the Richardson number, Ri. This is defined as Ri = $N^2/(\partial u/\partial z)^2$ where $N^2 = -(g/\rho)(\partial \rho/\partial z)$ and $\partial u/\partial z$ is the local shear, which Kent and Pritchard (1959) took as $(2/\pi)(U/H)$. There are very substantial differences in the Richardson number dependence of K_v chosen by different authors (Fig. 5), over and above differences in the diffusivity K_0 at Ri = 0 and in the choice between modeling the time development of $\partial u/\partial z$ or using an approximate formula.

The formulas of Kent and Pritchard (1959) and Blumberg (1977) were devised for use in estuaries, whereas that of Munk and Anderson (1948) came from a study of the oceanic mixed layer. Of particular interest in the context of this paper is the work of James (1977), who modeled the evolution of the ver-

Coastal Dynamics, Mixing, and Fronts

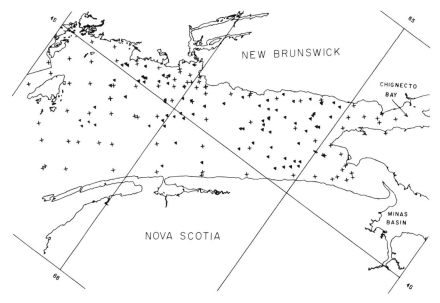

Figure 3. (b)

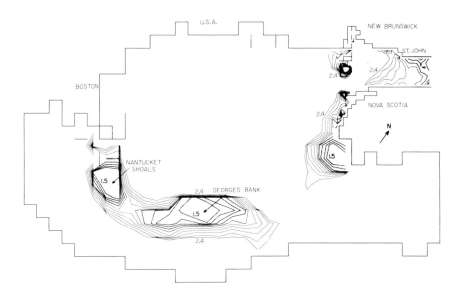

Figure 4. Contours of log (H/D) for the Gulf of Maine, with the depth H in meters and the tidal dissipation W m^{-2}. Only contours between 1.5 and 2.4 are shown; hydrographic data from the Bay of Fundy suggest that the areas with log $(H/D) < 1.9$ should remain well mixed throughout the summer (from Garrett et al., 1978).

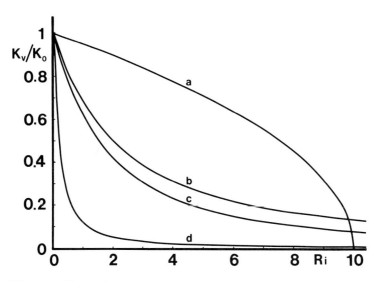

Figure 5. The reduction in the vertical eddy diffusivity K_v as a function of Richardson number Ri according to various authors. (a) $(1 - 0.1 \text{ Ri})^{1/2}$, Blumberg (1977); (b) $(1 + 0.3 \text{ Ri})^{-3/2}$, James (1977); (c) $(1 + 0.276 \text{ Ri})^{-2}$, Kent and Pritchard (1959); and (d) $(1 + 3.33 \text{ Ri})^{-3/2}$, Munk and Anderson (1948).

tical temperature structure in the Irish Sea in response to solar heating and mixing by winds and tides. He adopted the basic formula (common to many authors)

$$K_v = K_0 (1 + \sigma \text{Ri})^{-p}, \qquad (2)$$

and investigated the effects of varying σ and p. The reader is referred to his paper for details of K_0 and his approximation to $\partial u/\partial z$, which was not modeled. The important results of his work are that $\sigma = 0.3$ gave a transition from well-mixed to stratified conditions that compared well with observations (and hence with the H/U^3 criterion), whereas $\sigma = 0.1$ or 0.5 gave too mixed or too stratified a solution. He also found $p = 1.5$ to be appropriate, with $p = 1$ or 2 giving, respectively, too diffuse or too sharp a thermocline in the stratified region.

Similar formulas have been used for the vertical eddy viscosity A_v, which is required, not just for calculating the mean flow, but also if Ri is to be based on a calculation, rather than an estimate, of $\partial u/\partial z$. It is usually assumed that $A_0 = K_0$, and that $K_v/A_v < \text{Ri}^{-1}$, as is necessary on energetic grounds if the mixing is achieved by local conversion of kinetic to potential energy (Turner, 1973). Again there is a multiplicity of formulas, with parameters adjusted to give reasonable agreement between prediction and observation.

One wonders whether the basic assumptions, involving eddy diffusion and Richardson number dependence, are appropriate. More complicated models, which involve calculating the evolution in time and space of turbulent kinetic energy and a turbulent length scale, and relating mixing coefficients to these, are still being developed (Rodi, 1980; Blumberg and Mellor, 1980), but closure still requires a number of *ad hoc* assumptions as well as the determination of empirical constants from laboratory boundary layer data. The uncertainty in all these formulations seems to increase with increasing Richardson number, so that vertical mixing rates in highly stratified conditions are very poorly known.

In summary, while the vertical mixing rate due to the tides in the well-mixed region on Georges Bank is reasonably well established, the appropriate rates, parametric dependence, and physical processes in the stratified waters off the bank are still far from known. This ignorance not only limits our ability to predict vertical profiles of temperature and current but also, as we shall see, severely limits our investigation of other processes, such as those occurring at fronts.

One apparent shortcoming of shelf studies to date is the lack of direct measurements of eddy fluxes of heat of momentum; most "verification" of mixing formulas comes from comparing predicted and observed mean conditions. In the deep ocean direct measurement of temperature and velocity microstructure (see Gregg and Briscoe, 1979 for review) and direct measurement of Reynolds stresses in stratified conditions (Ruddick and Joyce, 1979) have shed considerable light on mixing rates. Although they are technically difficult, one hopes that these approaches will be adapted to Continental Shelf conditions.

4 Frontal Processes

The identification of H/U^3 as an important parameter in determining the extent of mixing of coastal waters in tidally energetic regions has led to increased interest in the frontal region between well-mixed and stratified conditions. To give one example, Pingree et al. (1975) reported much higher primary productivity in the frontal region than on the nutrient-limited stratified side of the front or in the light-limited well-mixed region. The explanation they proposed was that as the tides changed from springs to neaps (or after a storm), the front would advance into the mixed region and a phytoplankton bloom could occur in the freshly stratified nutrient-rich water.

However, James (1977) found, using the model referred to above, that a front should not in fact move much from springs to neaps due to the Richardson number dependence of vertical mixing. Water that becomes stratified at neap tides does not become mixed at spring tides. Observational support for the small movement of one particular front relative to the water was found by Simpson and Bowers (1979) from examination of satellite infrared imagery. Possibly the high productivity at a front is associated not so much with spring-neap variations as with a general cross-frontal exchange.

A calculation of the ageostrophic cross-frontal flow to be expected depends rather critically on the parameterization of vertical momentum transfer. James (1978) predicted the flow for a particular choice of eddy viscosity and its Richardson number dependence, using a numerical model, and Loder (1980a) applied a semianalytical model to Georges Bank. Both authors took a prescribed density field, rather than considering the full interrelationship of the density field and cross-frontal flow. Further work is obviously required, though the basic nature of the circulation is probably as shown in Figure 6. Viscous forces permit an ageostrophic cross-frontal flow, consistent with the flow of light water over denser water.

These models are for the flow averaged over a tidal cycle and may mask important horizontal fluxes, across the mean frontal position, that occur as the front moves to and fro with the tides, as suggested by Allen et al. (1980). Frontal models may have to be developed in some semi-Lagrangian frame, moving with the tides.

Cross-frontal flow can also occur at lower frequencies. Pingree (1979) has documented the baroclinic eddies that occur at some fronts on the continental shelf, and they are sometimes observable in infrared images of the front on the north side of Georges Bank (Fig. 2). One suspects that these eddies arise as a result of baroclinic instability of an along-frontal jet, although this flow is itself far from certain. Bigelow (1927) and later authors have suggested that the flow along the north side of Georges Bank (Fig. 1) is the geostrophic response, modified by turbulent exchanges, to horizontal density gradients, but even a geostrophic prediction depends on a knowledge of the depth of no motion, or surface slope. Loder (1980a), in the model referred to above, assumed that this is determined by a condition of zero net mass flux onto the bank and found that for a surface jet to exist the water on the bank must be lighter, as well as more mixed, than that off the bank. He pointed out that this is consistent with observations in late summer or fall and could be a consequence of the summer heat flux into water of variable depth.

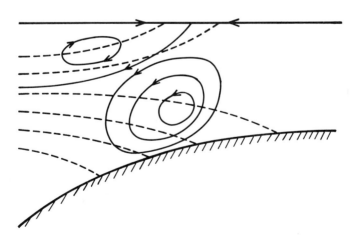

Figure 6. Schematic of the circulation in the plane normal to a front on one side of which the water is well mixed and also has a slightly lower density than the average over the same depth on the other side of the front. (---) Isopycnals, (—) Streamlines.

In summary, the mean flow and variability at fronts in coastal seas have profound implications for cross-frontal exchange, budgets, and biological processes. Many of the physical processes are understood qualitatively. A more quantitative understanding obviously depends on further observations, but also, for modeling efforts, on improved parameterization of turbulent fluxes of mass and momentum.

5 Tidal Rectification

As mentioned above, the summertime anticyclonic circulation around much of Georges Bank (Fig. 1), first documented by Bigelow (1927), may be a consequence of the horizontal density gradient induced by variable mixing. However, there are indications (Scarlet et al., 1979) that this circulation persists for much of the year, and that in some locations the mean flow is modulated on the time scales of variation of the semidiurnal tide, with variations that correlate with the variations in amplitude of the tide (Spiegel and Magnell, 1979).

This suggests that the clockwise gyre may, to some extent, be generated by rectification of the strong tidal currents. The basic theory for such a situation is due to Huthnance (1973) and has been extended in a variety of ways by Loder (1980b) for application to Georges Bank. From a mathematical point of view, a mean flow on the sloping sides of a bank is driven by horizontal gradients of the Reynolds stress, which varies due to the effects of the Coriolis force and bottom friction on the oscillatory tidal currents.

The basic physics, for the type of rectification that seems to be occurring on the sides of Georges Bank, is illustrated in Figure 7. On-off bank continuity leads to a larger x component of the velocity in the shallow region than in the deep region, hence a larger y component through the action of the Coriolis force, and a larger tidal ellipse. Assuming, for the moment, no mean Eulerian current, a

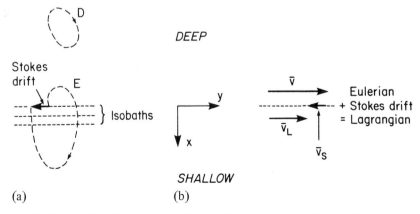

Figure 7. Schematic of the generation of (a) a Stokes drift and (b) a mean Eulerian flow due to cross-isobath tidal excursions (from Loder 1980b).

given water parcel moves in the negative y direction over the course of a tidal cycle, the average speed being the Stokes drift. Now the bottom friction on this water parcel does not average to zero over a tidal cycle, due to the larger velocity and larger bottom friction in the shallow region. To balance the forces we must add to the Stokes drift a (larger) mean Eulerian velocity. The net result is mean Eulerian and Lagrangian flows, with the former being the greater.

A key parameter in a mathematical theory of this process is the ratio of the cross-isobath tidal excursion L_e to the scale L of significant variations in water depths. Unless $L_e/L \ll 1$ one has to allow for the interaction between the tide and the mean flow which it generates (Loder, 1980b). For the north side of Georges Bank L_e and L are comparable; the predicted mean flow, assuming uniformity along the bank, is shown in Figure 8.

Loder's (1980b) theory applies only to the depth-integrated flow. The vertical structure of the flow, and particularly the important on-off bank circulation, is still uncertain even for unstratified conditions. Loder's (1980b) preliminary qualitative estimate is shown in Figure 8, but much remains to be done and the answers are probably rather sensitive to the parameterization of vertical mixing of momentum.

However, the important point about the predictions for the vertically integrated flow are that they are based on well-defined dynamical calculations, with a minimum of uncertain parameterization. The assumptions of along-bank uniformity and linear bottom friction could easily be relaxed in a numerical model, and work by Zimmerman (1978) and Loder (1980a) shows that it would not be necessary to have a resolution of the topography (i.e., a grid scale) much less than the tidal excursion.

6 Horizontal Mixing

A knowledge of horizontal diffusion rates on Georges Bank might be required for any estimate of the spread of a patch of plankton, fish larvae, or spilt oil. Unfortunately there are not really any simple, general formulas into which one can plug "external" variables such as details of the water depth, wind, and tidal currents.

Bumpus (1976) presents some interesting evidence for horizontal diffusion on Georges Bank, inferred from the spread of herring larvae, and a quantitative assessment by E G & G (1979) of the movement of drogued buoys provides estimates of horizontal diffusivities between a few $\times\ 10^2$ and over 10^3 m^2 s^{-1} for patches with a length scale of about 50 km.

An approximate indirect estimate of horizontal diffusion can be obtained from an examination of the horizontal temperature gradients that build up on Georges Bank in the course of the summer. As shown in Figure 2, the water over the shallow central area of the bank becomes warmer by September than the deeper, but still vertically well-mixed, waters nearby. However, Loder (1980a) has estimated that the heat content of the central water is only about half what

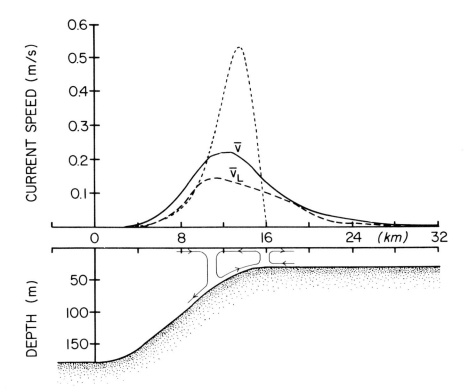

Figure 8. Predicted mean currents along the northwestern side of Georges Bank, and a schematic of the expected cross-isobath flow. (——) Eulerian mean current, (---) Lagrangian mean current, (···) Eulerian mean current ignoring the interaction between the tide and the mean flow (from Loder, 1980b).

it would be in response to the surface heat flux and finds that the heat loss could be accounted for by horizontal diffusion with a diffusivity of a few × 10^2 m^2 s^{-1} (Loder et al., 1982).

What are the mixing processes that give these large values of horizontal diffusion? One suspects that the strong tidal currents play some role, and the regular nature of tidal forcing makes tidal mixing one of the easier topics to investigate. Small-scale turbulence, which causes vertical mixing with a diffusivity of order 10^{-3} UH in water of depth H with a tidal current of amplitude U, presumably causes horizontal mixing at a comparable rate. This is very small for $U \cong 1$ m s^{-1} and $H \cong 30$ m. Of greater importance is the so-called shear dispersion in which a combination of vertical mixing and vertical shear of horizontal currents produces the same effect as longitudinal diffusion. For an oscillatory flow the effective longitudinal diffusion (or horizontal diffusion for the rotary tidal currents on Georges Bank) corresponds to a diffusivity of about UH (Bowden, 1965) with some uncertainty due to the dependence of the result on the tidal velocity profile and the rate of vertical mixing. The validity of this estimate depends on the vertical mixing time (of order $200H/U$) being much less than the tidal period, as is true on Georges Bank. If not, a vertical line of dye does little more than oscillate with the sheared velocity, and the horizontal dispersion is much reduced (Fischer et al., 1979).

This value, of order UH, is only about 30 m² s⁻¹ for Georges Bank, much less than the value of a few $\times\ 10^2$ m² s⁻¹ that seems to be required (and shear dispersion is hardly applicable to drogued buoys anyway!) We seem to require substantial mixing by large-scale eddies, and wonder whether the tides are responsible.

Zimmerman (1978, 1980) has shown how residual current eddies can be generated by tidal rectification over irregular bottom topography. Vorticity at the tidal frequency is produced both by stretching of vortex lines over the rough topography, and by the torque of bottom friction; and damping of this tidal vorticity combined with advection by tidal currents produces a field of residual eddies. The process, which embodies the same physics as the rectification theories of Huthnance (1973) and Loder (1980b), is most effective if the topographic length scale is comparable with the tidal excursion.

In an earlier paper, Zimmerman (1976) shows how the combination of oscillatory tidal currents and spatially random residual eddies leads to the separation of particles that are initially close to each other. For particle separations greater than the scale of the residual eddies, the dispersion may be described in terms of longitudinal and transverse diffusivities, with values depending on the eddy energy and on the scale of the eddies compared with the tidal excursion.

Combining the rather complicated formulas derived in these important contributions by Zimmerman leads to an estimate of a diffusivity of up to 10^3 m² s⁻¹ over the rough areas on top of Georges Bank, much less over the smooth. It certainly seems plausible that tidally generated residual eddies are largely responsible for the observed horizontal mixing on Georges Bank. However, given the nonhomogeneity of topography and the complex dependence of Zimmerman's (1978, 1980) results on the parameters of the problem, it might actually be best in any important region to develop an "eddy-resolving" tidal model, with a model resolution of only about half the tidal excursion being required (Zimmerman, 1978).

The main conclusion is that, at least in vertical well-mixed areas, progress is being made in the understanding and parameterization of horizontal mixing in shallow seas, although the theories discussed above need to be extended to include momentum as well as scalars. On Georges Bank tidal dispersion is probably dominant, but dispersion due to the interaction of oscillatory wind-driven currents with variable bottom topography can probably be predicted in much the same way, either statistically or in terms of a fine-scale model.

In stratified waters the processes and rates of horizontal mixing are not understood. In the surface layer one may want to appeal to Okubo's (1971) diffusion diagram to obtain a scale-dependent mixing rate, but the theoretical justification of this is in terms of homogeneous three-dimensional turbulence, which is hardly applicable. Needless to say, more work is required.

In the context of Georges Bank, most of the processes responsible for the exchange of water on and off the bank have been mentioned earlier. However, the role of remote forcing can also be important, with significant changes on the bank occurring occasionally due to intrusions of water from Gulf Stream rings (E G & G, 1979), or due to water swept in from the Scotian Shelf to the northeast.

7 Discussion

We have seen how partial answers may be given to the limited set of questions listed in Section 2, although I have not pursued the applications to Georges Bank to their conclusion, using the bank merely as a motivation for a discussion of the physics. There are essentially five levels of answer:

1. The qualitative, such as a vague unquantified statement that cross-frontal fluxes may occur at tidal frequencies;
2. Empirical quantitative, such as answers using horizontal mixing rates derived from drogue dispersion studies, without any understanding of the underlying physics or parametric dependence of the rates;
3. Semiempirical quantitative, such as an explanation of vertical uniformity in terms of the H/U^3 criterion, in which the parametric dependence is established, but the critical value of a dimensionless parameter is determined empirically;
4. Inconclusive theoretical, such as dynamical theories of frontal circulation with predictions that are rather sensitive to the parameterization of vertical mixing;
5. Conclusive theoretical, such as the various dynamical theories of tidal rectification.

Physical oceanographers are generally not happy to give answers of type 1 or 2 alone but are not often in a position to give a type 5 answer. Most of the work that is performed on the continental shelves, either in data interpretation or as input to the other studies, is of type 4, and I suspect we often miss type 3 answers (indeed the Simpson and Hunter (1974) H/U^3 criterion could have been proposed and tested decades earlier).

Very often a type 2, 3, or 4 answer is adequate, but there are bound to be occasions on which an improvement will be required. In some cases (as, perhaps, for horizontal mixing) one way of making progress will be through improved resolution of our models, but in other instances (as for vertical mixing) adequate resolution of all the important sub-grid-scale processes will never be possible.

Indeed the correct parameterization of vertical mixing emerges as one of the most central and least tractable problems of continental shelf physics, and is the weak link in studies of a number of other processes. I have suggested that progress may be aided by application of some of the techniques of deep-sea oceanography to shelf studies.

We must be careful in general not to separate studies of the physical oceanography of the continental shelves and deep oceans. Although much of what I have reviewed applies only to tidally dominated areas of the continental shelf, other areas which are strongly stratified and predominantly wind driven may have much more in common with circulation and mixing problems of the deep ocean.

It is impossible to be very specific about what physical oceanographers will do, or should be doing, in the next 50 years; but, as I have mentioned, vertical mixing seems to be one of the fundamental topics that needs more attention. I suspect we also have to learn more about the biological oceanography and fisheries studies which justify much of our efforts. Most physical oceanographers have an understanding that goes no deeper than "light + nutrients = fish," whereas biological oceanographers give us the impression that even if we presented them with temperature, salinity, and currents at all places for all times they would not be able to produce useful predictive models. Clearly there is a vast middle ground between these two disciplines waiting to be occupied.

On problems of environmental impact, where management answers are required in a short time, we often seem to indulge in excessive and rather mindless data collection, together with over-elaborate models, rather than make simple estimates based on a combination of common sense and the rather rough predictive formulas available to us. The time still seems far off when practical problems will involve no more than the application of engineering-type formulas. Instead, the interaction of basic and applied research will continue to be necessary, with all the associated pitfalls and challenges.

This brings me to my final point, as seen from a university perspective. There is a desperate need to attract bright young people into our field. Without them, any 50-year plan will be unworkable; with them any plan will continually evolve for the better.

Acknowledgments

I thank John Loder for discussion of most of the topics of this paper, and the Natural Sciences and Engineering Research Council for financial support.

References

Allen, C. M., J. H. Simpson, and R. M. Carson. 1980. The structure and variability of shelf sea fronts as observed by an undulating CTD system. Oceanolog. Acta 3, 59–68.
Bigelow, H. B. 1927. Physical oceanography of the Gulf of Maine. Bull. U.S. Bur. Fish. 40, 511–1027.
Blumberg, A. F. 1977. Numerical model of estuarine circulation. J. Hydraul. Div. Soc. Civil Eng. 103 (HY3), 295–310.
Blumberg, A. F., and G. L. Mellor. 1980. A coastal ocean numerical model. In: Lecture Notes on Coastal and Estuarine Studies, 1, Mathematical Modelling

of Estuarine Physics (J. Sündermann and K.P. Holz, Ed.). Springer-Verlag, Berlin and New York, pp. 203-219.
Bowden, K. F. 1965. Horizontal mixing in the sea due to a shearing current. J. Fluid Mech. 21, 83-95.
Bumpus, D. F. 1976. A review of the physical oceanography of Georges Bank. ICNAF Res. Bull. 12, 109-134.
Cohen, E. B. 1982. An energy budget for Georges Bank. In: Multispecies approaches to fisheries management advice. (M. C. Mercer, Ed.). Special Publ., Fisheries and Aquatic Sciences, Canada. In press.
Dietrich, G. 1951. Influences of tidal streams on oceanographic and climatic conditions in the sea as exemplified by the English Channel. Nature (London) 168, 8-11.
E G & G. 1979. Analysis Report. Appendix F, in: 10th Quarterly Progress Report, New England Outer Continental Shelf Physical Oceanography Program. E G & G, Environmental Consultants, Waltham, Mass., 122 pp.
Finn, D. P. 1980. Georges Bank: The legal issues. Oceanus 23(2), 28-38.
Fischer, H. B., E. J. List, R. C. Y. Koh, J. Imberger, and N. H. Brooks. 1979. Mixing in Inland and Coastal Waters. Academic Press, New York, 483 pp.
Garrett, C. J. R., J. R. Keeley, and D. A. Greenberg. 1978. Tidal mixing versus thermal stratification in the Bay of Fundy and Gulf of Maine. Atmosphere-Ocean 16, 403-423.
Gregg, M. C., and M. G. Briscoe. 1979. Internal waves, finestructure, microstructure and mixing in the ocean. Rev. Geophys. Space Phys. 17, 1524-1548.
Huthnance, J. M. 1973. Tidal current asymmetries over the Norfolk Sandbanks. Estuarine Coastal Mar. Sci. 1, 89-99.
James, I. D. 1977. A model of the annual cycle of temperature in a frontal region of the Celtic Sea. Estuarine Coastal Mar. Sci. 5, 339-353.
James, I. D. 1978. A note on the circulation induced by a shallow-sea front. Estuarine Coastal Mar. Sci. 7, 197-202.
Kent, R. E., and D. W. Pritchard. 1959. A test of mixing length theories in a coastal plain estuary. J. Mar. Res. 18, 62-72.
Loder, J. W. 1980a. Secondary tidal effects in tidally-energetic shallow seas, with application to the Gulf of Maine. Ph.D. Thesis, Dalhousie University, Halifax, Nova Scotia.
Loder, J. W. 1980b. Topographic rectification of tidal currents on the sides of Georges Bank. J. Phys. Oceanogr. 10, 1399-1416.
Loder, J. W., D. G. Wright, C. Garrett, and B.-A. Juszko. 1982. Horizontal exchange on central Georges Bank. Can. J. Fisheries and Aquatic Sciences. In press.
Munk, W. H., and E. R. Anderson. 1948. Notes on a theory of the thermocline. J. Mar. Res. 7, 276-295.
Okubo, A. 1971. Oceanic diffusion diagrams. Deep-Sea Res. 18, 789-802.
Pingree, R. D. 1979. Baroclinic eddies bordering the Celtic Sea in late summer. J. Mar. Biol. Assoc. U.K. 59, 689-698.
Pingree, R. D., and D. K. Griffiths. 1978. Tidal fronts on the shelf seas around the British Isles. J. Geophys. Res. 83, 4615-4622.
Pingree, R. D., P. R. Pugh, P. M. Holligan, and G. R. Forster. 1975. Summer phytoplankton blooms and red tides along fronts in the approaches to the English Channel. Nature (London) 258, 672-677.
Riley, G. A. 1979. Summation. Second Informal Workshop on the Oceanography of the Gulf of Maine and Aadjacent Seas. Dalhousie University, Halifax, Nova Scotia, May 14-17, 1979.
Riley, G. A. 1980. Biological processes on Georges Bank. Unpublished.

Rodi, W. 1980. Mathematical modelling of turbulence in estuaries. In: Lecture Notes on Coastal and Estuarine Studies 1, Mathematical Modelling of Estuarine Physics. (J. Sündermann and K.-P. Holz, Ed.). Springer-Verlag, New York and Berlin, pp. 14-16.

Ruddick, B. R., and T. M. Joyce. 1979. Observations of interaction between the internal wavefield and low frequency flows in the North Atlantic. J. Phys. Oceanogr. 9, 498-517.

Scarlet, R. I., B. A. Magnell, D. Frye, C. N. Flagg, and J. B. Andrews. 1979. Physical oceanography of Georges Bank. EOS, Trans. Am. Geophys. Union 60, 278.

Simpson, J. H., and D. Bowers. 1979. Shelf sea fronts' adjustments revealed by satellite IR imagery. Nature (London) 280, 648-651.

Simpson, J. H., and J. R. Hunter. 1974. Fronts in the Irish Sea. Nature (London) 250, 404-406.

Simpson, J. H., D. G. Hughes, and N. C. G. Norris. 1977. The relation of seasonal stratification to tidal mixing on the continental shelf. A Voyage of Discovery. Deep-Sea Res. (Suppl.), 24, 327-340.

Spiegel, S. L., and B. A. Magnell. 1979. Interaction between low-frequency currents and tidal currents on the northern flank of Georges Bank. EOS, Trans. Am. Geophys. Union 60, 279.

Turner, J. S. 1973. Buoyancy Effects in Fluids. Cambridge University Press, Cambridge, 367 pp.

Zimmerman, J. T. F. 1976. Mixing and flushing of tidal embayments in the western Dutch Wadden Sea-II: Analysis of mixing processes. Neth. J. Sea Res. 10, 397-439.

Zimmerman, J. T. F. 1978. Topographic generation of residual circulation by oscillatory (tidal) currents. Geophys. Astrophys. Fluid Dynam. 11, 35-47.

Zimmerman, J. T. F. 1980. Vorticity transfer by tidal currents over an irregular topography. J. Mar. Res. 38, 601-630.

Shoreline Research

Orrin H. Pilkey

1 Looking Back To See Where We Are Going

The decade of the 1970s was an era of unprecedented advancement in our understanding of shorelines and coastal processes. Important advances on all scales of studies ranged from sophisticated instrumentation of the surf zone (to measure wave pressure, currents, and suspended sediment) to large long-range studies of the origin and history of barrier island systems. The 1970s saw a peaking of the environmental movement for which the shoreline became a major focus of concern—so much so that 1980 was declared the year of the coast.

During the 1970s the postwar trend of the rush to the American shore continued and even accelerated. More and more Americans purchased coastal property, often right on the beach; 75% of all Americans live within a few hours' drive of either an oceanic or a great lake beach. The coast and its complexities and problems achieved great visibility; this visibility brought with it increased opportunities, increased need, and increased funding for shoreline and coastal zone research.

Perhaps more than any other group of American marine scientists, those involved in the study of beach and coastal processes find themselves in the midst of an environmental, political, and economic crisis. The crisis is shoreline development converged upon by a rising sea level. The post-1930 sea level (Fig. 1) jump is causing widespread erosion. We can fairly say that the environmental crunch at the beach in the 1980s and beyond will make the crunch of the 1970s seem like a picnic. Shoreline scientists will be involved because they will be increasingly called upon to offer expert testimony and advice concerning practical

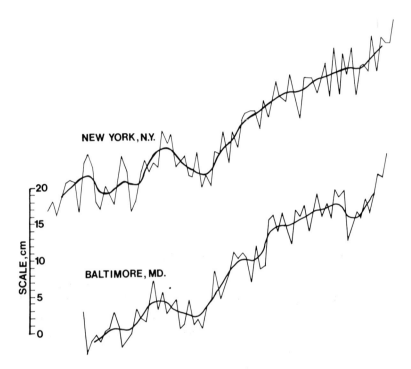

Figure 1. The sea-level rise in two American cities (modified from Hicks and Jones, 1974).

problems at the shore; they will also be involved because coastal process funding will come from agencies directly concerned with "man vs. shoreline" problems.

Recently the National Academy of Sciences (Marshall, 1980) has expressed strong concern over the possibility that the rise in sea level is related to the production of excess CO_2 by fossil fuel combustion—the greenhouse effect. There is a strong likelihood that the present sea level rise, which accelerated in the 1930s (Hicks, 1972, 1981), may have accelerated further in the 1970s (Emery, 1980). If the highly likely scenario of continued acceleration of sea level rise continues (Gornitz et al., 1982), research into the processes of the shoreline should be among the better funded branches of marine sciences. This is because rising sea level means shoreline migration, called erosion (Fig. 2) if a building is threatened, will continue and probably accelerate.

This is not to imply that those who study coastal and beach processes are the only marine scientists involved in environmentally critical areas. However, a row of falling beach cottages owned by prominent and wealthy people is a far more visible and pressworthy crisis than, for example, a polluted estuary.

The profound lack of understanding that rising sea level directly impacts our coastal development has resulted in a recent reduction in federal funding for programs monitoring the sea-level rise. The probable reason for the lack of federal interest in the sea-level rise phenomenon is that the rise rate is very small. Perhaps 1 foot a century is the present rate of rise (Emery, 1980). What many do not understand is that the horizontal translation of the shoreline in a rising sea is basically controlled by the slope of the land surface (Pilkey et al., 1978). This means a 1 foot a century rise will produce a retreat of 500 to 1500 feet during the same century along coastal plain-bordered continents (Fig. 3).

2 What Is The Shoreline?

The shoreline is first, in the minds of men, a boundary. It is a political, biological, physical boundary, but it rarely coincides closely with boundaries of major physiographic or geological components of the earth's crust. (For example, the shoreline does not represent the boundary between oceanic and continental crust, nor does it usually occur at the shelf-slope break.) The shoreline is also an energy buffer, a dynamic zone of interaction between land and sea processes. Most important of all, from the standpoint of marine sedimentology studies, the shoreline is a filter of material going to the ocean basins. Different combinations

Figure 2. "Erosion" at Holly Beach, Louisiana (photograph by Steve Heron).

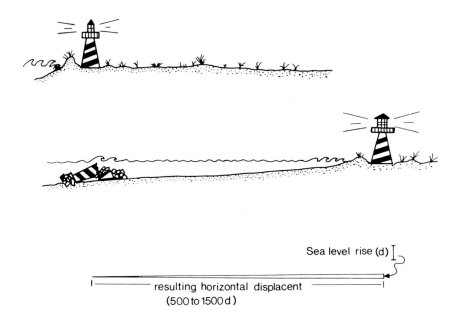

Figure 3. The relation between sea-level rise and horizontal shoreline retreat on coastal plain coasts.

of sea-level changes, river discharges, and coastal types release varying sediment types to the sea, even from identical source areas. Understanding the boundary filter is important because all marine sediments are influenced. It is an important factor in all models of marine sedimentation, shallow or deep. The most recent summary of our level of understanding of various coastal sedimentary environments is by Davis and Ethington (1978).

For example, the Southern and Mid-Atlantic shoreline is deeply embayed over the very flat surface of the Atlantic Coastal Plain. The estuaries are not now filled, so most mud and all sand is deposited in upper estuaries, far inland from the open-ocean shoreline. Fines escape only in relatively small amounts and only in time of flood. Sands (and much mud) are released to the marine environment only by the process of shoreface bypassing (Swift, 1976). Such bypassing and consequent release of stored river materials occurs when former upper estuary deposits are overrun by the transgressing shoreline, responding to a rising sea level. As a consequence of the present boundary filter conditions, relatively small amounts of material are released to the sea and deep-sea sedimentation rates are low. The calcareous fraction of shelf sediments is dominated by beach or nearshore shells; some open-shelf shells are found in beach sediments.

A large literature on the classification of shorelines and beaches is summarized in King (1972). For our purposes, we can simply divide shorelines into two broad types: barrier island and nonbarrier. The United States Atlantic and Gulf coasts south of New England are bordered by the broad low-lying coastal plain. The barrier islands along these shorelines formed basically because nature likes straight shorelines (Swift, 1975, 1976). As the sea level rose after the last ice age, coastal plain valleys became coastal plain estuaries. The ridges between the estuaries became headlands. Waves attacked the headlands and longshore transport carried the material laterally. Since longshore currents are wave derived, the transported sand could not be carried into the estuaries where the wave energy was dissipated. Instead the sand built parallel spits out into the mouth of the estuary. Meanwhile the rising sea level was flooding behind the spits and eventually detached islands formed (Fig. 4). The continued existence of barrier islands during the Holocene Era was due to the rising sea (Fig. 5). During a time of a static or falling sea level such islands probably do not exist.

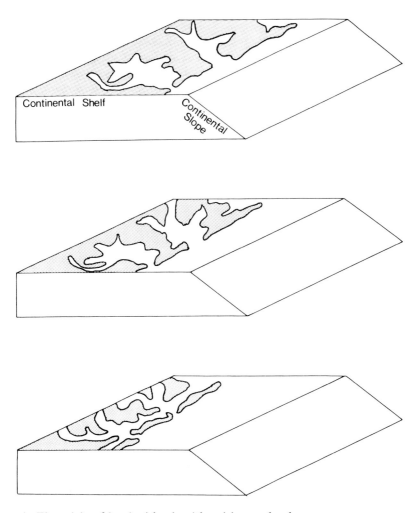

Figure 4. The origin of barrier islands with a rising sea level.

Figure 5. A migrating barrier island. This is Conch Bar, north of Mispillion Inlet, in Kent County, Delaware. The island is migrating back into salt marsh. Mosquito canals dug on the marsh in the 1930s can now be seen on the frontside of the island. (Photograph by Evelyn Maumeyer.)

On steeper coasts such as our Pacific and New England shores, barrier islands are minor features or not present at all. Barrier spits across the mouths of rivers do form for the same reason that barrier islands form but these probably do not migrate continuously with a rising sea level as do the coastal plain barriers. The rest of the nonbarrier shoreline owes its configuration simply to the continuing battle between breaking storm waves and the materials of the shoreline. A factor affecting the outline of the nonbarrier coast is tectonic, the degree to which and the rate at which a coastal area is rising or subsiding. Each of these major types of shorelines responds uniquely to the forces of the sea, and understanding this response is the major goal of shoreline scientists the world over.

3 Coping With the Sea-Level Rise

In the spring of 1981, a small conference of coastal geologists convened at Skidaway Institute of Oceanography, Savannah, Georgia for the purpose of drafting a position paper. This paper, "Saving the American beach: a position paper by concerned coastal geologists," was signed by a number of American coastal geologists. Perhaps better than any other available literature, this document sums up the present state of the American shoreline and our needs for the future (Fig. 6). A fundamental turnaround in shoreline management and in the direction of coastal research is called for. Following is a summary from the Skidaway document.

New approaches to the management of the American shoreline are urgently needed to preserve our recreational beaches for future generations. Approximately half of the 10,000 miles of the "lower 48" American shoreline facing the open ocean is under development pressure. Well over 2,000 miles are considered by the U.S. Army Corps of Engineers to be in a state of critical erosion. Erosion is occurring along almost all of the U.S. coast and when shoreline retreat collides with shoreline development, a state of "critical erosion" is achieved. Shoreline retreat is due to many causes but a major one is rising sea level and indications are that the rise will continue for the foreseeable future.

The usual response to critical erosion on America's shore is stabilization; halting of shoreline retreat by engineering means. Such stabilization of America's shore has been successful in increasing the length of life of buildings built adjacent to the beach. However, stabilization in the long run (50 years ±) and sometimes in a much shorter time frame has resulted in severe degradation of the recreational beach area. Dollar costs of halting shoreline retreat by stabilization are very high. Replacement of the beach by pumping in new sand costs about 1 million dollars or more per shoreline mile each time it is done and it must be carried out repeatedly; commonly in 3 to 10 year intervals. Another approach, the building of seawalls, costs between $100 to $600 per linear open ocean shoreline foot. Combining these cost figures with the 2,000 mile figure of critically eroding shoreline gives some idea of the magnitude of the potential economic crisis on the American shoreline if we continue to stabilize.

American taxpayers are paying huge sums of money to temporarily protect the private property of a relative few. Furthermore this practice commonly leads to the ultimate destruction of a highly valued public recreational area.

Stabilization costs can be justified for major coastal cities or harbor entrances (Chicago, Galveston, Miami Beach, Coney Island, the Columbia River entrance, for example), but stabilization of most American shores is not justifiable in the broader scope of national interests. Numerous projects involving public and private money along virtually all developed coastal and lake shores presently threaten most of America's recreational shoreline.

The following summarizes our views on stabilization of America's open ocean shorelines.

1. People are directly responsible for the "erosion problem" by constructing buildings near the beach. For practical purposes, there is no erosion problem where there are no buildings, farm or crop lands.
2. Fixed shoreline structures (breakwaters, groins, seawalls, etc.) can be successful in prolonging the life of beach buildings. However, they almost always accelerate the natural rate of beach erosion. Resulting degradation of the beach may occur in the immediate vicinity of structures or it may occur along adjacent shorelines, sometimes miles away.

Figure 6. A narrowed beach in front of a revetment on Lake Erie. The groins in the background are trapping no sand because no sand is moving laterally in this shoreline system. Narrowed recreational beaches are increasingly part of the American beach scene.

3. Most shoreline stabilization projects protect property, not beaches. The protected property belongs to a few individuals relative to the number of Americans who use beaches. If left alone, beaches will always be present, even if they are moving landward.
4. The cost of saving beach property by stabilization is very high. Often it is greater than the value of the property to be saved, especially if long range costs are considered.
5. Shoreline stabilization in the long run (10 to 100 years) usually results in severe degradation or total loss of a valuable natural resource, the open ocean beach.

6. Historical data show that shoreline stabilization is irreversible. Once a beach has been stabilized, it will almost always remain in a stabilized state at increasing cost to the taxpayer.

The consequences of responding to rising sea level by shoreline stabilization are so serious that we urge immediate measures to explore totally new approaches to shoreline management. Such approaches may even involve drastic and unpopular measures such as assuming that buildings adjacent to the beach are temporary or expendable. Equally important, the new approach to shoreline management must incorporate the very significant advances in geologic understanding of shoreline processes that have occurred during the last decades. In the past the American public has been largely unapprised and unaware of the long range environmental and dollar costs of shoreline stabilization. There is a critical and immediate need for the public to know the direction in which American shoreline management is leading.

Coastal studies have a bright funding future; the basis for this optimistic prediction is the burgeoning coastal crisis. Both basic and applied research will benefit from this increase in support, but the emphasis will likely be on applied problems. If the National Academy's findings about the sea-level rise come to be, the next 50 years could see the study of coastal processes become the dominant branch of marine science.

4 Anticipated Directions of Research

One of the problems in understanding and applying coastal processes to practical situations has been our failure to recognize local shorelines as being a small component in a much larger system. The system may involve the inner continental shelf, local rivers and estuaries, beaches for miles in either direction, and adjacent land areas. Such large-scale studies will require large individual grants (perhaps of an International Decade of Ocean Exploration, or IDOE, nature) involving teams of scientists of various specialties including geologists, physical oceanographers, and biologists.

The role of the continental shelf as a sink or source of sediment is poorly understood; the role of the physiography of individual inner shelf segments in controlling wave characteristics and orientations is usually unknown. Barrier island studies have emphasized single islands, accompanied by vague speculation concerning the role of adjacent islands; even on the state political level (Delaware and New Jersey) the need for studying island systems has been recognized.

The documentation of the sea-level rise must be done on an international level with the cooperation of scientists from many countries; the rise must be observed in many types of shoreline tectonic situations and in a wide range of latitudes. This must be combined with monitoring of the progress of the greenhouse effect. This will include observation of the West Antarctica Ice Sheet, a mass of ice particularly vulnerable to degradation.

In recent years, studies of the past have definitely held the spotlight in advancing our understanding of sediment types and sediment structures of the shore zone. A case in point is the widespread discovery of hummocky cross stratification in ancient sandstones (e.g., Bourgeois, 1980). This is probably a structure caused by storms in the lower surf zone (Clifton, personal communication). The difficulty of obtaining information on sedimentary structures in the surf zone is obvious, although Hill and Hunter (1976) and Howard and Reineck (1981), among others, have made important contributions.

One of the bright spots of present day research is the increasing emphasis on instrumentation of the surf zone. Simultaneous measurements of wave, current, and water characteristics and suspended sediment load are leading to a fine tuning of our understanding of shore processes. Recent studies by Douglas Inman and his group at Scripps Institute of Oceanography exemplify this trend. Unquestionably surf zone instrumentation is a major research direction for the coming decades.

The many directions in which the small- or local-scale studies lead are exemplified by the papers in the Society of Economic Paleontologists and Mineralogists Special Paper No. 24 (Davis and Ethington, 1976). Although already slightly outdated, this volume *(Beach and Nearshore Sedimentation)* shows the future. The studies reported include various types of surf zone instrumentation measuring suspended sediment and sediment transport; various theoretical approaches to wave modeling, relating shelf and shoreline processes of sedimentation processes and wave forces; weather patterns and coastal processes; the role of organisms in surf zone sedimentation; and various studies of nearshore sedimentary structures and facies relationships.

One of the greatest needs in terms of small- or local-scale studies is the documentation of a catastrophe—a close look at the beach during a storm. Coastal scientists have long been aware that on many shorelines everything happens during the storm, and that fair weather conditions have little bearing on processes of island migration, coastal retreat, inlet migration and formation, cliff retreat, etc. (Hayes and Boothyroyd, 1969; Komar, 1976). Obviously, storms are difficult to document; in the future, instrumentation or other types of studies of storm conditions will be a highly fundable type of coastal study requiring new and imaginative approaches and devices.

5 Man and the Shoreline

Our society's greatest need in terms of coastal research concerns the relationship between man and the shoreline. An interesting social problem at present is, in my opinion, hindering the progress of applied coastal research (from a geological view). This is the conflict between the engineering approach and the geological approach. If a beach community leader asks an engineer for a solution to his or her erosion problem, a solution is immediately forthcoming. If the same individual asks a geologist, he or she is likely to be confronted with a shrug of the shoulders and the statement that there is no solution. The engineer is correct for a short range; the geologist is correct for a longer range (30 to 50 years or more).

A prominent coastal engineer was quoted recently in a newspaper: "In our national studies as well as in my own extensive coastal engineering experience in this field, I have never heard of engineering or geological studies which show that such revetments damage or diminish the beach."

Basically the statement implies that revetments mustn't cause erosion. Such a statement is geologically absurd, but from a short-range viewpoint it may be much closer to the truth.

The conflict is highlighted by the fact that engineers are more quantitative in their approach to shoreline processes. Basically, engineers are forced to quantify highly complex oceanographic processes in order to arrive at design criteria for construction purposes. Geologists tend to be critical of the assumptions used in the computations where many variables are not included.

Much research effort has been expended in the study of the interaction of coastal processes and coastal engineering structures. What has not been specifically sought in this research is quantification of the "bad side" of stabilization. For example, replenished beaches disappear at rates that probably are at least 10 times that of the natural beach they replace. A thorough documentation of natural vs. artificial beach erosion rates would form a basis for predicting long-term costs of stabilization. Furthermore, how much more rapidly will the artificial beach be expected to erode as the sea level continues to rise and the beach becomes increasingly out of equilibrium? The rate and extent of shoreface steepening in front of seawalls under various wave climate conditions remains unknown.

Geologists and engineers alike continue to study and to apply coastal processes as they occur on pristine undeveloped beaches. Factors in beach evolution such as storm response, sources of sand, and rates of transport must change drastically as stabilization of a shoreline proceeds. Engineers must be able to predict not only what changes will happen to the beach but also how those changes will affect shore processes. The whole spectrum of detailed scale or local studies discussed above will certainly be applied to heavily developed shorelines in the coming decades.

A specific example of a research need on heavily developed barrier islands is determining the role of the backside of the island in storm response. Finger canals and bulkheads on the backside of New Jersey beach communities have funneled flooding waters returning to the sea, causing new inlets to form right through developed areas. The role of the lagoon on island or barrier spit response to storms has long been recognized but has never been formally investigated.

An area where a great deal of imagination and creativity will be called for is the response to the sea-level rise. The Corps of Engineers presently has a demonstration program including the testing of new shoreline stabilization devices. Although the new devices tested often show a great deal of engineering imagination, to date the demonstration program has basically not even admitted that the sea level is rising. The experimental devices trap sand and block wave energy just like older type structures but do not respond to the sea-level rise.

The following statement was made by the author of a prominent textbook in coastal engineering: "The subject of sea level hardly warrants a comment. [Problems caused by rising sea level] . . . will be solved by future generations." The coastal scientists of the next 50 years *are* the future generation.

The ultimate state of coastal stabilization has been termed New Jerseyization (Fig. 7). Destabilization of a New Jerseyized shore could be termed de-New Jerseyization.

The retreat from the shore, believed by many to be inevitable in the coming decades, will require a great deal of input from coastal researchers. The basic problem will be one of recommending how to get an island or beach back into equilibrium with the sea level as well as with other factors in the dynamic equilibrium of beaches or islands. If seawalls and buildings are removed, how will the steepened shoreface system of a long-stabilized island respond? How far will a shoreline retreat before coming back into equilibrium? How far back do we have to move?

Undeniably a major problem with shoreline stabilization schemes has been the acceptance of short design lives for shore structures. The problem is that the impact of the structures on the beach environment has been viewed in a similar short time frame.

In the coming decades shoreline scientists and engineers will be under increasing pressure to make long-range predictions concerning stabilization plans. Inherent in such predictions is recognition of the destabilizing effect of the sea-level rise.

Figure 7. A New Jersey beach scene showing several generations of coastal engineering structures on the beach.

6 Conclusion

We begin the next half-century of shoreline studies from a plateau of solid but general understanding of many important aspects of coastal processes. Because of the expected continuation and acceleration of the sea-level rise, increasing conflict between man and erosion can be expected. This fact virtually assures greatly increased funding for studies of coastal processes. The next generation of marine scientists and engineers concerned with shoreline problems can expect to be at the center of an important sociological, political, and economic controversy as we begin our retreat from the shore. The next 50 years of coastal studies will be most exciting.

Acknowledgments

William Neal and Douglas Glaeser furnished many ideas for this paper. Douglas Glaeser criticized the manuscript in detail. I am grateful to my friends and associates in the geological and engineering community for many conversations on the topic of this paper.

References

Adams, J. W. R. 1981. Florida's beach program at the crossroads. Shore and Beach 49(2), 10-14.
Bourgeois, J. 1980. A transgressive shelf sequence exhibiting hummocky stratification: The Cape Sebastian sandstone (Upper Cretaceous), Southwestern Oregon. J. Sed. Petrol. 50, 681-702.
Davis, R. A. (Ed.). 1978. Coastal Sedimentary Environments. Springer-Verlag, New York, 420 pp.
Davis, R. A., and R. L. Ethington. 1976. *Beach and Nearshore Sedimentation*. Soc. Econ. Paleon. and Mineral, Spec. Publ. No. 24, Tulsa, Okla., 187 pp.
Emery, K. O. 1980. Relative sea levels from tide-gauge records. Proc. Natl. Acad. Sci. (U.S.) 77(12), 6968-6972.
Gornitz, V. O., S. Lebedeff, and J. Hansen. 1982. Global sea level trends in the past century: Science 215, 1611-1614.
Hansen, J., D. Johnson, A. Lacis, S. Lebedeff, P. Lee, D. Rind, and G. Russell. 1981. Climate impact of increasing atmospheric carbon dioxide. Science 213, 957-966.
Hayes, M. O., and D. D. Boothyroyd. 1969. Storms as modifying agents in the coastal environment. In: Coastal Environments. N. E. Massachusetts. Geol. Dept., Univ. of Mass., Amherst, Mass., 290-315.
Hicks, S. D. 1972. On the classification and trends of long period sea level series. Shore and Beach 40(1), 20-23.
Hicks, S. D., and E. C. Jones. 1974. Trends and Variability of Yearly Mean Sea Level. NOAA Tech. Memo No. 13, 39 pp.
Hicks, S. D. 1981. Long-period sea level variations for the United States through 1978. Shore and Beach 49(2), 26-29.
Hill, G. W., and R. E. Hunter. 1976. Interaction of biological and geological processes in the beach and nearshore environments, Northern Padre Island, Texas. In: Beach and Nearshore Sedimentation (R. A. Davis and D. D. Ethington, Eds.). SEPM, Tulsa, Okla., Spec. Publ. No. 24, pp. 169-187.

Howard, J. D., and H. E. Reineck. 1981. Depositional facies of high energy beach to offshore sequence: comparison with low energy sequence. Bull. Am. Assoc. Petrol. Geol. 65, 871-829.

King, C. A. M. 1972. Beaches and Coasts (2nd ed.). St. Martin's Press, New York, 569 pp.

Komar, P. D. 1976. Beach Processes and Sedimentation. Prentice Hall, Englewood Cliffs, N.J., 129 pp.

Leatherman, S. P. 1981. Barrier beach development: a perspective on the problem. Shore and Beach 49(2), 2-9.

Marshall, E. 1980. By flood, if not by fire, CEQ says. Science 221, 463.

O'Brien, M. P. 1980. Editorial: Let's Look at the Record. Shore and Beach 48(3), p. 2 and p. 42.

Pilkey, O. H., Jr., W. J. Neal, and O. H. Pilkey, Sr. 1978. From Currituck to Calabash: living with North Carolina's barrier islands. North Carolina Science and Technology Research Center, Research Triangle Park, N.C., 228 pp.

Skidaway Inst. of Oceanography. 1981. Saving the American Beach: a position paper by concerned coastal geologists: Unpub. 18 p.

Swift, D. J. P. 1975. Barrier-island genesis: evidence from the central Atlantic shelf, eastern U.S.A. Sediment. Geol. 14(1), 1-43.

Swift, D. J. P. 1976. Coastal sedimentation. Chapter 14 in: Marine Sediment Transport and Environmental Management (D. J. Stanley and D. J. P. Swift, Eds.). John Wiley and Sons, New York.

Watson, T. L. 1979. Coastal erosion, some causes and some consequences: with special emphasis on the state of Florida. Shore and Beach 47(2), 7-12.

The Ocean Nearby: Environmental Problems and Public Policy in the Next Fifty Years

Evelyn Murphy

1 Introduction

No part of the ocean is more affected by people than the coastline. And, in the next 50 years, no part of the ocean is more likely to bear the brunt of man's destructive and indifferent treatment.

To anticipate the next 50 years is an awesome task. Think for a moment: who in 1930 foresaw Congressional actions that led to a Coastal Zone Management Act; a Fisheries Management and Conservation Act; a Marine Protection, Research and Sanctuaries Act? We had some idea then of the economic importance of ocean resources. In 1920, Congress passed the Federal Mineral Leasing Act; by 1930, offshore deposits of oil and gas off the California coast were being developed. But in 1930, this nation was only beginning to awaken to the vast economic opportunities of the ocean.

And, 50 years ago, federal ocean policies and programs were oriented toward the use of surface waters for transport. Federal agencies regulated commerce, revenue collection, harbor and port development, and obstacles to navigation. National policy expanded from focusing almost exclusively on the ocean's surface to the current panoply of authorities that encompass ocean space, resources, and activities, including the lands that lie beneath the ocean surface.

What more public policy, therefore, could be needed for the next 50 years? Some people will argue that the legislation passed in the 1960s and 1970s is sufficient to set the course of ocean policy and programs for decades to come; that we need only implement existing laws to manage the oceans wisely and well. Perhaps. But I think not.

And I ask you to consider what I believe will be the trouble spot of the oceans in the next 50 years—the coastline.

2 People Along the Coast

As we enter the 1980s nearly four out of five Americans live within 100 miles of the ocean or the Great Lakes. That means that fifty million people live on the Atlantic Coast; eleven million people on the Gulf Coast; and twenty-three million people on the Pacific Coast.

By the end of this decade it is predicted that 75% of the American people will live within 50 miles of these shores.

Think what that means to a coastline. It means power plants to provide electricity for the homes, offices, and industries that these people rely upon—power plants discharging millions of gallons of cooling waters daily into coastal bays and tidelands, waters warmer than most coastal marine life can tolerate. By 1995, over 100 additional fossil fuel and nuclear plants are scheduled to go into operation in coastal waters—this, on top of the several thousand plants already discharging cooling waters along the coast. (In Massachusetts, for example, 80% of the state's energy facilities are located along the coast.) The ocean provides an abundance of free cooling waters for power plants, and coastal states find this an irresistible attraction in making decisions about the location of energy plants.

It means housing and schools and hospitals and industrial and commercial development; it means vacation homes along with recreational development, such as marinas, boat ramps, and related parking facilities; it means harbor and channel dredging; it means infilling lands and marshes and estuaries when existing lands seem used up.

And, not least important, it means the discharge of wastes from all these activities into coastal waters—human wastes, industrial wastes, dredge spoils, and runoff. (The Hudson and East Rivers of New York City, for instance, now receive 250 million gallons of raw sewage each day.) Over a billion gallons of treated sewage is dumped along our coastline every day, too.

Add to these staggering figures the fact that the Army Corps of Engineers issues permits for 300 to 400 million cubic yards of dredge spoil to be dumped annually and you begin to get some sense of the magnitude of the daily disruption we cause in coastal waters.

The encroachment of people on the coastal ecology brings additional pollution. New England now experiences an average of one oil spill a day. These may not be spills of the magnitude of the *Argo Merchant,* or the AMOCO *Cadiz,* but the effects from the spill of the barge *Florida,* over 10 years ago, can still be seen. Spills of any size deposit highly toxic hydrocarbons into coastal waters, thereby causing long-lasting harm to marine life throughout the water column and on the ocean floor.

Offshore oil and gas drilling, with routine discharges of drilling muds and frequent spills in loading vessels, also contribute to the pollution of America's coastline.

The damage to coastal marine life from all these activities is documented only in part; but what documentation exists points to staggering losses. Louisiana loses wetlands at the rate of 16.5 square miles a year, and nationwide, 100,000 acres of wetlands are lost annually. Between 1947 and 1980, an estimated 740,000 acres of important estuarine habitat were lost to coastal development. Millions of acres of shoreline are closed to shellfishing due to persistent, low-level pollution.

Overall, along the entire coastline of the United States, day after day, the pressures caused by sheer numbers of people seeking to live and work near the ocean is resulting in enormous losses of coastal marine life.

So far, however, I have only presented one dimension of the anticipated pressures along the coastline—pressures emanating from land-based activities pushing for access to coastal waters.

The scientific advances of the last decade, sponsored in large part by the International Decade of Ocean Exploration, have opened up the ocean for economic exploitation. All forecasts point to the next 50 years as the emergence of an enormous variety of economic activity in the ocean—deep-sea mining for scarce minerals; continued exploitation of oil and gas reserves offshore; increased fishing for commercial species, underutilized species, and species yet to be commercially recognized (such as krill).

Clearly, we will look more and more to the ocean as a source of energy in renewable resources—tidal power; current power; wind power; ocean thermal energy conversion; even perhaps kelp beds for incineration into methane gas.

And, as the world becomes increasingly one large marketplace, the transport of cargo—raw materials, intermediate products, and final products—will require increasing use of ocean surface waters as transportation lanes.

These ocean activities will bring great pressure on coastal resources—for more dock space; for more dredging; more use of coastal lands as staging areas for offshore operations—taking materials and staff to offshore rigs; and bringing back materials, staff, and products to land.

3 The Coastal Ecology: A Matter of Principles

We are all aware of the significance of coastal marine life in the food chain, as spawning grounds, nurseries, and essential life support systems for fish and wildlife.

Many scientists and naturalists have written eloquently about the ecological value of bays, estuaries, salt marshes, lagoons, and the variety of nature's resources that constitute the coastal ecology. I want to focus, therefore, not on the ecology of the coast, but rather on the public policy needed to protect coastal ecology in the face of growing pressures from human activities.

If we start with specific public policy—regulatory structures—we will likely differ on many fronts: on what amount of what pollutants in what quantities over what period of time in what water temperatures, and so on, will be toxic to what species of marine life at what time in its life cycle.

Every public regulator nowadays feels the irony of the following situation: on any specific issue, highly regarded scientists will testify on various sides of the regulatory spectrum, presenting credible cases using data and assumptions they deem worthy; and, in the face of inconsistencies, contradictions, yet seemingly plentiful data, the regulator almost always yearns for just a few more scientific facts and data on which to rest a decision.

Four principles seem essential:

First, given the crucial nature of coastal resources for marine and wildlife, public policy must protect the ecological vitality of these resources.

Second, with the pressures on coastal resources expected to grow in the future, public policy must become increasingly protective, too.

Third, in instances of uncertainty about the extent of pollution from proposed activities, public policy must be biased in favor of protecting marine life first, and permitting development afterward.

Finally, economic pressures along the coastline must be recognized and dealt with realistically, not by prohibiting the use of the coastline, but by directing activities that require access to coastal waters into already developed areas, thereby stopping the loss of any more coastal marine resources.

Many environmentalists will argue that so much loss has been incurred that restoration and reclamation of natural resources must be part of national public policy. On the other hand, industrial interests will argue that we have already sacrificed enormous economic opportunities for protection unnecessarily; and that we cannot sacrifice economic progress for vague, nonessential environmental protection.

This is neither the time nor the issue on which to recreate the antagonisms of a decade ago between environmentalists and business. Ample scientific evidence exists to persuade all people that coastal natural resources are fundamental to man's existence. Achievement of nondegradation assumes, first of all, that we know the base of coastal resources with which we start.

We do not know that base. Despite the major legislation and programs of the last several decades, there is widespread disagreement as to whether the decline of coastal resources has been stayed. Most of that disagreement hinges on a lack of any comprehensive inventory of coastal resources.

4 Coastal Zone Management

A single piece of Congressional legislation constitutes this nation's effort to protect the natural resources in coastal waters—the Coastal Zone Management Act of 1972. Eight years after passage, this statute was scheduled for reauthorization this year. In the Congressional hearings for reauthorization conflicting views of the program's effectiveness emerged.

Almost every public interest group testified that the Act was too weak to stop the decline of coastal ecology. Testimony by the Natural Resources Defense Council (NRDC) characterized the views of most environmentalists:

> NRDC believes that the coastal zone management program is not living up to

its promise. It is not ensuring the protection of our valuable coastal resources. It is not effectively addressing the problems which Congress recognized when it passed the Coastal Zone Management Act of 1972. States have failed to utilize their full powers to effectively protect coastal resources and direct necessary coastal development to appropriate areas . . . .

NRDC believes that these failures are attributable to inherent weaknesses in the statute, as well as to inadequate federal and state implementation . . . .

In sum, Sarah Chasis, staff attorney for NRDC, concluded that, based on NRDC's evaluation of a dozen state coastal-management programs:

It's just not working . . . the coastline is in worse shape now than it was when Congress passed the Coastal Act in 1972.

Federal officials disagreed. Mr. Robert Knecht, the director of the Coastal Zone Management Program since its inception, highlighted the progress in protection of coastal resources in his testimony:

Many states are developing special management programs for fragile areas. Puerto Rico has proposed management recommendations for 14 natural areas.

During the past year, Hawaii has designated ten sites under a nature area reserve system and established one new marine life conservation district, to protect uniquely valuable coastal resources.

California has improved dramatically the protection of wetlands under the jurisdiction of the San Francisco Bay Commission.

Mr. Knecht cited many examples of accomplishment in protection, in the management of coastal development, and in increasing recreational access to the shoreline. He reminded Congress that all 35 eligible states and territories have participated in the program; and that 19 coastal states covering 68% of the nation's coastline are implementing coastal management programs that fully meet the requirements of the law.

How can two different groups—public interest activists and public officials—both intent on protecting coastal resources, differ so markedly about what is happening along the coast? The answer is, in large measure, that there is no comprehensive inventory of coastal resources from which to measure the program's accomplishments and effectiveness.

Data collected by the federal Office of Coastal Zone Management are intended to portray the accomplishments of the program. Understandably so, this office is an advocate for its existence; and in the year of Congressional reauthorization, the Office must put its best work forward.

But the fact remains: public interest groups and public agencies alike rely on anecdotal, piecemeal data to make their case about the well-being of our coast. And this is foolish. Coastal resources are too valuable, the pressure on these resources too great, and our technological capabilities too sophisticated to continue to predicate coastal management policy for the next 50 years on such impressionistic arguments.

5 A Proposal

Therefore, I ask you, scientists and technologists, to create an inventory of coastal natural resources. This country needs a complete, comprehensive account of the natural resources of our coast. This must include the resources in states that do not participate in the Coastal Zone Management Program—as yet a voluntary program—as well as those that do. Like the census, the inventory could be constructed in great detail every 10 years with on-site scientific teams measuring and assessing natural resources. At intervals within the decade, high-resolution aerial photography, perhaps from satellites, could update the inventory.

Second, I ask and urge you to build the body of knowledge about the effects of pollution on coastal resources. So much still needs to be understood about the effects of industrial sludges, organic and inorganic wastes, and human wastes on marine biota.

These are the challenges to the ocean science community in future decades; and these are the needs for leadership in public policy and public management.

We have begun to live in an era of limits—limits to natural resources and limits to financial resources. These circumstances demand the setting of priorities. And, I would argue, with regard to the ocean, the priority must be the wise and prudent management of coastal natural resources. That, in turn, depends upon a much more scientifically sophisticated inventory of resources and the effects of people's use of these resources.

Part II
Regional-Scale Oceanography

v

Acoustics and Ocean Dynamics

J. Walter Munk

1 Sound's the Thing

In September 1944, I was dispatched to Woods Hole Oceanographic Institution (WHOI) to learn something about wave recording. Earlier in the year, Ewing and Worzel had listened aboard the *Saluda* at 26°N, 76°W to 4-lb charges being detonated in the sound channel at distances up to 900 miles. They had heard, for the first time, the characteristic signature of a SOFAR (Sound Fixing and Ranging) transmission building up to its climactic finale (Fig. 1):

bump   bump   bump   bump   bump   bump bumbumb.

In the words of the authors, "The end of the sound channel transmission was so sharp that it was impossible for the most unskilled observer to miss it" (Ewing and Worzel, 1948). What's more, this was one of the rare cases where the experiment had confirmed a previously worked out theory. (We oceanographers are not accustomed to such luxury.) Ewing and Worzel spoke even then of the probability of transmissions over 10,000 miles. The oceans were as transparent to sound as they were opaque to electromagnetic radiation.

Having thus confirmed long-range sound transmission to a hydrophone at the sound axis suspended from a ship, Ewing and Worzel went on to see if comparable results could be obtained with the hydrophone lying on the bottom at the prerequisite depth. For this purpose, they established a station at Eleuthera Island in the Bahamas in October 1944. I have included copies of two records as originally presented (Fig. 1).

The following few years saw an amazing sequence of discovery here at Woods

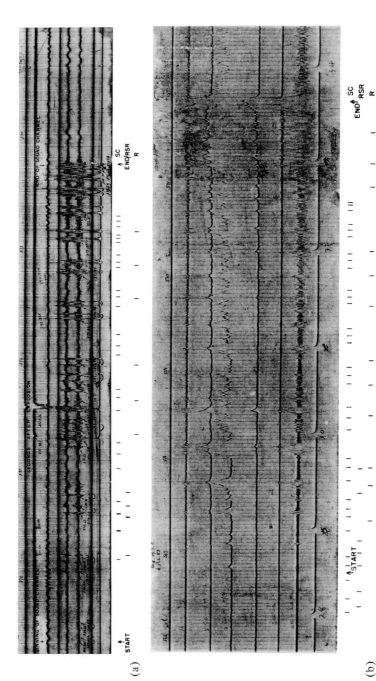

Figure 1. (a) Shot 43 recorded aboard the *Saluda* on April 3, 1944. Charges were exploded at 4000-ft depth and a range of 320 nautical miles. Times are labeled for 370, 371, . . . 374 seconds following the explosion. (b) Shot 42 recorded at Eleuthera on January 28, 1945 (the cable failed on January 29) at a range of 462 nautical miles. Times are labeled 566, 567, . . . 572 seconds.

Hole, led by the trio of Vine, Hersey, and Schevill. Columbus Iselin had come back from the West Coast with the news of the discovery of the deep scattering layer by Raitt, Eyring, and Christensen. Martin Johnson (who had studied the diurnal migration of copepods) had immediately recognized its biological origin. On a cruise on the *Atlantis* the first evening out, Hersey noticed the echo recording paper darkened at 200 fathoms and then shoaled after sunset to merge with the outgoing ping (the first evidence that the Atlantic followed the Pacific). On the same cruise, Hersey and Bergstrom recorded echoes from $\frac{1}{2}$-lb TNT charges that came from layers beneath the sea floor. Schevill was able to attribute the so-called fish noises (consisting of whistles and squeals punctuated with ticking and clucking, reminiscent of an orchestra tuning up) to white porpoises. Liebermann demonstrated that the anomalous sound absorption in sea water was due to ionic relaxation involving traces of magnesium sulfate. The fact that the "afternoon effect" (a dramatic decrease in the range of sonar targets) was due to the warming of the surface waters (and not the after-lunch lethargy of sonar operators) was independently suggested by Steinberger and Iselin and by a West Coast group consisting of Roger Revelle, Harald Sverdrup, and Richard Fleming. And finally, Vine, Hersey, and Schevill recognized that, in deep water, sound from a shallow source beneath the duct must focus again at the same depth. This was the discovery of the convergence zone (it was then called skip-distance focusing).

All this took place in 1946 to 1948, with most of the work classified. When these important discoveries were eventually reported in the open literature after some $1\frac{1}{2}$ decades, the authorship bore little or no relation to the people who had done the pioneering work.

A growing acoustic community under Navy sponsorship set out to exploit the sound channel for submarine detection. The attendant development of underwater technology under the leadership of Bell Laboratories could be termed miraculous. But this was classified work and had little impact on oceanography.

In the 1950s, physical oceanographers turned their attention to the large-scale steady ocean circulation, its response to wind and thermohaline forcing, and its modification by the effects of the Earth's rotation. This work on "DC" (or zero-frequency) oceanography was of little use to the acousticians, who became increasingly concerned with the extreme variability of acoustic signals transmitted through the oceans. Fadeouts and phase jumps are the rule, not the exception, of long-range sound transmission. And these implied an ocean variability of small scale and high frequency.

And so the communities of physical oceanographers and ocean acousticians drifted apart.[1] The acousticians were driven to invent their own ocean. Some of them thought of the ocean as a transmission line with imperfections, the imper-

---

[1] But there was little or no disconnect between acousticians and geologic oceanographers. Early discoveries of strong reflectors beneath the seafloor ($\sim$1948) were soon followed by the development of the precision echo sounder ($\sim$1950), the continuous seismic profiler (1956), acoustic arrays (early 1960s) and finally the side scan sonars. Each of these developments had played a profound role in marine geology. The relation between acoustics and marine biology was more tenuous.

fections expressed in terms of space and time correlations of ocean finestructure. More often than not the finestructure was taken as homogeneous and isotropic, with a $-5/3$ spectrum characteristic of homogeneous isotropic turbulence in the inertial subrange. We now know that the ocean finestructure is neither homogeneous nor isotropic and has little to do with turbulence in the usual sense.

2 Ocean Finestructure

By the 1960s the oceanographers were turning their attention to fine-scale features. Even toward the end of the war years, when the bathythermograph (BT) came into its own, with temperature-depth profiles scratched on literally hundreds of thousands of smoked glass BT slides, people had occasionally noticed a steppy finestructure superimposed on the gross profile. (This was usually attributed to stylus "stiction" and the instruments suitably repaired.)

The finestructure (down to a meter scale) has by now been well established, owing largely to the stubborn evolution of the CTD (Conductivity, Temperature, and Depth) by Neil Brown, Karl Schleicher, and Al Bradshaw. Free-fall instruments sinking slowly and employing tiny rapid-response transducers have resolved the microstructure down to the centimeter scales (here the pioneering work is by John Woods and by Charles Cox and his collaborators Michael Gregg and Tom Osborn). At this scale (the dissipation scale) molecular processes first make themselves directly felt, and even smaller scales are suppressed exponentially. There have been parallel developments in the measurement of velocity finestructure (Tom Sanford) and microstructure (Tom Osborn). The time is drawing near when we shall record from a single free-fall apparatus the entire finestructure and microstructure, down to the molecular dissipation scales, of temperature, salinity, and currents and hence of the buoyancy frequency and of the Richardson number.

I have previously referred to the extreme variability in long-range acoustic transmissions, and how this related to ocean finestructure. The finestructure is dominated by the variable straining due to the ever-present internal waves. We had learned a great deal about the scales of internal wave motion from the "site D" moorings that Woods Hole (under the leadership of Fofonoff and Webster) had set in the western North Atlantic in the late 1960s. The time was ripe for the acoustic and physical oceanographers to take up where they had left off twenty years earlier.

3 MIMI Transmissions

Meanwhile, Steinberg and Birdsall (1966) were conducting their pioneering sound-transmission experiment across the Straits of Florida as part of the Navy's acoustic program. They discovered tides were an important factor. (This was not surprising to oceanographers, who were quite accustomed to tidal components in all their measurements.) There were some difficulties in the interpretation associated with shallow-water effects, and subsequent efforts (in which they were joined by Kronengold, Clark, and others) were shifted to a 1250-km path from Eleuthera to Bermuda (Clark and Kronengold, 1974). The Eleuthera site was very close to where Ewing and Worzel had placed their bottom hydrophone in 1945.

An essential feature in these experiments (called MIMI for the Miami-Michigan participation) was that they gave continuous observations over many months, and this opened the way for a meaningful geophysical interpretation. In essence, the experiment consisted of transmitting a 406-Hz signal and recording the relative phase and intensity of the received signal using a perfectly synchronized 406-Hz oscillator. The resulting time series of acoustic phase $\phi(t)$ and intensity $\iota(t)$ are dominated by occasional fadeouts and phase jumps, which are the result of interference among the many paths from source to receiver. Here we must distinguish the multipath phase $\phi(t)$ and associated phase rate $\dot\phi(t)$ from the corresponding single path $\phi_i(t)$ and $\dot\phi_i(t)$. The single path $\langle\dot\phi_i^2\rangle$ can be inferred from the measured multipath $\langle\dot\phi^2\rangle$ under reasonable assumptions. The result is $\langle\dot\phi_i^2\rangle = 1.6 \times 10^{-5} \sec^{-2}$.

Now this parameter depends on the fluctuations of sound velocity along the transmission path and can be calculated if certain statistical properties of these fluctuations are known. Using an internal wave spectrum (based entirely on temperature, salinity, and current measurements from ships and moorings) and performing these calculations leads to the result $\langle\dot\phi_i^2\rangle = 2.7 \times 10^{-5} \sec^{-2}$ for the 1250-km transmission (Munk and Zachariasen, 1976). There are no loose parameters here, and the data sets are entirely independent—one acoustic, the other based on traditional oceanographic parameters. It was this rough agreement that first convinced me that acoustic fluctuations and ocean finestructure had something to do with one another (and gave an oceanographic interpretation of the measured "decorrelation time" of a few minutes for long-range transmission). This work has since been greatly expanded by members of the JASON group (Flatté et al., 1979), and led to a fairly complete interpretation of the acoustic fluctuation statistics measured by A. W. Ellinthorpe on the Azores Fixed Acoustic Range (Buehler, 1979). There remain some outstanding problems, of course.

Long-range transmissions are ultimately limited by the signal degradation associated with ocean finestructure. But we can say now that this degradation is surprisingly mild, and that with proper signal processing at low frequencies the 10,000-mile transmission ranges envisioned by Ewing can be attained. This brings us to the possibility of exploring large-scale ocean structure by acoustic remote sensing, which is my principal subject. But first let me show the results of a remarkable set of measurements by Gordon Hamilton nearly 20 years ago (Fig. 2). Precisely located and timed SOFAR charges were fixed at axial depth off Antigua, using the hydrophone array of the Atlantic Missile Range. From the axial cutoff of the received signal at the Bermuda and Eleuthera field stations, travel times were ascertained to within 30 msec. These are found to vary by fractions of a second over time scales of a few months; the maximum change was by

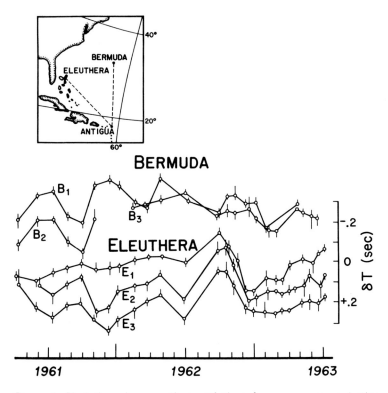

Figure 2. Variations in acoustic travel time from a source at Antigua to three Bermuda hydrophones and to three Eleuthera hydrophones (from Hamilton, 1977).

500 msec in 3 months. Hamilton remarks on the lack of correlation between the fluctuations for the Antigua-Bermuda and Antigua-Eleuthera paths (Bermuda and Eleuthera are separated by more than 1000 km). Fluctuations at the three individual Eleuthera hydrophones (60-km separations) are clearly correlated, but even here there are significant differences. This should have told us then and there of an intense ocean variability on a 100-km scale, with characteristic times of the order of months. We now call this mesoscale variability; the discovery and study of this important variability was to occupy the physical oceanographers throughout the 1960s and 1970s.

The work of Hamilton was done to satisfy specific needs of the Air Force missile test program, and was not published for $1\frac{1}{2}$ decades. As a matter of general interest, at the same time George Shor and Jack Nafe fired 200-lb depth charges offshore from Perth, Australia, and this then was recorded by Hamilton in Bermuda 10 dB above noise level. Yes, there is a great circle route; the range is 10,000 miles, reminiscent of the comment by Ewing and Worzel and about as far as one can go in the oceans.

4 Demise of Zero-Frequency Oceanography

The classical physical oceanographer cast his Nansen bottles, computed densities, contoured dynamic heights, and from these he inferred the geostrophic currents which he published on charts of ocean "circulation." It was generally agreed that, as time went on with more and more casts available, the precision and detailing of the charts would continue to improve. This pleasant viewpoint was rudely interrupted in 1959 as a result of the *Aries* measurements, when John Swallow acoustically tracked some neutrally buoyant midwater floats, giving direct measurements of motion at depth. These measurements revealed a variable current structure with kinetic energy exceeding that of the mean motion by one or two orders of magnitude!

The subsequent MODE (Mid-Ocean Dynamic Experiment) and POLYMODE efforts have clearly established the importance of the mesoscale ocean structure. Figure 3 is a sketch prepared at Woods Hole which dramatizes the revolutionary change which has taken place in our view of the ocean circulation. We can think of the classical view as a climatological mean taken over several years. But at any one moment a snapshot of the ocean circulation is dominated by the mesoscale eddies which we can regard as the ocean weather. The mesoscale scale differs from the atmospheric counterpart: storms have dimensions of 1000 km and time scales of a few days. Mesoscale eddies (the ocean storms) have dimensions of around 100 km and time scales of 100 days.

I believe that the monitoring of ocean weather is as important to the wise use of the oceans as is the monitoring of storms to atmospheric research. But the ocean problem is much more difficult on account of the small spatial scales and long time scales. This is where acoustics might be useful.

5 "Bermuda Square" Demonstration Experiment

The essential idea of remote acoustic monitoring can be illustrated by an experiment to be performed early in 1981 by R. Spindel and D. Webb, WHOI; C. Wunsch, Massachusetts Institute of Technology; T. Birdsall, University of Michigan; and by our group at Scripps Institution of Oceanography. We shall measure travel times from each of four sources to each of five receivers moored at the periphery of a 300 km × 300 km square (Fig. 4).

In the event that a cold cyclonic (say) eddy drifts through the square, then any of the acoustic ray paths through the eddy will be delayed (since sound speed increases with temperature), whereas the other ray paths will exhibit normal travel times. A typical delay time is $\frac{1}{4}$ sec, within easy reach of the precision of the measurements. The extent to which different source-receiver combinations (out of a total of 20) are affected tells us something of the location and strength of the eddy.

The problem is more complicated in that there are many multipaths between

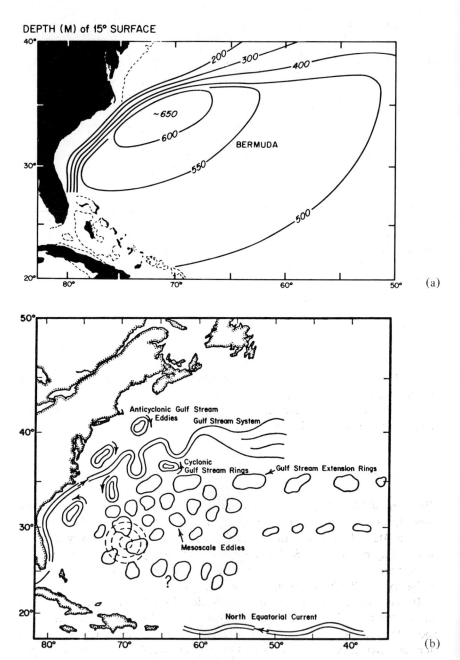

Figure 3. (a) A sketch of the climatic mean circulation in the northwest Atlantic. The contours surround the subtropical gyre; the western intensification (climatological Gulf Stream) corresponds to the region of crowded contours. (b) Sketch of a snapshot of Stream lines, showing a meandering Gulf Stream with eddies on either side.

Acoustics and Ocean Dynamics 117

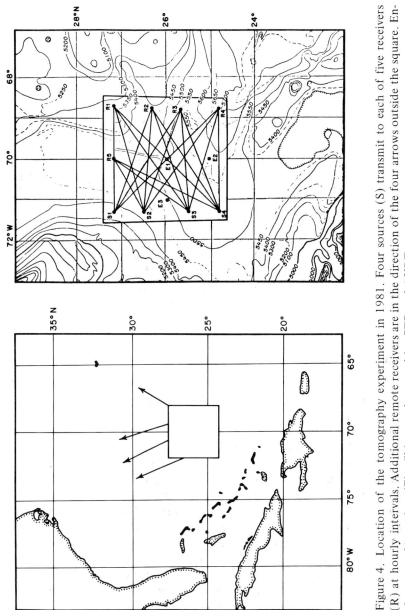

Figure 4. Location of the tomography experiment in 1981. Four sources (S) transmit to each of five receivers (R) at hourly intervals. Additional remote receivers are in the direction of the four arrows outside the square. Environmental moorings (E) will be supplemented by CTD and AXBT surveys.

any one source-receiver pair. Figure 5 shows a typical North Atlantic ray diagram for a 1000-km range. Rays with turning points near the surface and bottom come in early; flat rays that stay near the axis come in late. (The axial "SOFAR arrival" is missing because the source is not at the axis.) If the multipaths can be resolved, one can get information about the depths of the eddy structure. For example, a very shallow eddy would affect only the early steep rays; an eddy extending below the axis would affect all arrivals.

Figure 6 shows the result of a recent experiment. Here 14 multipaths have been resolved over a long range. The principal advance since the work of Ewing and Worzel and of Hamilton is that we can work with resolved multipaths and are not limited to the final SOFAR cutoff. (In fact we avoid the final stages of unresolved arrivals by placing the source and receiver moorings well beneath the axis.) In this way we can obtain the depth information. We no longer use the Ewing-Hamilton explosive sources which are not too well timed and have non-reproducible bubble pulses. Nor do we use the MIMI CW transmissions which are incapable of resolving multipaths. Rather we transmit at a low power level (~ 1 watt) over long periods (~ 100 sec) a pseudorandom sequence which has the property of a uniquely sharp autocovariance. Arrival times are measured by correlating the received sequence with a replica of the transmitted sequence, giving a correlation spike for each multipath arrival. Here we are limited by the relatively narrow transmitter bandwidth $\Delta f \approx 20$ Hz and corresponding coarse resolution $1/\Delta f = 0.05$ sec, which is, interestingly enough, equivalent to the precision of the cutoff used by Ewing and Worzel in 1944.

6 Inverse Problem

The transmission from each of the four sources is recorded at each of the five receivers. For 10 resolved multipaths we then have $4 \times 5 \times 10$ travel times. The method of using these 200 numbers measured once an hour to derive the best possible field of sound speed (temperature) as a function of x,y,z,t is a problem of inverse theory. In general, the problem is underdetermined and the solutions are ambiguous. One can remove the ambiguity by imposing restraints. One such restraint is to seek the least wiggly solution consistent with the measurements and allowing for the precision of the measurements (Munk and Wunsch, 1979).

Suppose one uses the acoustic travel times to produce ocean weather maps at weekly intervals. Carl Wunsch has been thinking about the interesting question of whether, and to what extent, one should use hydrodynamic theory to aid in updating the maps from one week to the next. There was a time just after computers were introduced into weather prediction when the predictive skill actually diminished. We now know that this was the result of not paying enough attention to the extrapolation (by theory and experience) of past data. We want to avoid such errors. This is an interesting example of the interplay of acoustics and ocean dynamics.

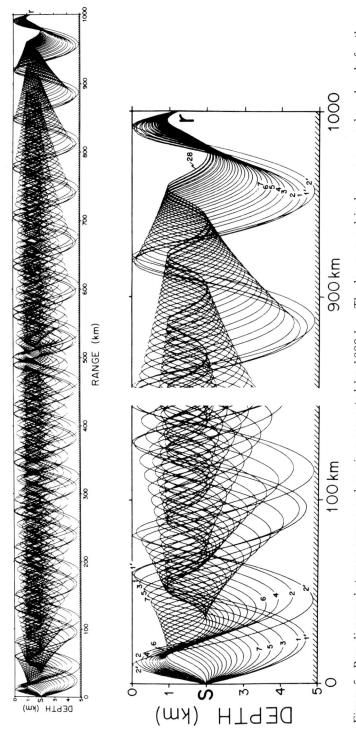

Figure 5. Ray diagram between source and receiver separated by 1000 km. The lower panel is drawn on an enlarged scale for the first and last 100 km. Only rays which tilt downward at the receiver are plotted.

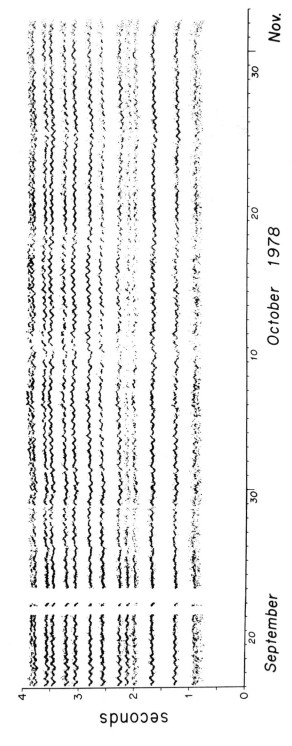

Figure 6. Relative travel times over a 900-km path, according to Spiesberger et al. (1980). About 14 multipaths are resolved, and these arrive within about 3 sec. The paths are quite stable over a 2-month period. The small wiggles are the result of propagating alternately with and against the M_2 tidal current.

Worcester (1977) has carried out reciprocal transmissions over a 25-km path, with sources and receivers colocated. The differential travel times furnished very precise information on current components along the acoustic paths. (Worcester's installation can be regarded as a stretched acoustic current meter.) There is no fundamental obstacle to performing reciprocal transmissions over the scales of our demonstration experiment. A comparison of the temperature field and the current field could provide an estimate of the small (but important) departure from geostrophy. This was one of the goals of MODE, but so far the ageostrophic component has not been measured.

One could ask: what are the advantages of this roundabout way of estimating the variable ocean structure between moorings as compared to direct measurements at the moorings? There are several considerations: m moorings consisting of r receivers and s sources yield $r \cdot s$ pieces of information, rather than $r + s = m$ from "spot" measurements; spatially integrated measurements (such as travel times) filter out the small scales which contaminate conventional spot measurements; we can avoid the difficulties of mooring in intense currents by placing the moorings to the sides or beneath the strong currents.

We have called this method "ocean acoustic tomography" because of its resemblance to medical tomography, where the interior of a part of the body is obtained from reconstruction of views at many different angles. One advantage of the oceanographic problem as compared to the medical problem is that the interior of our physical space is actually accessible without damage. In the long run this may turn out to have been a liability, for it has delayed indirect methods which (in our prejudiced view) are made inevitable by the magnitude of the ocean monitoring task.

7 Monitoring an Ocean Basin

Can the method be extended to the dimension of an ocean basin? The required number of moorings is proportional to the linear dimension, so for a 1000-km square we would need three times the nine moorings to be deployed in the 1981 experiment. This is a large number, but not an absurd number. Further, following a suggestion by John Clark, one can replace some of the moored sources and receivers with a few towed instruments whose position is precisely determined by satellite fixes. We think that in this way the total effort can be kept at a reasonable level.

Our view is that the acoustic remote sensing should be in concert with satellite remote sensing. In particular, satellite altimetry can provide global coverage of sea-surface topography with excellent spatial resolution. This is the ocean equivalent to surface weather maps. Perhaps the principal function of acoustic remote sensing is to provide a downward extension of the surface features, just as we need to augment the surface weather maps with measurements aloft.

I need to add a word of caution. Acoustic tomography is in an experimental stage. If we should prove successful, it is still just one of a number of modern developments for sensing the oceans, and it would have to be used in proper balance with those developments.

8 Scatterometry and Other Acoustic Applications

I have emphasized tomography here because of my personal involvement. Acoustic techniques have proved useful in many other ways in recent work. Some of these have been summarized in a recent National Oceanographic and Atmospheric Administration Workshop on OARS (Ocean Acoustic Remote Sensing) (Murphy and Schulkin, 1980).

Pinkel (1975) has measured the backscatter of a narrow high-frequency beam transmitted horizontally (refraction curves it downward) from the bottom of the platform *Flip*. By measuring the Doppler shift as a function of range r he obtains the current component $u_{ray}(r)$. This is like towing a current meter at 3000 knots! The work has given a new view of internal wave dynamics.

Marshall Orr uses a downward beam from a surface vessel that is underway at slow speed. Scattering horizons related to the density structure are vertically distorted, presumably by internal waves. The scatterers are mostly biological; they

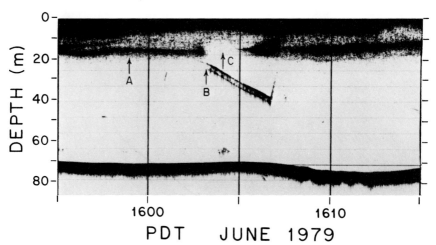

Figure 7. Backscatter within the water column of a narrow 357-kHz acoustic beam pointing downward from an anchored vessel. Scatterers are concentrated in a layer at 15 m depth (A). As a previously launched slowly sinking (8 cm/sec) instrument comes into view (B), the biological scatterers are found to have moved out of the acoustic beam (C) until the instrument has descended 15 m beneath the scatterers. I am indebted to M. Orr (1980) for this remarkable recording.

are more plentiful at night, and in one instance (Fig. 7) have maneuvered to escape a descending instrument (a harmless Gregg microstructure recorder). Orr's technique can prove very useful in monitoring particle concentration on ocean dump sites (Fig. 8).

Upward-looking sonars (inverted echo sounders) can give a measure of the heat content of the water column. A bottom-based ambient noise device has been used to infer wind speed and rainfall. The directional surface wave spectrum has been measured from surface scattering of an upward beam. Doppler current shear profilers can be used from ships underway. And although echo sounders have been used by commercial fishermen for half a century, a quantitative stock assessment by acoustic means is a recent development. The management of Peruvian anchoveta and Barents Sea copelin is now guided principally on the basis of data derived from acoustic surveys.

This gives a flavor of the present high level of activity of acoustic ocean monitoring. But perhaps the most influential application to ocean dynamics has been the work of Rossby et al. (1975). This is an extension of the Swallow float, which started the mesoscale revolution; now the floats are monitored from shore stations at ranges up to several 1000 km, for years. This has provided a Lagrangian view of the ocean to which we were quite unaccustomed: float drifts change in a jerky, sometimes almost discontinuous fashion.

9 Will We Use the Oceans Wisely? The Next 50 Years

We are now taking up where we left off in the 1950s. What role will acoustics play in the oceanography of the future? I believe it will be as vital as the role played by optics, radar, and radio in the management and study of the atmosphere and the space beyond. Perhaps the key to the potential contribution by acoustic remote sensing is that it can provide a sampling density in space and time that is adequate to the complexity of ocean structure.

I have puzzled about the reasons for the relative lull between the pioneering discoveries of the postwar years and the present accelerating pace. The Navy has always played a vital and positive role in ocean acoustics, but problems of security have necessarily placed some distance between the Navy-oriented ocean acoustic community and the remaining oceanographic community, to the detriment of both.

I should like to end this essay with two proposals:

1. Someone (Hersey, Vine, Schevill, perhaps) who participated in the development of the underwater sound technology should give a firsthand account of this era before it is too late. This should include not only the development of a remarkable technology, but also the history of Navy-civilian interaction from which much could be learned for future benefit.

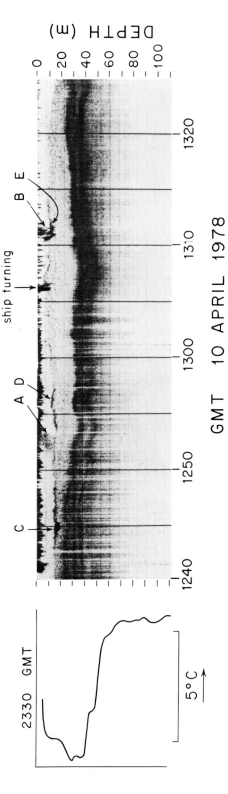

Figure 8. An acoustic recording during a chemical waste disposal on Deep Water Dumpsite 106. The particulate phase of the chemical waste can be seen distributed in the upper 15 m of the water column (A, B) and accumulating onto a density interface (C, D, E). An XBT profile taken about 10 hours later shows the structure of the springtime water column (from Orr and Baxter, 1980).

2. We need a civilian acoustic array to serve as an observatory for ocean studies, at a location and in a mode of operation that does not interfere with the Navy mission. This ocean acoustic observatory can serve two broad purposes: one is to develop all aspects of acoustic oceanography, the other is to serve as a receiving antenna for acoustically telemetered underwater observations from remote stations.

Acknowledgments

I am deeply grateful to the Office of Naval Research for having supported my work throughout my career (and with no strings tied), up to and including the present efforts in remote acoustic sensing. This essay is based on experience gained during this work. Much of the information was learned from Brackett Hersey. I am grateful to Roger Revelle for critical comments.

References

Buehler, B. G. 1979. Volume propagation experiments at the Azores Fixed Acoustic Range. Naval Underwater Systems Center, Technical Report 5785.
Clark, J. G., and M. Kronengold. 1974. Long-period fluctuations of CW signals in deep and shallow water. J. Acoust. Soc. Am. 56, 1071-1083.
Ewing, M., and J. L. Worzel. 1948. Long-range sound transmission. Geol. Soc. Am. Memoir 27.
Flatté, S. M., R. Dashen, W. H. Munk, K. M. Watson, and F. Zachariasen (Eds.). 1979. Sound Transmission Through a Fluctuating Ocean. Cambridge University Press, Cambridge, 299 pp.
Hamilton, G. R. 1977. Time variations of sound speed over long paths in the ocean. Paper presented at the International Workshop on Low-Frequency Propagation and Noise. Woods Hole, Massachusetts, Oct. 14-19, 1974, pp. 7-30.
Munk, W. H., and C. Wunsch. 1979. Ocean acoustic tomography: a scheme for large scale monitoring. Deep-Sea Res. 26A, 123-161.
Munk, W. H., and F. Zachariasen. 1976. Sound propagation through a fluctuating stratified ocean: theory and observation. J. Acoust. Soc. Am. 59, 818-838.
Murphy, J. R., and M. Schulkin. 1980. Ocean acoustic remote sensing. Paper presented at the NOAA workshop in Seattle, Washington, January 21-24, 1980. Washington Sea Grant, Div. of Marine Resources, University of Washington.
Orr, M. H. 1980. Remote acoustic detection of predator-prey interaction, biological response to oceanographic instrumentation, and the response of biological organisms to fluid processes. J. Fish. Res. Bd. Can. In press.
Orr, M. H., and L. Baxter, III. 1980. Seasonal dependence of the vertical and horizontal dispersion of particles formed or released during the disposal of industrial chemical waste or sewage sludge. Proc. Second International Ocean Dumping Symposium. In press.
Pinkel, R. 1975. Upper ocean interval wave observations from FLIP. J. Geophys. Res. 80, 3892-3910.

Rossby, T., A. D. Voorhis, and D. Webb. 1975. A quasi-Lagrangian study of mid-ocean variability using long range SOFAR floats. J. Mar. Res. 33, 355–382.

Shockley, R. C., J. Northrop, and P. G. Hansen. 1982. SOFAR propagation paths from Australia to Bermuda: Comparison of signal speed algorithms and experiments. J. Acoust. Soc. Am. 71(1), 51–60.

Spiesberger, J. L., R. C. Spindel, and K. Metzger. 1980. Stability and identification of ocean acoustic multipaths. J. Acoust. Soc. Am. 67, 2011–2017.

Steinberg, J. C., and T. G. Birdsall. 1966. Underwater sound propagation in the Straits of Florida. J. Acoust. Soc. Am. 39, 301–315.

Worcester, P. F. 1977. Reciprocal acoustic transmission in a midocean environment. J. Acoust. Soc. Am. 62, 895–905.

Oceanic Biology: Lost in Space?

James J. Childress

1 Introduction

The topic which I have been asked to address in this presentation is the past and future of studies in oceanic biology. My first reaction to this request was that the topic was a staggeringly large one, for which I was only slightly qualified. My second reaction was that when scientists reach the stage of lecturing on what they see in crystal balls, they are generally over the hill creatively and have little to contribute directly to the living religion of science. John Teal reassured me, however, that he thought I was qualified and that I could proceed to have fun with the topic. So with this disclaimer, here is my shot into the darkness.

First I would like to define my area of concern as the waters covering about 62% of the earth's surface with depths of 1 km or more (Fig. 1a). For the purposes of this presentation, the euphotic zone is also excluded from the oceanic realm, since I know little about these surface layers. What is left is simply the largest volume known to be occupied by living organisms (Fig. 1b), excluding the atmosphere which has no permanent aerial inhabitants. This immense realm is studied by few biologists; in fact, most biological oceanography takes place embarrassingly close to shore and certainly not in oceanic waters. There are many ways to illustrate the biological neglect of oceanic organisms, but my favorite is to examine invertebrate zoology texts. The typical text gives such major groups of oceanic crustaceans as the copepods, mysids, and euphausiids a fraction of a page apiece while spending many pages on such quantitatively trivial, but predominantly shallow-living groups as barnacles. In short, biological oceanography is largely anthropocentric.

Even within ocean biology one finds intellectual constructs which are pro-

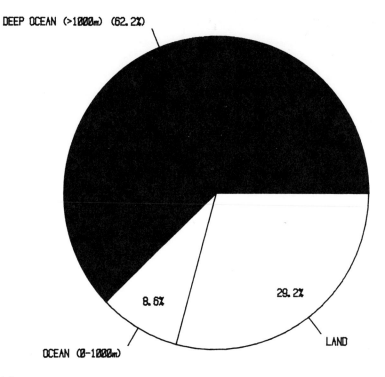

(a)

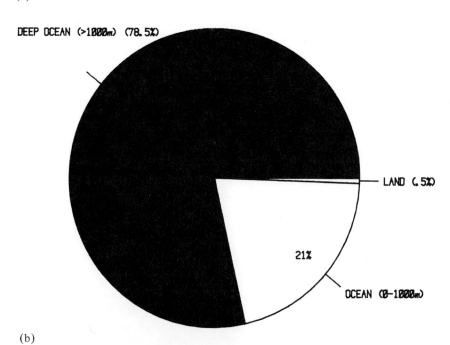

(b)

foundly limiting to oceanic studies. One of my colleagues has frequently expressed concern that our interests overlap too much and that we should be careful not to study similar organisms for fear of jeopardizing our funding. His fears are not unrealistic, since one agency which funds him has clearly told me that they will only support one investigator working on the metabolic rates of oceanic animals. Is there only room for one person studying the metabolism of terrestrial animals? Another point of view which I have heard often is that there are very few important questions in oceanic biology. Therefore, it is appropriate that few people work on them.

2 The Alien Life

Remember that our topic is the biology of organisms occupying the largest volume known to be inhabited. Not only is it the largest, it is also far and away the most alien of known major habitats. It is dark and cold; many of its inhabitants make their own light; and most of this habitat is characterized by remarkable stability and low food levels. Organisms within this realm exist at high hydrostatic pressures. Most of the pelagic inhabitants of this realm spend their entire lives in midwater completely out of contact with the bounding surfaces.

How alien are these characteristics? In talking to nonbiologists, and sometimes to biologists, I often find them completely amazed at the idea of organisms living their entire lives out of contact with surfaces (in midwater) and in total darkness except for the light produced by the organisms themselves. A geologist described being told by a membrane biochemist that life could not exist in the deep sea, because of pressure effects on membranes; the biochemist was right, of course—surface living animals can't survive in the deep sea, nor can deep sea animals survive at the surface. Further, this environment and its inhabitants are alien in many other ways.

If only for this alienness, biology as a whole has much to learn from oceanic organisms. However, there are even anthropocentric reasons for wanting to know more about oceanic biology. Disposal of both low- and high-level nuclear wastes in the ocean poses biological questions which we cannot now answer, and a variety of other human interactions with the oceanic biota are becoming interesting.

Yet, as I suggested before, oceanic biology has remained a small backwater in biology. There has been progress; however, even the practitioners of this science have often seemed unaware of its exciting potentials. I believe that the field and its workers, by and large, lack a vision of oceanic biology. This lack of larger

---

Figure 1. (a) Fraction of earth's surface covered by ocean greater than 1000 m deep (Kossinna, 1921). (b) Volume of major habitats on earth. Calculated from Kossinna (1921) assuming that the volume occupied by life on land is 50 m thick.

vision has led in many ways to a stagnation in this field, a stagnation often cultivated by the major oceanographic institutions. In short, we have a field which is, in the jargon of the times, "lost in space."

3 The Roots of Oceanic Biology

As we are all aware, early Europeans held that the oceanic realm was devoid of life (Schlee, 1973). Gradually, as it became apparent this was not so, oceanic exploration gained considerable impetus. This interest was fueled both by the strangeness of the organisms found and by the hope that evolutionary "missing links" might be found. This sort of relatively haphazard exploration continued through the 1960s. This work described the broad outlines of oceanic communities and many of the morphological and taxonomic characteristics of oceanic animals. The impression was that we knew a good deal about the organisms of this realm. At the same time, with the ascendance of population biology to the forefront of ecology, came an emphasis on hypothesis testing in ecology. In fact, within ecology in general as well as biological oceanography, the hypothesis testing approach has almost come to be the definition of good science. This is the sort of thinking that has led to the push away from exploration, the push to do "real science."

This has had at least two effects on oceanic biology. The first is that exploratory work has become increasingly difficult to get funded. The second is that since in most of oceanic biology we simply don't know enough to formulate substantial testable hypotheses, this criterion tends to shut off necessary exploration and produce trivial hypothesis testing—that is, to destroy the field. The antidote, as far as I am concerned, is to remember that science is a spirit of inquiry into the nature of the world; and that scientific method, logic, statistics, and gas chromatographs are tools of scientists. I have read that the scientific method works very well for motorcycle mechanics, but this does not make a motorcycle mechanic a scientist.

Proof that exploration is not intellectually dead in oceanic biology comes from the fact that much of the most significant progress in recent years has come from exploration. This has been both exploration of the physical ocean and exploration of biological phenomena within the ocean.

4 The Latest Findings

In addition to the already considerable variety of known oceanic habitats, several new ones have recently been found. The deep-sea spreading center hydrothermal vent communities discovered by Ballard and other geologists provide a habitat and community radically different from any known previously (Corliss et al., 1979). The findings of Felbeck, Cavanaugh, Rau, Somero, Arp, and Jones clearly point to the probability that the vestimentiferans and some other major vent animals may, with the use of symbiotic bacteria, derive food energy from the vent water hydrogen sulfide, fix carbon dioxide into organic material, and perhaps even fix dissolved nitrogen (Felbeck, 1981; Cavanaugh, 1980; Cavanaugh et al., 1981; Rau and Hedges, 1979; Rau, 1981; Arp and Childress, 1981; and Jones, 1981). These findings not only open up an existing avenue for study in the deep sea but also change our perspective on many environments. As Cavanaugh (1980) has suggested, wherever sulfide and O_2 are present one should suspect the possibility of these symbioses. This sort of symbiosis may explain the mystery of the food supply of pogonophorans in general.

Another very different vent community was discovered by Eric Barham in 1966 and rediscovered by Peter Lonsdale (1979) along faults off Southern California. I and one of my students, Alissa Arp, dove last week on Lonsdale's site on the San Clemente fault at a depth of 1800 m. This habitat lacks the dramatic outflow of water which characterizes spreading center vents; in fact, a wide variety of typical deep-sea nonvent life grown on and around the vestimentiferans and clams found there (Fig. 2a). Yet preliminary observations by Cavanaugh, Felbeck, Mickel, Arp, myself, and Jones all suggest that the same kind of autotrophic metabolism is in use here. These kinds of habitats and symbioses may prove to be widely distributed in the deep sea even if they are highly localized in a given location.

Yet another deep-sea community only recently gaining attention is the group of pelagic animals living in association with the bottom but not on it. Karen Wishner (1980a, 1980b) has documented the small creatures living there. I have worked on larger organisms in San Clemente Basin and East Cortez basin off Southern California. There is a tremendous biomass near the bottom (up to 5 mg dry wt m^{-3}) and most of it consists of the pelagic sea cucumber *Scotoanassa* (Fig. 2b; Barnes et al., 1976). The biomass of the near bottom pelagic community in these basins is much denser than the overlying pelagic communities (about 0.02 mg dry wt m^{-3} at 1000 m above the bottom). Undoubtedly there are other different habitats and communities waiting to be found in the ocean. Many of the known ones also need further exploration.

In addition to these findings of different habitats and communities, many exciting findings have come out of explorations of a more quantitative nature. Sanders and Hessler's discovery of the large number of kinds of critters in the deep sea has provided the basis for a great deal of discussion over the years (Sanders and Hessler, 1969). Jannasch and co-workers took advantage of perhaps the most important work ever done with a deep submersible in the *Alvin* lunch study, in which a sandwich was recovered from the deep sea (Jannasch et al., 1971). These observations clearly suggested that microbial community growth and metabolism was quite low at depth.

My own finding of much lower metabolic rates in individual bathypelagic animals came as a complete surprise to me and was the opposite of what I expected (Childress, 1969, 1971, 1975). Yet I made the first measurements only because the animals were available.

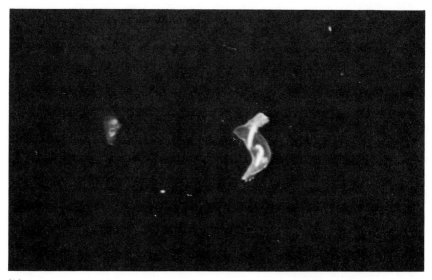

(b)

Figure 2. (a) Vent area on San Clemente Fault at a depth of 1750 m (Lonsdale, 1979). Vestimentiferan tube worms in the foreground. (b) Near-bottom pelagic sea cucumber (*Scotoanassa* sp). This individual was photographed from *Alvin* in the San Clemente Basin off Southern California at a depth of about 2000 m.

George Somero's finding of the low catalytic efficiency of pressure-adapted enzymes was quite surprising (Somero and Siebenaller, 1979). The low turnover rates and higher energies of activation of these enzymes are the opposite of what is found in shallow-living cold-water species. Sam Taylor and I (Childress et al., 1980) have recently been investigating the growth rates, ages, reproduction, and energy budgets of the bathypelagic fishes. The surprising picture which emerges here is that these deep-living fishes grow faster and live shorter lives than do the shallower living, vertically migrating fishes. Further, the bathypelagic fishes apparently defer reproduction until the end of their lives, reproducing only once with rather low fecundity. This life history pattern is so unusual that it forces a major reconsideration of the selective forces acting upon animals in the bathypelagic realm, as well as a reconsideration of the life history consequences of stable, food-poor environments.

Another quite unexpected observation is Ken Smith's recent finding of an apparent daily rhythm in the metabolism of a fish living at 1200 m (Smith and Laver, 1981). In measurements of three individual *Cyclothone acclinidens* living at 1200 m off Southern California, Smith found 3- to 6-fold increases in O_2 consumption rates during nighttime hours. These data raise important questions about both the cueing of such behavior and its significance.

In summary, we still know so little about oceanic biology that exploration is an extremely productive and important way to proceed. This is not a flaw in our science but a statement of how much remains to be studied.

5 Exploring the Future

As we now turn to the future with a spirit of exploration, I want to speak of the role of the major oceanographic institutions. These institutions, through their resources and activities, dominate the field of oceanography in a way that few other fields of science are dominated. The temptation is always present for these institutions to seek their own goals of institutional growth, security, and defense at the expense of the community at large. Giving in to such temptations would be a terrible blow to oceanography, since much of the creative force in the field comes from its periphery, where individuals trained outside oceanography are drawn to oceanographic problems. The health of such a multidisciplinary field as biological oceanography depends upon maintaining and increasing the access of those outside the core of oceanography to the tools of trade. I hope that the major institutions will take these community responsibilities more seriously in the future. A quote from Jim Morrison of "The Doors" may be appropriate here: "When you make your peace with authority, you become an authority."

My prescription for the future of oceanic biology is to provide committed, innovative individuals with the "space," that is, the resources needed for them to work productively. In particular I believe we need further exploration, both for new kinds of oceanic habitat and creatures as well as quantitative exploration of a variety of phenomena. I believe that it will be particularly valuable to compare diverse habitats since this will provide a much greater range of any given variable than trying to follow short-term changes in a single habitat. All of this is, of course, dependent on the larger world of human concerns and politics. I would like to close this with one of my favorite poems by Wallace Stevens, "Of Mere Being":

OF MERE BEING

The palm at the end of the mind,
Beyond the last thought, rises
In the bronze distance,

A gold-feathered bird
Sings in the palm, without human meaning,
Without human feeling, a foreign song.

You know then that it is not the reason
That makes us happy or unhappy.
The bird sings. Its feathers shine.

The palm stands on the edge of space.
The wind moves slowly in the branches.
The bird's fire-fangled feathers dangle down.

References

Arp, A. J., and J. J. Childress. 1981. Blood function in the hydrothermal vent vestimentiferan tubeworm. Science 213, 27-40.
Barnes, A. T., L. B. Quetin, J. J. Childress, and D. L. Pawson. 1976. Deep-sea macroplanktonic sea cucumbers: suspended sediment feeders captured from deep submergence vehicle. Science 194, 1083-1085.
Cavanaugh, C. M. 1980. Symbiosis of chemoautotrophic bacteria and marine invertebrates. Biol. Bull. 159, 457.
Cavanaugh, C. M., S. L. Gardiner, M. L. Jones, H. W. Jannasch, and J. B. Waterbury. 1981. Prokaryotic cells in the vestimentiferan pogonophore *Riftia pachyptila* Jones; possible chemoautotrophic symbionts. Science 213, 340-342.
Childress, J. J. 1969. The respiratory physiology of the oxygen minimum layer mysid *Gnathophausia ingens*. Ph.D. dissertation, Stanford University, Palo Alto, Ca., 142 pp.
Childress, J. J. 1971. Respiratory rate and depth of occurrence of midwater animals. Limnol. Oceanogr. 16, 104-106.
Childress, J. J. 1975. The respiratory rates of midwater crustaceans as a function of depth of occurrence and relation to the oxygen minimum layer off Southern California. Comp. Biochem. Physiol. 50A, 787-799.
Childress, J. J., S. Taylor, G. Cailliet, and M. H. Price. 1980. Patterns of growth, energy utilization and reproduction in some meso- and bathypelagic fishes off Southern California. Mar. Biol. 61, 27-40.

Corliss, J. B., J. Dymond, L. I. Gordon, J. M. Edmond, R. P. von Herzen, T. D. Ballard, K. Green, D. Williams, A. Bainbridge, K. Crane, and T. H. van Andel. 1979. Submarine thermal springs on the Galapagos Rift. Science 203, 1073–1083.

Felbeck, H. 1981. Chemoautotrophic potentials of the hydrothermal vent tube worm, *Riftia pachyptila* (Vestimentifera). Science 213, 336–338.

Hopkins, Jerry and D. Sugarman. 1980. No One Here Gets Out Alive. Warner Books, New York, 387 pp.

Jannasch, H. W., K. Eimhjellen, C. O. Wirsen, and A. Farmanfarmian. 1971. Microbial degradation of organic matter in the deep sea. Science 171, 672–675.

Jones, M. L. 1981. *Riftia pachyptila* Jones: Observations on the vestimentiferan worm from the Galapagos Rift. Science 213, 333–336.

Kossinna, E. 1921. Die Tiefen des Weltmeerls. Berlin Univ., Institut F. Meereskunde. Veröff, N.F., A. Geogr.-natururss. Reihe, Heft 9, 70 pp.

Lonsdale, P. 1979. A deep-sea hydrothermal site on a strike-slip fault. Nature (London) 281, 531–534.

Rau, G., and J. I. Hedges. 1979. Carbon-13 depletion in a hydrothermal vent mussel: Suggestion of a chemosynthetic food source. Science 203, 648–649.

Rau, G. H. 1981. Carbon-13/Carbon-12 in vestimentiferan worm and bivalve tissue from hydrothermal vents: Further evidence of non-photosynthetic food sources. Science 213, 338–340.

Sanders, H. L., and R. R. Hessler. 1969. Ecology of the deep-sea benthos. Science 163, 1419–1424.

Schlee, S. 1973. The Edge of an Unfamiliar World. E. P. Dutton, New York, 398 pp.

Somero, G. N., and J. F. Siebenaller. 1979. Inefficient lactate dehydrogenases of deep-sea fishes. Nature (London) 282, 100–102.

Smith, K. L., Jr., and M. B. Laver. 1981. Respiration of the bathypelagic fish *Cyclothone acclinidens*. Mar. Biol. in press.

Stevens, Wallace. 1972. Opus Posthumous. Alfred A. Knopf, New York, 301 pp.

Wishner, K. F. 1980a. Near-bottom sound scatterers in the Ecuador Trench. Deep-Sea Res. 27A, 217–223.

Wishner, K. F. 1980b. The biomass of the deep-sea benthopelagic plankton. Deep-Sea Res. 27A, 203–216.

Eddies and the General Circulation

H. Thomas Rossby

1 Introduction

The advances in our knowledge of the ocean circulation in the last two decades have been tremendous. In the past we perforce viewed the ocean as a steady-state system of large-scale wind-driven currents and a weaker thermohaline abyssal circulation. Now our attention is drawn to the omnipresent, often very energetic fluctuations of ocean currents that occur over a wide spectrum of length and time scales. The variety of oceanic motion that falls under the general rubric of eddy motion is extraordinary. Indeed, we now wonder whether the presumption of a steady-state circulation is valid or merely reflects our inability to understand the very long time-scale secular changes of the oceans.

The study of eddies, as a generic reference to oceanic variability, is a rapidly evolving discipline in oceanography and one that frequently encompasses all aspects of marine science, not just physical oceanography. We cannot possibly give justice to the many exciting efforts that are presently underway. The reader is referred to the forthcoming treatise entitled *Eddies in Marine Science* edited by A. R. Robinson (1981) for a detailed and comprehensive survey. There is also an issue of *Oceanus* (1976) devoted to ocean eddies including the extraordinary Gulf Stream rings. Rhines (1977) offers an excellent and detailed review, entitled "The Dynamics of Unsteady Currents."

There are probably many ways eddies influence the general circulation. Certainly they help to dissipate the larger scale flows in an energy cascade to smaller and smaller scales. But they may also interact constructively with each other to create mean flows (negative viscosity) by means of Reynold stresses. Another role of certain types of eddy motion may be to redistribute water masses by means of small-scale dynamics having the appearance of a large-scale flow. We will touch upon this below. Finally, there are "virtual" eddies which arise due to the spatial variability of steady systems, such as the meandering of currents or the passage of rings.

In this overview we explore eddy motions as evidenced by the motions of neutrally buoyant drifters known as SOFAR (sound fixing and ranging) floats (Rossby et al., 1975). We look first at the open ocean (subtropical gyres) where the eddy kinetic energy levels are low, then the Gulf Stream recirculation region (the POLYMODE Local Dynamics Experiment) where the eddy activity is very pronounced, and finally we speculate on the behavior of several SOFAR floats that got entrained into the Gulf Stream.

SOFAR floats are freely drifting subsurface instruments which are tracked over great distances (>1500 km) by means of a natural acoustic wave guide in the oceans known as the SOFAR channel. Every 8 hours each instrument transmits a low-frequency acoustic signal. From the time of arrival of this signal at several receivers the position of the float can be determined to within a few miles (Rossby et al., 1982a).

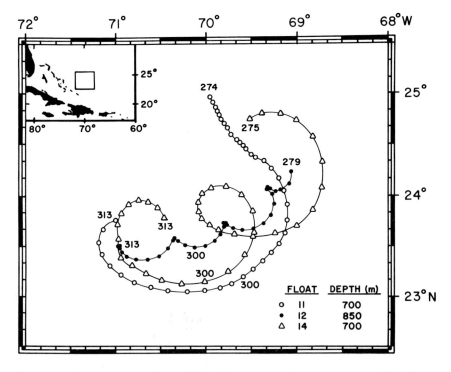

Figure 1. Trajectories of three SOFAR floats launched in October 1976. Daily positions and time in yeardays are indicated. The small map shows the area relative to the Bahama Islands.

2 The Open Ocean

We begin with the trajectories of two float groups, each of which revealed the presence of a small-scale, thin baroclinic lens containing water of a type that can only be found thousands of kilometers from where the eddy was located.

The first one is the Mediterranean eddy, a lens about 600 m thick and 100 km in diameter (McDowell and Rossby, 1978). Several SOFAR floats were set in this eddy, and three of these exhibit striking anticyclonic motion with time (Fig. 1). One of the floats continued this orbital motion for over 8 months. Hydrographic stations (salinity, temperature, depth STD profiles) revealed an extraordinary salinity excess anomaly of about $0.2°/_{oo}$ at $10°C$, which is about 17 standard deviations outside the normal T-S curve for the area, shown in Figure 2. The salinity is so high that there can only be one area the water could have come from—the Mediterranean outflow. In Figure 5a we show the eddy in relation to the distribution of salt throughout the North Atlantic. Figures 5b and 5c show the corresponding salinity distributions at greater depths (McDowell, 1981).

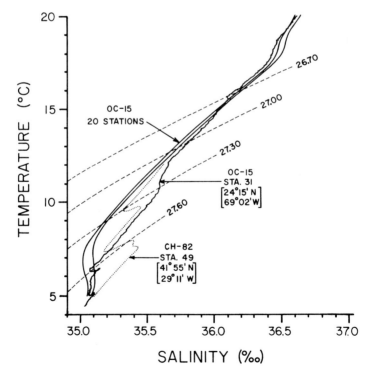

Figure 2. Temperature-salinity curve from an STD cast inside the eddy. The very high salinities in the lower main thermocline indicate that the water in this eddy must have come from the region of the Mediterranean outflow.

Attention was drawn to the other eddy by the trajectory of a 700-m float that underwent almost monochromatic anticyclonic oscillations with a $3\frac{1}{2}$-day period for over 2 months before the float was retrieved (Fig. 3). A brief STD survey revealed a small negative salinity anomaly, but a striking O_2 minimum (Fig. 4). The combination of temperature, salinity, and oxygen suggests that this water originates from somewhere along the low-O_2 tongue that emanates from

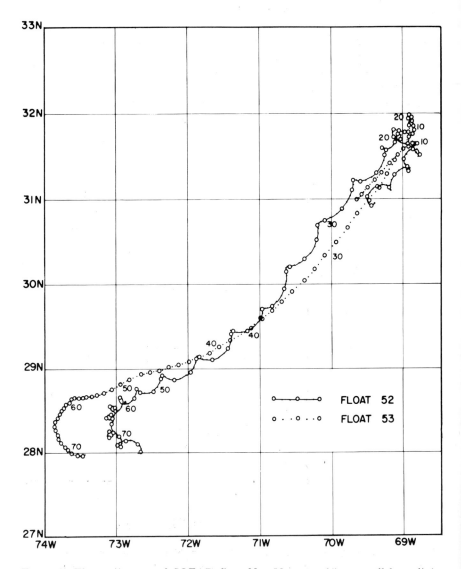

Figure 3. The trajectory of SOFAR float No. 52 trapped in a small baroclinic lens of water. The path is well described by a low-frequency translation and an orbital motion with a period of $3\frac{1}{2}$ days.

Africa between 15° and 20°N (Fig. 6a) (Rossby et al., 1982b). Figures 6b and 6c show the corresponding oxygen distributions at greater depths.

The zonal orientation of both the salinity and oxygen distributions in Figures 5 and 6 is striking. Even at depths where the mean flow patterns might be expected to differ from the main thermocline circulation, the zonal extension is still apparent. The f/H contours (Gill, 1970), especially near the Mid-Atlantic Ridge, along which a steady circulation might be patterned, are far from zonal, as shown in Figure 7. This suggests the possibility that the distributions of salt and dissolved oxygen are not maintained by large-scale "steady dynamics" but by small-scale processes which do not "feel" the bottom.

How these eddies made their way across the ocean is presently a matter of conjecture. Are they passively advected by the larger scales, or are they actively propelled westward on a β plane, drawing upon their available potential energy to sustain their locomotion? Either way they must be several years old, indicating that they are very stable and able to resist erosion by and dilution with the surrounding waters. We are not aware of any atmospheric counterpart to these eddies.

To further dramatize the zonality of these distributions, Figure 8 gives a schematic view of a number of midocean tongues, all of which are oriented to the west except for the Antarctic Intermediate Waters (AAIW), which, as Wüst (1936) pointed out, moves north, primarily along the western boundary of the Atlantic basin.

One suspects that if other interior or eastern boundary tracer sources were available these, too, would have tonguelike distributions. The classical core

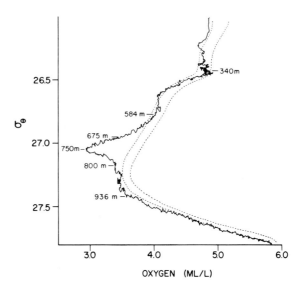

Figure 4. Oxygen-sigma θ diagram for an STD cast inside the low-oxygen eddy. Compare this to the typical O_2-σ_θ curve indicated by the dashed lines.

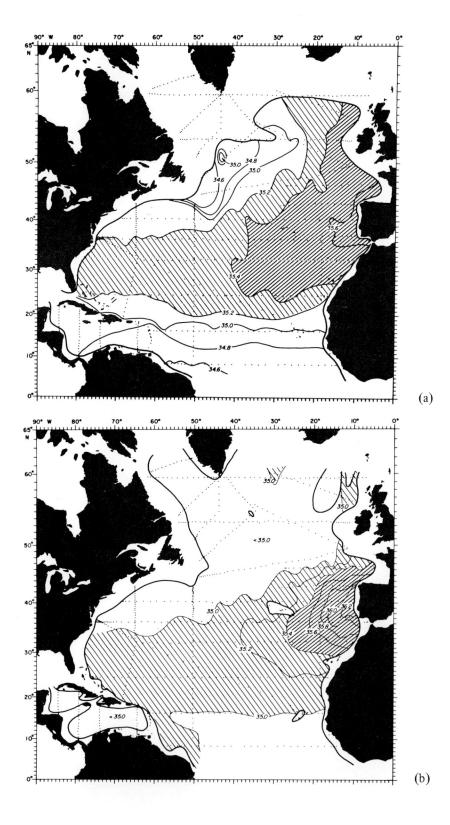

method would give us an ocean moving *en masse* from east to west, clearly an absurd situation. The point here is an old one: tongues are not simply advective but are also functions of boundary conditions, source distributions, and diffusive mechanisms. In addition to these well-known considerations, the two lenses here suggest a new, specific mechanism for the active transport of material in rather undiluted form westward (nearly) along constant latitude lines. This question is given added relevance by the fact that there is some evidence of an opposing, eastward flow between 20°N and 30°N in the western North Atlantic (Reid, 1978; Riser and Rossby, 1982).

Should this hypothesis be borne out by further studies, we would in effect have a small-scale mechanism for the large-scale redistribution of oceanic tracers. If so, we must be careful how we use observed property distributions in diagnostic models, since these can be maintained by small-scale dynamics not repre-

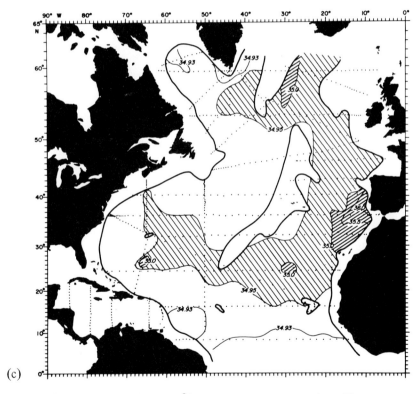

Figure 5. (a) Distribution of salt in °/oo on the $\sigma_\theta = 27.3$ surface. The approximate depth and temperatures are 800 m and 9°C. Note the position of the Mediterranean eddy relative to its parent water mass. (b) Distribution of salt in °/oo on the $\sigma_\theta = 27.8$ surface, 1500 m deep (the temperature is $\sim 4°C$). (c) Distribution of salt in °/oo on the $\sigma_\theta = 27.9$ surface, 3000 m deep (temperature $\sim 2.5°C$). These three charts were prepared by McDowell (1981) from the IGY surveys (Fuglister, 1960).

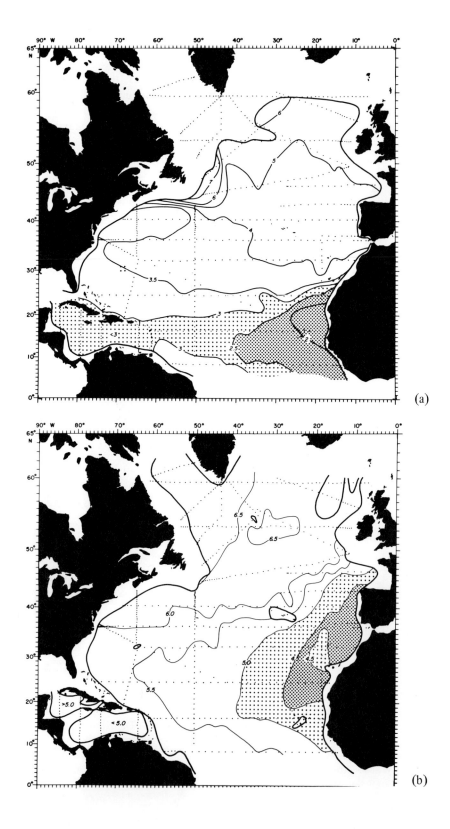

sented in the usual advective/diffusive calculations. A simple calculation will show that only a few Mediterranean eddies a year are required to provide a salt flux comparable to that, given the observed salinity gradients, by classical eddy diffusion. This could explain why they have not been observed in the past. Interestingly, the structure of the tongues in low-kinetic-energy regions suggests how pollutants might spread out from "point sources" (at least if they are not on the bottom).

Of equal importance to determining tracer fluxes is the transport of water itself. Clearly this is a low-frequency or mean-flow question and one where much averaging will be required to gain statistical significance. It is with these questions in mind that we have started a program to develop a low-cost deep-sea drifter that can be manufactured in large quantities. Each one, after extended submergence, surfaces and is relocated by means of satellite. Ensemble averaging

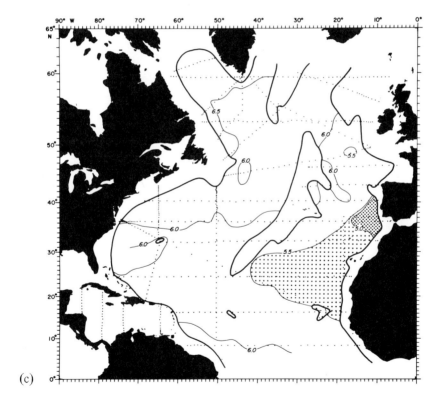

Figure 6. (a) Distribution of dissolved oxygen (ml/l) on the $\sigma_\theta = 27.3$ surface. Note the position of the low-O_2 eddy relative to its parent water mass. The sharp zonal orientation may partly be due to the North Equatorial Current. (b) Distribution of dissolved oxygen (ml/l) on the $\sigma_\theta = 27.8$ surface. (c) Distribution of dissolved oxygen (ml/l) on the $\sigma_\theta = 27.9$ surface. These three charts were prepared by McDowell (1981) from the IGY surveys (Fuglister, 1960).

such vectors over an appropriate region should provide statistical significance and, depending upon the deployment strategy, render the results less sensitive to the source distributions that were discussed above. These large atoms might be thought of as the physical oceanographer's version of marine geochemistry!

3 The Gulf Stream Recirculation Region

As one moves closer to the Gulf Stream, the eddy activity increases rather rapidly. The POLYMODE program in general and the Local Dynamics Experiment (LDE) in particular were designed to obtain quantitative information on the dynamical properties of the eddies themselves, their interactions, and their role in and relationship to the mean flow. While a wide variety of data of excellent quality have been collected by many investigators, most analyses have only begun. In this survey we limit ourselves to a few observations made by the SOFAR float program.

Perhaps the most striking result to date is the manifestation of potential vorticity conservation by about ten SOFAR floats at 1300-m depth. As the clus-

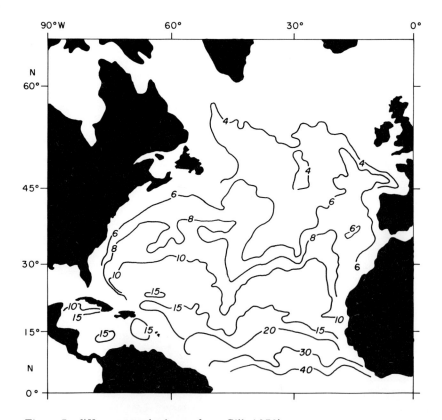

Figure 7. f/H contours (redrawn from Gill, 1970).

ter oscillates back and forth between the NE and SW, it rotates first anticyclonically (clockwise), then cyclonically, and then anticyclonically again in direct response to the latitudinal changes in the effect of the earth's rotation. (This may be the simplest, most explicit geophysical demonstration of the conservation of angular momentum to date.) Figure 9 shows a series of panels in 12-day steps of the float cluster, both in geographical coordinates as well as relative to each other by subtracting off their common low-frequency translation. In a recent paper, Price and Rossby (1982) show that excellent quantitative agreement with theory emerges if vortex stretching by the sloping bottom is included.

In the main thermocline there is conspicuous eddy activity on much longer time scales. This is borne out by the strikingly different fates of two groups of 700-m SOFAR floats, set 2 months apart in 1978 (Fig. 10, a and b). The first

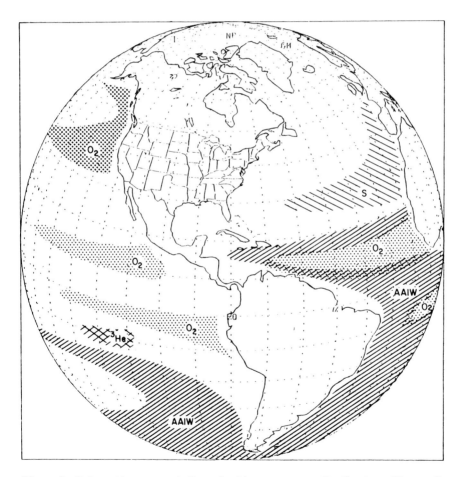

Figure 8. Schematic representation of mid-ocean tracer distributions. The small ^{3}He tongue in the South Pacific is taken from Lupton (1979). Surface and bottom distributions are not included.

cluster (May 1978) broke off rather quickly into groups to the west and southwest; of the second cluster (July 1978) all but one or two floats took off at varying speeds to the east, two of them going 700 km in less than 3 months. This suggests a low-frequency (annually forced?) zonal oscillation of the upper ocean. Owens et al. (1982) comment on the possibility of a vernal appearance of a zonal jet flowing to the west. Whether this jet can be thought of as a structural component of the Gulf Stream recirculation is not clear, but it does seem possible. One of the earlier SOFAR floats was rapidly swept westward by this jet from 34°N 66°W to 33°N 74.2°W in 50 days, equivalent to a mean speed at 700 m of more than 17 cm/sec. Although the float did not enter the Gulf Stream at the time, one is tempted to speculate that the jet is a significant tributary to the increase in the Gulf Stream in the springtime. Clearly there is much to be learned.

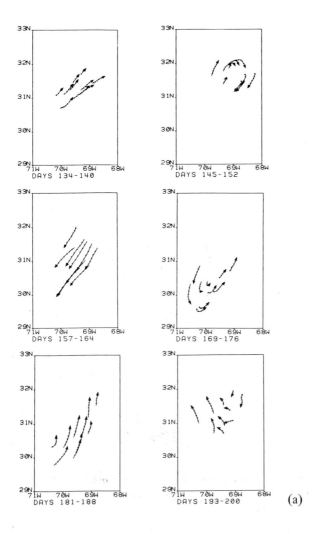

(a)

Eddies and the General Circulation 149

Several of the floats from the first setting in the LDE came into close proximity of the Gulf Stream within weeks. A striking aspect of these is the number of brushes the floats make against the Gulf Stream without getting swept away. For over $1\frac{1}{2}$ float years there were nine contacts, before one of these floats got caught (Figs. 11 and 12). Another float, after a brief stay off San Salvador, got caught in the Gulf Stream near 31°N and was swept north and east to 36°N 70°W where it broke off to the north and slowly drifted to the west in the slope water region for 50 days. It then rejoined the Gulf Stream and continued rapidly to the east (Fig. 13). Two other floats farther to the east, both at 1200 m, had earlier skipped north across the Gulf Stream and drifted westward.

What do these skipping events portend about lateral exchange processes in the Gulf Stream? Certainly the warm surface waters of the Gulf Stream suggest a narrow, continuous flow, which, although highly baroclinic in the downstream

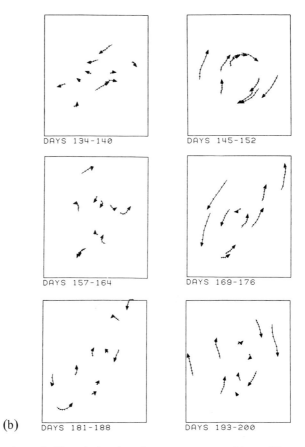

Figure 9. Covering one and one-half periods of a planetary-topographic oscillation in 12-day steps, (a) shows the geographical or "planetary" position of ten 1300-m floats, and (b) shows their position relative to the cluster's center of gravity.

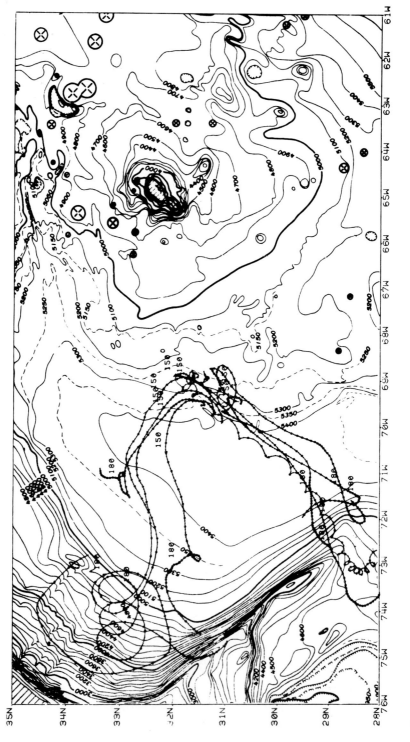

Figure 10. (a) A set of 700-m SOFAR floats launches in May 1978. Note how these drift rapidly to the west. Time in year-days.

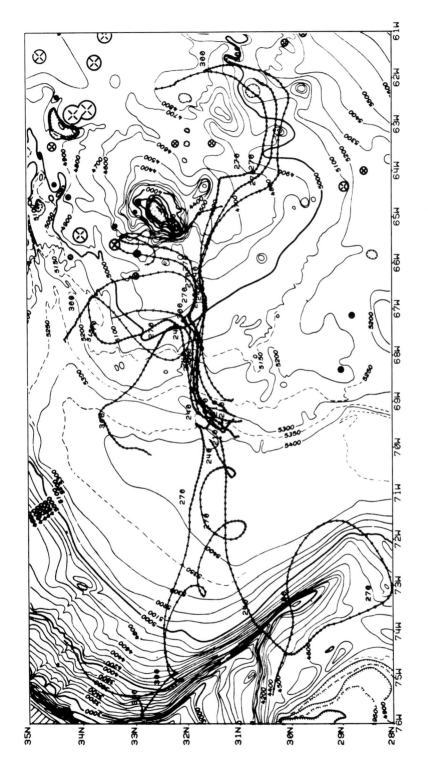

Figure 10. (b) A set of 700-m SOFAR floats launched in July 1978, 2 months later. These drift to the east instead!

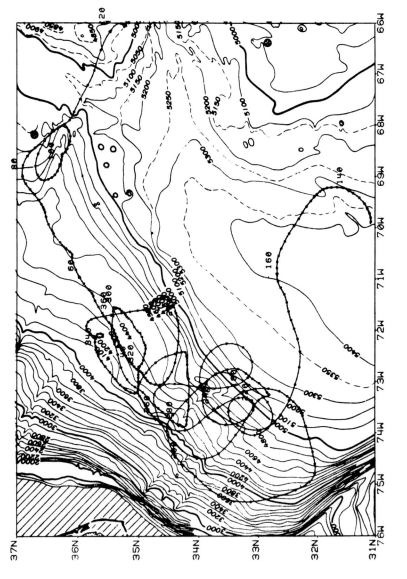

Figure 11. The trajectory of float 54.

Eddies and the General Circulation 153

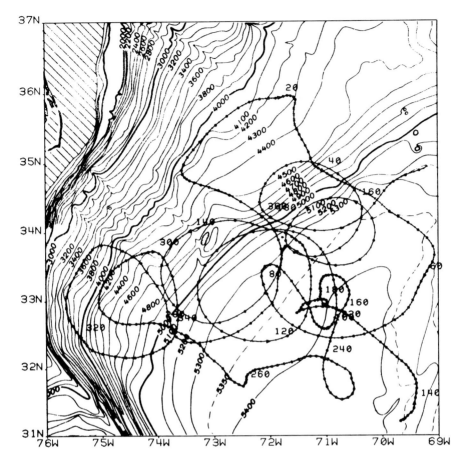

Figure 12. The trajectory of float 55. This float did not get swept away by the Gulf Stream.

direction, meanders coherently like an oversized hose. Getting floats across such a hose as effortlessly as indicated in Figure 12 poses a problem. Indeed, one might ask whether these observations are merely artifacts of floats, their property of constant pressure (rather than constant density). We can look at historical data for help. We consider the famous "Gulf Stream '60" data set (Fuglister, 1963) and examine the distribution of dissolved oxygen on the O_2 minimum surface ($\sigma_\theta = 27.0$; $\theta \sim 10°C$). Figure 14 shows the depth of this surface, almost 900 m to the south and shoaling rapidly across the Gulf Stream (and in a cold ring) to 300 m depth on the Slope Water side. Figure 15 shows that the distribution of oxygen is about 3.5 ml/l everywhere, except very close to the shelf break to the north. The striking feature here is the absence of any indication of a Gulf Stream! This is not what one might expect of an intense

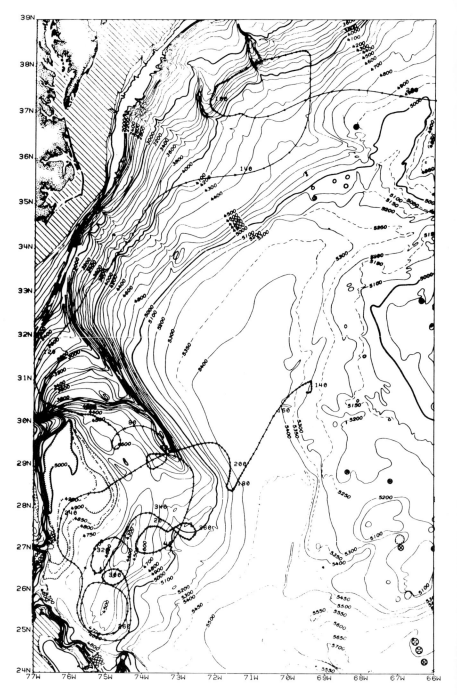

Figure 13. The trajectory of float 59. Note how the float doubles back while it is in the Slope Water region.

Eddies and the General Circulation 155

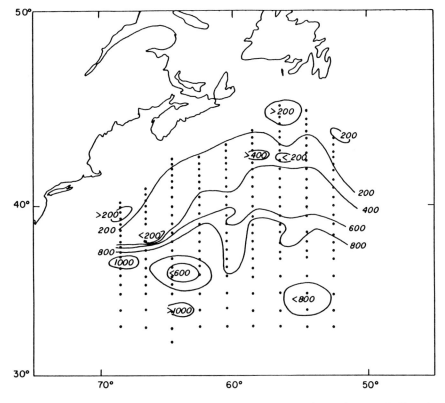

Figure 14. The depth (in m) of the oxygen minimum layer ($\sigma_\theta = 27.0$) estimated from the Gulf Stream '60 survey (Fuglister, 1963).

boundary current separating the Sargasso Sea from the Slope Water region or of a current advecting low O_2 water into this region from the Gulf of Mexico.

Actually this distribution was well known by the 1930s (Rossby, 1936; Seiwell, 1937) and was noted again more recently by Gatien (1976). The point here is that unless the waters being advected from the Gulf of Mexico have the same O_2 levels at this σ_θ (and this is not the case), there must be considerable mixing across the Gulf Stream.

To visualize the problem of floats slipping across the Gulf Stream we consider schematically four types of lateral motion and indicate what the motion of a fluid particle and a float might be like. In all cases the density structure is coherent in the vertical and the downstream flow strongly baroclinic.

> Small-scale (random) isentropic motion. This is no doubt always present, but it is difficult to see how bold float movement can be accounted for this way. The particle and float trajectories are presumably uncorrelated (Fig. 16).

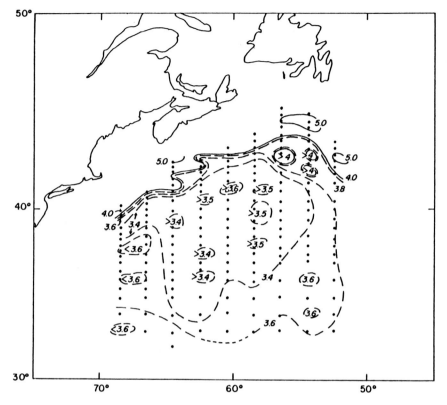

Figure 15. The distribution of dissolved oxygen (ml/l) on the oxygen minimum layer.

Barotropic meandering (hose). No vertical velocity, only lateral translation. Both particles and floats remain the same depth; thus floats cannot cross the Gulf Stream (Fig. 17).

Baroclinic meandering (phase propagation). Uplifting of isotherms leads to pattern propagation, but not bulk translation. In this case a fluid particle is brought closer to the surface and accelerated downstream until it is "left behind" on the Slope Water side (Fig. 18). In short, the Gulf Stream crosses the float, not vice versa.

Large-scale isopycnal flow. In this case there is a fluid flow along isopycnal surfaces (even across a stationary pattern). A float moves according to the local horizontal component of flow (Fig. 19). Evidence of such motion may be found in the entrainment and subduction of shelf waters by the Gulf Stream (Fig. 20).

In some sense all of these processes as well as the larger scale recirculation are active, but where and why each of these contributes to the observed distribu-

Eddies and the General Circulation

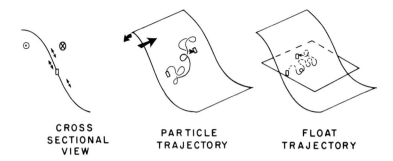

Figure 16. Schematic representation of small-scale (random) isentropic motion. The three panels show the cross-stream structure (left), and perspective views of particle motion (center) and constant pressure float motion (right) respectively.

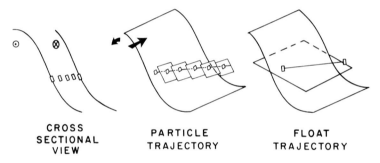

Figure 17. Schematic representation of barotropic meandering. No vertical displacement is allowed so particles, such as floats, may only move laterally.

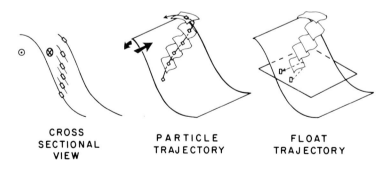

Figure 18. Schematic representation of baroclinic meandering (phase propagation) due to the vertical displacement of isotherms. The left and center panels give a cross-stream and perspective view of particles being uplifted and the right panel shows the corresponding float trajectory.

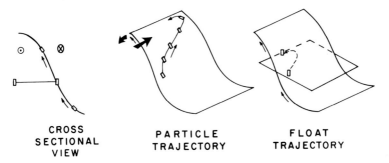

CROSS SECTIONAL VIEW   PARTICLE TRAJECTORY   FLOAT TRAJECTORY

Figure 19. Schematic representation of isopycnal flow. The density pattern can be stationary while particles slide up (or down) along isopycnal surfaces, shown in the left and center panels. A float moves according to the local component of horizontal motion.

tions of O_2 and others will be major questions to be explored in the coming years.

4 Transient Events

In this survey we have deliberately taken a mechanistic approach in portraying a variety of time-dependent events. This reflects the view that eddies as such not only command considerable interest in their own right but are also a key to our understanding of the general circulation. To employ a parable, it is difficult to know how a machine operates without knowledge of its parts and their function, even more so if it is a nonlinear system.

At the same time, however, it is important to learn how to identify and to distill the essential properties of the eddy field—to parameterize them. This is necessary for the correct representation of the aggregate effect of the smaller scale processes in studies of the larger scales. Given the remarkable variety of eddy phenomena that is coming to light, we may need to develop additional ways of parameterizing them; perhaps to include not only second-order statistics, but also aspects of the processes themselves.

We conclude with an illustration of this point. Consider the Gulf Stream. (For the sake of the argument, we shall assume that even while it meanders, it is made up of a narrow, continuous, and persistent baroclinic flow.) Suppose one wishes to summarize an observable property, such as thermocline depth or velocity. A set of random independent observations will transform a very specific structure into a broad region of enhanced variability—so-called "virtual eddies." This transformation of a low-frequency or even steady signal in natural coordinates into one of great temporal variability in geographical coordinates is of course well known, and we do it because we have little choice; the bulk of our historical data has been collected this way. Yet as we come to recognize these

Figure 20. Infrared satellite image of the Gulf Stream region off Cape Hatteras. Note the entrainment and probable downstream subduction of shelf waters near the north wall.

limitations, we may need to develop statistical sampling schemes for discrete systems, such as the Gulf Stream or Gulf Stream rings (Kim and Rossby, 1979) that are more discriminating than the classical Gaussian definitions of means and standard deviations.

In summary, whether we are dealing with little eddies, rings, or fronts, the ocean seems to be made up of transient phenomena more discrete and long-lived than hitherto recognized. And, as we have sought to illustrate here, some eddies may give rise to property distributions suggestive of an apparent mean flow, on the one hand, and a mean field can create regions of apparent enhanced eddy activity, on the other. Understanding this, and the overall role of eddies in the general circulation system, poses a formidable and exciting challenge for future research.

Acknowledgments

The research discussed in this overview is funded by the Office of the International Decade of Ocean Exploration at the National Science Foundation. Their continuing support is most gratefully acknowledged. The SOFAR float program is one of several collaborative projects that together make up the U.S. POLYMODE program under the leadership of A. Robinson and H. Stommel.

The other major components of the field program are the extended local dynamical studies with moored current meters under B. Owens and J. Luyten; the intensive hydrographic program under J. McWilliams and B. Taft; and the geographical studies under P. Niiler and C. Wunsch. While much of the analysis has yet to be completed, there is little doubt that the effort has been worthwhile and exciting and the data obtained have been of excellent quality, technically as well as scientifically. Indeed, the caring, the attention to detail, and sense of excellence that has characterized POLYMODE is most impressive and by itself a major contribution to its success.

There are many people who have contributed very substantially to the research program using SOFAR floats, but I wish to specially mention Mr. Douglas Webb, whose skill, imagination, and perseverance have brought his development of the SOFAR float to working reality.

References

Fuglister, F. C. 1960. Atlantic Ocean atlas, temperature and salinity profiles and data from the International Geophysical Year of 1957-1958. Woods Hole Oceanographic Institution, Atlas Series 1, 1-209.

Fuglister, F. C. 1963. Gulf Stream '60. In: Progress in Oceanography, Vol. 1 (M. Sears, Ed.). Pergamon Press, New York, pp. 265-273.

Gatien, M. G. 1976. A study in the slope water south of Halifax. J. Fish. Res. Bd. Can. 33(10), 2213-2217.

Gill, A. 1970. Contours of "hcosecθ" for the world's oceans. Deep-Sea Res. 17, 823-824.

Kim, K., and T. Rossby. 1979. On the eddy statistics in a ring-rich area: A hypothesis of bimodal structure. J. Mar. Res. 37(1), 201-213.

Lupton, J. E., and H. Craig. 1979. A major ^{3}He source on the East Pacific Rise. Unpublished manuscript.
McDowell, S. E. 1981. Isopycnal hydrography and mixing in the North Atlantic Ocean. Submitted.
McDowell, S. E., and T. Rossby. 1978. Mediterranean water: an intense mesoscale eddy off the Bahamas. Science 202, 1085-1087.
Owens, W. B., J. R. Luyten, and H. L. Bryden. 1982. Moored velocity measurements during the POLYMODE Local Dynamics Experiment. J. Mar. Res. In press.
Oceanus. 1976. Special issue on ocean eddies. Oceanus 19(3), 88 pp.
Price, J., and T. Rossby. 1982. Observations of a barotropic/topographic wave. J. Mar. Res. In press.
Reid, J. L. 1978. On the middepth circulation and salinity field in the North Atlantic Ocean. J. Geogr. Res. 83, 5063-5067.
Rhines, P. B. 1977. The dynamics of unsteady currents. In: The Sea, Vol. 6 (E. D. Goldberg, Ed.). Pergamon Press, New York, pp. 189-318.
Riser, S., and T. Rossby. 1982. Lagrangian structure and variability of the western North Atlantic circulation. Submitted.
Robinson, A. R. (Ed.). 1982. Eddies in Marine Science. Springer-Verlag. In preparation.
Rossby, C. G. 1936. Dynamics of steady ocean currents in the light of experimental fluid mechanisms. Pap. Phys. Oceanogr. Meteorol. 5(1), 43 pp.
Rossby, T., A. D. Voorhis, and D. Webb. 1975. A quasi-Lagrangian study of midocean variability using long range SOFAR floats. J. Mar. Res. 33(3), 355-382.
Rossby, T., J. Price, and D. Webb. 1982a. The spatial and temporal evolution of a cluster of SOFAR floats in the Polymode Local Dynamics Experiment (LDE). Submitted.
Rossby, T., S. Riser, and S. McDowell. 1982b. On the origin and structure of a small-scale lens of water observed in the North Atlantic thermocline. In preparation.
Seiwell, H. R. 1937. The minimum oxygen concentration in the western basin of the North Atlantic. Pap. Phys. Oceanogr. Meteorol. 5(3), 24 pp.
Wüst, G. 1936. Atlas a. Schichtung und Zirkulation des Atlantischen Ozeans. Ergebn. Deutscher Atl. Expedition METEOR. Bd. VI, Atlas.

Radioactive Tracers in the Sea

Hugh D. Livingston and William J. Jenkins

1 Introduction

The oceans are a critical component of our environment. Physically, they modify climate, not only on a regional and local scale, but on a planetary scale as well. A significant fraction of the intrahemispheric heat flux is caused by oceanic circulation. The high productivity regions of the ocean represent an important source of food. The oceans not only exert a long time-scale control over the geochemical cycling of many elements, but they play an increasingly important role in the disposal of mankind's unwanted byproducts.

All of these effects are related to the ocean's ability to transport material, whether by simple fluid flow or by some more complex pathway. An understanding of the processes responsible for transport in the oceans is necessary for predictions of mankind's impact on the global environment, and to rational policy decisions concerning future activities.

Over the past several decades, new substances produced by human activities have entered the atmosphere and hydrosphere and are being transported throughout these systems through a number of physical, chemical, and biological pathways. Observations of the evolution of the oceanic distribution and evolution of these tracers can provide unique information about the magnitudes and natures of the processes at work. Because the time history and geographical character of the introduction of these substances range markedly, in addition to their differing respective environmental behaviors, each tracer illuminates a different portion of the spectrum of oceanic transport processes.

Anthropogenic radionuclides are of particular utility to oceanographers: first, their radioactivity allows their measurement to exceedingly low concentrations, and therefore under large dilutions; second, the half-life represents a built-in clock—either by diminution of the species, by accumulation of a daughter product, or by the presence or absence of secular equilibrium with the parent species; and third, the release of these substances to the environment tends to be relatively well documented and regulated.

The two major sources of anthropogenic radionuclides are the atmospheric nuclear weapons tests of the 1950s and 1960s and the reprocessing of nuclear fuel. The fallout from nuclear weapons tests generally produced a geographically large-scale input of radiotracers to the ocean which are subsequently being redistributed mostly by planetary-scale processes, particularly thermohaline circulation and abyssal mixing. On the other hand, the nuclear fuel reprocessing byproducts are being introduced on a more local scale and are thus affected by smaller scale processes. Within these two groups are a variety of tracers with differing time histories, spatial inputs and biogeochemical natures, all of which provide complementary and valuable information on oceanic processes.

The early days of measurement were characterized by sparse and sometimes unreliable data, as the sophisticated techniques and technology evolved for the measurement of infinitesimal amounts of material in the face of large volume sample logistics and severe contamination problems. Unfortunately the fallout transient evolved more rapidly than large-scale sampling efforts, so that the earlier part of the fallout pulse injection into the ocean was not as well documented as one would have liked. Fallout nuclide measurement programs initially consisted of individual laboratory efforts, with relatively sporadic sampling. The early 1970s saw the advent of the first large-scale synoptic tracer measurement program (GEOSECS—Geochemical Ocean Sections Survey) and the field advanced into a new era. As the data base increased, the interpretation expanded from the purely descriptive to include more quantitative, diagnostic models. The role of the tracer models at present appears to be one of constraining or quantifying parameterizations which may or may not be physically realistic. The application of multiple tracer distributions to these simple diagnostic models not only tests the consistency of the parameterization, but, in a broad sense, the validity of the models themselves. As the data base further increases, more and more sophisticated models are brought to bear on the tracers. However, we recognize that the model tracer distributions are as strongly affected by boundary and initial conditions (such as the spatial and temporal nature of the input function), as they are by interior dynamics. There is a strong need, therefore, to carefully evaluate and document the pathways, time histories, and spatial patterns of the inputs of the transient radionuclides (and for that matter, any other anthropogenic substances) if the promise seen in these tracers is to be realized.

2 Input of Artificial Radioactivity to the Oceans

Although, as reviewed recently by Templeton (1980), artificial radioactivity has entered the oceans from a large and varied number of quantitatively small sources (such as low-level nuclear waste disposal, accidents involving nuclear materials, or discharges from coastally located nuclear power reactors), the major input has come from two sources. These are the fallout from atmospheric nuclear weapons tests and the discharges associated with nuclear fuel reprocessing. The inputs from these sources have been very different, in terms of the times and places of introduction and of their quantitative and qualitative factors. Knowledge of the nature of these factors is basic to an understanding of the nature and distribution of artificial radioactivity now present in the oceans.

By far the largest release of bomb-produced radioactivity to the atmosphere, and subsequently to the oceans via nuclear fallout, followed the massive series of weapons tests in the 1950s and early 1960s (Zander and Araskog, 1973). It has been estimated (Joseph et al., 1971) that 194 megaton equivalents of fission products were produced by atmospheric weapons testing in the period between 1945 and 1963. These were estimated to have introduced 21 MCi of ^{90}Sr and 34 MCi of ^{137}Cs to the globe—these nuclides being major, longer lived fission products. The time pattern of the entry of radioactive fallout into the oceans can be thought of as broadly similar to that of its arrival at points on land. An excellent record of the delivery of fallout radioactivity can be seen in the series of measurements of the quarterly deposition of ^{90}Sr on New York City (Toonkel, 1980). These show that fresh input since the middle of the 1960s has been trivial and that the delivery to the earth's surface came in essentially two pulses— a minor pulse between 1955 and 1959 and a major pulse between 1962 and 1965. Because most tests took place in the Northern Hemisphere, it received a much larger proportion of fallout from the atmosphere. The distribution of delivered fallout has followed a distinct pattern of maximum delivery at midlatitudes, decreasing to minimum values at the equator and the poles. The pattern of globally distributed plutonium which was observed in soil samples collected in 1970 typifies hemispheric and latitudinal delivery of fallout radioactivity (Hardy et al., 1973).

The contemporary geographical oceanic distribution of fallout radionuclides is still strongly dependent on these input patterns. Figure 1 shows the distribution pattern of ^{137}Cs inventories in the Atlantic and Pacific Oceans during the period 1972-1974. North Atlantic inventories increase from the equator toward mid-latitudes; but, as will be discussed later in relation to the importation of labeled bottom water from the Greenland Sea Overflow, the high latitudes show increased inventories. In the Pacific, in contrast, ^{137}Cs inventories decrease from mid-latitudes toward both equatorial and high-latitude regions.

Recent work (Bowen et al., 1980) has shown that the Pacific distribution of weapons testing fallout radionuclides cannot be completely understood unless account is taken of the input, in addition to that from global fallout, that resulted from close-in fallout of debris from United States Pacific nuclear tests. This input, representing tropospheric fallout subsequent to ground or near-ground level tests, was largely centered in the equatorial western North Pacific, around the Marshall Islands, from 1952 to 1958. There may have been some later input of this type from some tests around Christmas and Johnson Islands. This source was recognized (Miyake, 1963) to have effects on wide geographical areas in the western North Pacific, observable for ^{137}Cs in the low latitudes of the western Pacific by excesses found there compared to similar North Atlantic

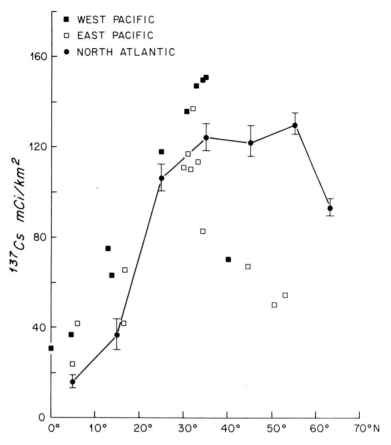

Figure 1. ^{137}Cs inventories in the North Atlantic and North Pacific Oceans, 1972 to 1974.

latitudes (Fig. 1; Bowen et al., 1980). This effect is much more pronounced for nuclides such as plutonium which lack rare gas precursors (Bowen et al., 1980). Figure 2 shows, plotted against latitude, North Pacific plutonium inventories from GEOSECS stations (Bowen et al., 1980) and from a series of soils from a network of sites believed to have received only global fallout. The low to mid-latitudes, especially at western North Pacific locations, are clearly characterized by amounts of plutonium in considerable excess of that expected from global fallout and explainable readily only in terms of the extra input from close-in fallout.

Until recently, discharges of artificial radioactivity to the oceans associated with the reprocessing of spent nuclear fuel appeared to represent relatively minor, and hence rather highly localized, perturbations of the fallout distributions. This was true of the early years of discharges to the Irish Sea from the United Kingdom reprocessing plant at Windscale (Joseph et al., 1971; their Table 18). Over the last decade, these releases have become quantitatively much

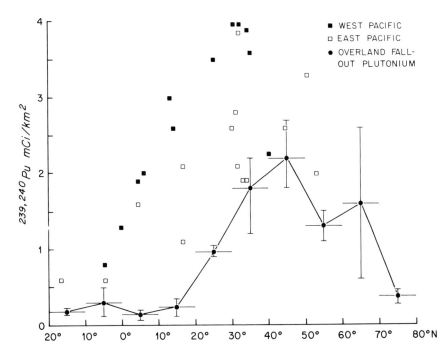

Figure 2. ^{239,240}Pu inventories in the North Pacific Ocean 1972 to 1974 versus those expected from global fallout from atmospheric nuclear testing.

larger and attention has been drawn to the potential application of the dispersal of radiotracers in these releases for oceanographic studies at high latitudes (Livingston et al., 1982). To date, published studies of Windscale-derived radiotracers have been confined to coastal areas around the United Kingdom, including the North Sea and its outflow in the Norwegian Coastal Current (Kautsky, 1977; Mitchell, 1977; Livingston and Bowen, 1977). There have also been releases from the French fuel reprocessing plant at Cap de la Hague on the north coast of France. These releases are moved through the English Channel and North Sea to join those coming from Windscale (Kautsky, 1977). However, the magnitude of these releases have to date been quite small relative to those from Windscale; hence the latter has provided the major source term for high-latitude studies of the entry of fuel reprocessing releases to the world ocean. Table 1 compares the total alpha and beta activities released by La Hague and Windscale from 1972 to 1976 (Luykx and Fraser, 1978) and shows the Windscale releases to have dwarfed those of La Hague during this time period. A substantial fraction of beta activity released from Windscale has been ^{137}Cs, which has been used extensively in coastal studies cited above. The significance of increase in input of ^{137}Cs over the last decade becomes clear when it is noted that, of the 603 kCi released from 1957 to 1977, 581 kCi were released subsequent to 1969 and 448 subsequent to 1974 (Livingston et al., 1982). The very recent entry of this tracer from this source into the ocean offers a unique

Table 1. Comparison of radioactivity release from Windscale and Cap de la Hague 1972-1976

	Total beta activity released (kCi)		Total alpha activity released (kCi)	
	Windscale	La Hague	Windscale	La Hague
1972	140	11.6	3.86	0.0031
1973	127	13.7	4.90	0.0036
1974	207	25.3	4.57	0.0270
1975	245	31.9	2.31	0.0133
1976	183	19.3	1.61	0.0099

opportunity for tracer studies using it and other significant radionuclides released from Windscale.

In the strictest sense, a conservative tracer is a tracer which undergoes absolutely no chemical or biological alteration. In truth, then, there are no truly conservative tracers in the oceans, since all substances (even the noble gases) undergo some degree of reaction or adsorption. However, there are a number of radiochemical species which range from "essentially" conservative to "mostly" conservative. These species, in order of increasing reactivity, are listed in Table 2.

Krypton-85, because it is a noble gas, is effectively a conservative radioactive tracer. Its half-life of 10.3 years is well suited to studying thermocline structure, and the time constant of atmospheric increase (about 5 years; Pannetier, 1970) may make it useful for gas exchange studies on a regional (as opposed to local) scale. This tracer, although introduced by nuclear fuel reprocessing, behaves as a "global fallout" tracer in that its delivery (atmospheric concentrations) is geographically uniform (Farges et al., 1974; Telegadas and Ferber, 1975). Whereas atmospheric ^{85}Kr concentrations have been well studied, the paucity of this gas in sea water necessitates extraction from large water samples (several hundred liters); so few oceanic ^{85}Kr measurements have been made. The data obtained to date are not sufficient yet for detailed interpretation, but what are available (Schröder, 1975) appear both promising and consistent with other tracers.

Tritium (^{3}H) can effectively be regarded as a conservative tracer; as hydrogen,

Table 2. Conservative tracers

Species	Half-life (years)	Source*
^{85}Kr	10.3	NFR
^{3}H	12.45	NWT
^{90}Sr	28.8	NWT/NFR
^{137}Cs	30.2	NWT/NFR
^{134}Cs	2.06	NFR
^{14}C	5730	NWT

*NFR, Nuclear fuel reprocessing; NWT, nuclear weapons tests.

it exists as part of the water molecule itself. Its half-life is similar to that of ^{85}Kr, being 12.45 years, so that it appears well suited for thermocline studies, and it has been exploited toward that end (Rooth and Östlund, 1972). Tritium differs from other global fallout products in that it is strongly bound to the hydrologic cycle, with water vapor exchange and continental runoff contributing to the complex delivery pattern of tritium to the oceans (see Weiss et al., 1978). Nonetheless, the time history of surface water concentrations has been modeled effectively (Dreisigacker and Roether, 1978, for the North Atlantic), and these "source functions" can be used for quasidiagnostic tracer models of intermediate and deep water formation. Truly prognostic tritium tracer models, however, must ultimately use the boundary fluxes rather than the surface water concentrations.

For tritium, both the time constant of its delivery to the oceans and its half-life make it a decade time-scale tracer. Circulation and mixing which occurs on time scales of the order of years or longer and on regional and planetary scales are most strongly illustrated by the evolution of the distribution of tritium in the oceans. A graphic example is the tritium section (Fig. 3) obtained during the GEOSECS 1972 expedition in the western Atlantic (Östlund et al., 1977). The global delivery pattern is reflected in the surface water concentrations (most notably the interhemispheric asymmetry and polar enhancement: cf. Weiss et al., 1979), whereas the apparent ventilation of the deep and intermediate water masses is seen at the northern end of the section. The southward penetration of abyssal tritium in the western boundary jet is shown in Figure 4 at 30°N (Jenkins and Rhines, 1980). Coupled with current-meter data, estimates can be made of material fluxes and dilution, providing important constraints on future tracer models.

The daughter product of tritium, ^{3}He, is a stable inert tracer which provides additional information. For relatively short time scales (less than a decade), the two tracers may be coupled to obtain ^{3}H/^{3}He "ages," which may be interpreted as a gaseous isolation age and extend to time scales as short as a few months (Jenkins and Clarke, 1976). This technique has been employed to obtain oxygen utilization rates (Jenkins, 1977) and gas exchange rates (Weiss et al., 1978). For longer time scales, the complementary nature of their boundary conditions (tritium is enhanced while ^{3}He is depleted by contact with surface waters) leads to consistency checks on mixing and circulation models. This has been utilized to quantify the relative contributions of diapycnal and isopycnal mixing in a subtropical main thermocline (Jenkins, 1980).

Strontium-90 and ^{137}Cs, with half-lives of 28.8 and 30.2 years, respectively, were produced in global fallout in a remarkably constant ratio of 1.45 ^{137}Cs/^{90}Sr (Bowen et al., 1974). Because of the similarity of their half-lives, this ratio will change with time very slowly (with a half-life of ~600 years). Although neither element is perfectly conservative in sea water, comparison of sediment inventories with those in overlying water columns indicates that their residence times against scavenging are in excess of several hundred years (Noshkin and Bowen, 1973). Thus it appears that on time scales comparable to the time since

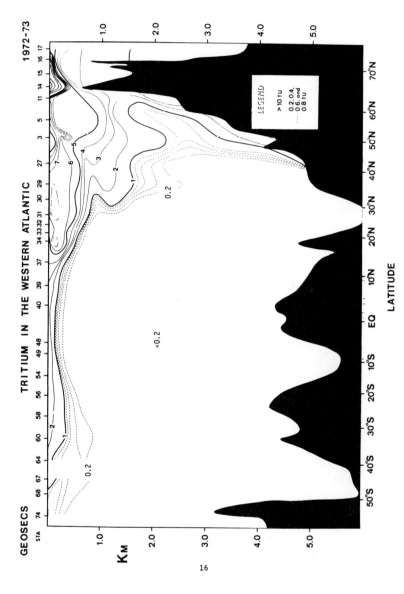

Figure 3. Distribution of tritium in the western North Atlantic Ocean from the GEOSECS expedition in 1972 (from Östlund et al., 1977).

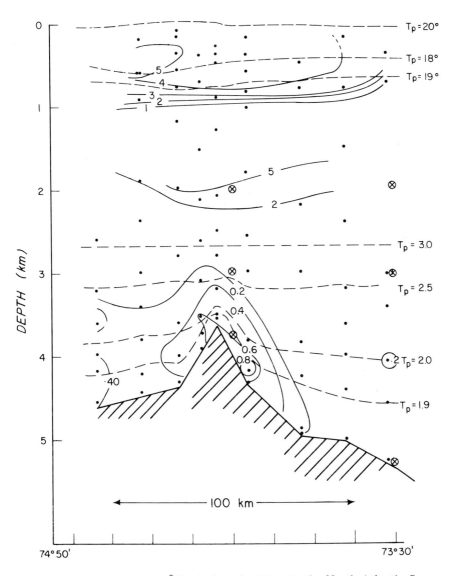

Figure 4. Penetration to 30°N of abyssal tritium in the North Atlantic Ocean western boundary undercurrent. (—) tritium isopleths; (---) isotherms; ⊙ current meter position (after Jenkins and Rhines, 1980).

the bomb tests ^{90}Sr and ^{137}Cs are effectively conservative. The oceanic distributions of ^{137}Cs and ^{90}Sr appear to be similar to that of tritium (Bowen and Roether, 1973; Bowen et al., 1980) but with differences which may be related to a close-in fallout component in the former two nuclides and to the vapor exchange input which is available only for tritium. A north-south section for ^{137}Cs approximately along 180° in the Pacific (Fig. 5) shows features which are

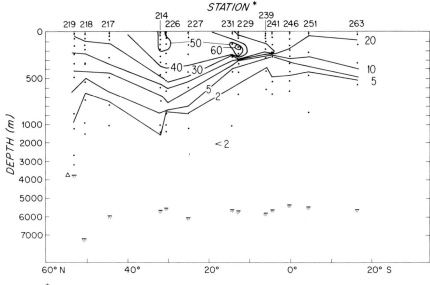

Figure 5. Distribution of ^{137}Cs in 1973 to 1974 along a north-south GEOSECS section between 170°W and 170°E in the western Pacific Ocean. (^{137}Cs: disintegrations per minute per 100 kg sea water.)

consistent with expectations: higher surface water concentrations north of the equator, a generally massive downward penetration in the subtropical gyre, and undetectable concentrations in the deep water. The ^{137}Cs isopleth generally parallels isopycnals on a regional scale, but on a larger scale they cross significantly, implying geographically isolated areas of downward mixing coupled with laterally heterogeneous isopycnal transport (Bowen et al., 1980).

Strontium-90 and ^{137}Cs are different from tritium in that significant quantities (relative to their fallout inventories) of these nuclides are being introduced to the oceans on a more local scale by discharge of waste from nuclear fuel reprocessing plants. The plant at Windscale, U.K., had released more ^{137}Cs by 1976 than the current fallout inventories of these isotopes north of 60°N in the Atlantic. Moreover, the ^{137}Cs/^{90}Sr ratio in the discharge has been generally different from the fallout value of 1.45. Both the absolute amounts and relative ratios have varied with time and have been carefully documented, so there are characteristic time markers and dilution indicators.

The plume of radionuclides from Windscale and the French reprocessing plant at Cap de la Hague has been exploited to study more local-scale transport (Livingston et al., 1980) in the North Sea and input regions of the Arctic Ocean. Additional and potentially very useful information is gained from the measurement of ^{134}Cs/^{137}Cs ratios. The shorter lived isotope (^{134}Cs has a 2.1-year half-life) provides an elapsed time indicator for regional-scale circulation.

The half-life of ^{14}C (5730 years) makes it attractive as a tracer of the planetary scale thermohaline circulation, and the apparent depletion of natural ^{14}C in the abyssal Pacific leads to estimates of the ventilation of the abyss (Munk, 1966). Correction for chemical effects (Craig, 1969) leads to somewhat different estimates, but to a first order, the problem has been solved. Evidence of temporal variations in the natural ^{14}C production rate (Stuiver and Quay, 1980) can, however, produce significant perturbations on model results for shorter timescale processes (Broecker and Peng, 1980).

The introduction of ^{14}C-poor carbon dioxide from the burning of fossil fuels tends to reduce the relative ^{14}C activity (^{14}C/^{12}C, the Suess effect), but by far the most significant perturbation was the production of about 10^5 moles of ^{14}C by nuclear weapons tests. The invasion of this bomb ^{14}C into ocean surface waters has been used to make estimates of the globally averaged CO_2 exchange rate (Munnich and Roether, 1967; Oeschger et al., 1975). This information is important in the construction of oceanic CO_2 uptake models.

A synoptic picture of the bomb ^{14}C distribution in the western Atlantic obtained during the GEOSECS expedition (Östlund and Stuiver, 1980), bears a striking resemblance to the tritium distribution shown in Figure 3. Application of simple vertical diagnostic models to the data by Oeschger et al. (1975) and others yielded quantitative estimates of the invasion rate of anthropogenic CO_2 into the ocean. Although such models resulted in unrealistically high vertical eddy diffusivities ($\gtrsim 1$ cm^2 sec^{-1}; Garrett, 1979), they probably are first-order accurate in the overall vertical transport—regardless of pathway. For example, the surface ocean Suess effect compares favorably to coral records (Broecker et al., 1980). Estimates of equatorial upwelling by such models (Broecker et al., 1978) and North Atlantic deep water formation rates (Broecker and Peng, 1980) are two examples of how such models and data can provide independent information about large-scale processes that are not directly observable.

This kind of application serves to outline the nature of the kind of information that such tracers give. The oceanic distributions of tracers provide ensemble averaged Lagrangian information: they are affected by processes that are not readily measurable and average over many scales and types of fluid motion. By the same token, information is lost in the averaging process, so that a direct inversion of the observed tracer distributions cannot lead us unambiguously to the processes themselves. Rather, the approach has been one of developing and testing simple diagnostic models in an attempt to obtain information which will eventually represent constraints or tests on more physically realistic models.

Nonconservative radionuclides (Table 3) are those whose chemistries or reactivities in sea water are such that they are moved by processes other than, or additional to, those advective and diffusive processes which transport conservative tracers. They generally include those radionuclides whose reactivities have usually been shown to have relatively high concentration factors for marine phytoplankton or high distribution coefficients for sediments or suspended materials (Duursma and Gross, 1971). As a group, nonconservative radionuclides are usually transition metals (such as ^{55,59}Fe, ^{60}Co, and ^{106}Ru), lanthanides (^{141,144}Ce and ^{147}Pm), or actinides (^{228,230,232}Th, ^{238,239,240,241}Pu, ^{241}Am, and ^{242,244}Cm). Nonconservative behavior in an oceanic water column results in the

Table 3. Nonconservative radionuclides

Species	Half-life (years)	Principal present source*
^{239,240}Pu	24,000, 6540	NWT
^{238}Pu	88	NWT, Snap 9-A satellite
^{241}Am	458	NWT
^{242,244}Cm	0.44, 17.6	NFR
^{55}Fe	2.6	NWT
^{95}Zr/^{95}Nb	0.18, 0.10	NFR
^{60}Co	5.2	NWT/NFR
^{106}Ru	1.0	NFR
^{147}Pm	2.6	NWT
^{144}Ce	0.78	NFR

*NWT, Nuclear weapons tests; NFR, nuclear fuel reprocessing.

nonconservative radionuclides separating and moving vertically away from conservative radionuclides. Their reactivities cause them to become associated with marine particulates and consequently moved in association with the local particle flux. This vertical separation from conservative nuclides has been described for such fallout-produced nonconservative tracers as ^{144}Ce and ^{147}Pm by Sugihara and Bowen (1962) and studies of vertical and horizontal studies of both these and of ^{95}Zr-^{95}Nb, ^{103,106}Ru, and ^{141}Ce have been summarized by Volchok et al. (1971).

Few of the major bomb-produced nonconservative tracers have half-lives long enough to allow them to be used to study nonconservative transport processes in the world ocean as opposed to the coastal or near-surface ocean. The time scales of these processes are such that tracers with half-lives of less than the range of 1 to 2 years have only limited use, whereas those with half-lives in the range of tens, hundreds, or thousands of years have long-term potential for nonconservative process studies. A group of nuclides with these properties are the transuranics, notably various plutonium isotopes and ^{241}Am. They have been demonstrated to exhibit nonconservative behavior in the oceans (Bowen et al., 1980; Noshkin and Bowen, 1973; Hetherington et al., 1976a,b) and their environmental behavior has been the subject of international symposia (IAEA, 1976).

A recent account (Bowen et al., 1980) of the distributions of fallout radionuclides in the Pacific at the stations of the GEOSECS program serves as an ocean-wide demonstration of the contrast between conservative radionuclides, traced by ^{137}Cs (Fig. 5), and nonconservative radionuclides traced by ^{239,240}Pu (Fig. 6). The distributions of ^{239,240}Pu were shown to be displaced vertically downwards from those of ^{137}Cs and to extend all the way to the ocean floor, through a water column in which ^{137}Cs was only measurable in the upper 1000 to 2000 m. Notable features of the ^{239,240}Pu distributions are the ubiquitous subsurface maxima at about 500 m; and the increase in bottom water concentrations found at most stations north of 25°N and west of 160°W.

Although ^{241}Am (Livingston and Bowen, 1976) and ^{55}Fe (Labeyrie et al.,

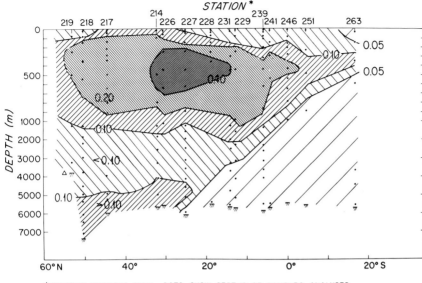

Figure 6. Distribution of ^{239,240}Pu in 1973 to 1974 along a north-south GEOSECS section between 170°W and 170°E in the western Pacific Ocean. (^{239,240}Pu: disintegrations per minute per 100 kg sea water.)

1976) have been shown to move more rapidly vertically than ^{239,240}Pu, the extent of the separation in open-ocean water columns of at least ^{241}Am from ^{239,240}Pu is not great. Although the distribution of ^{241}Am relative to ^{239,240}Pu is controlled, in part at least, by its production *in situ* by decay of ^{241}Pu ($t_{1/2}$ = 14.9 years), the fact that its water column profile shows no really measurable separation from that of ^{239,240}Pu suggests that the rate at which it is being removed, while greater than that for Pu removal, may indeed be relatively slow. In fact, fractionation between the two radioelements may well be greatest in the euphotic zone where ^{241}Am depletions appear significant.

The euphotic zone seems clearly the region where the strong separation of conservative and nonconservative tracers takes place. Biological processes of uptake and sinking of biogenic particles are thought to be major vectors of transport of reactive radionuclides (Noshkin and Bowen, 1973). This is analogous to the emphasis which has been laid on the major role of such processes on the biogeochemistry of ^{210}Pb (Nozaki et al., 1976) and the removal from the euphotic zone of various trace metals (Bruland, 1980).

Reactive radionuclides discharged in low-level nuclear fuel reprocessing wastes to shallow coastal waters have been shown to be scavenged quickly from the water column and to accumulate in sediments near to the points of release (Hetherington et al., 1976a, b). Such studies have so far not been sufficiently detailed to address the questions of how quantitative has been the transfer of discharged reactive radionuclides to local sediments, and what is the extent and nature of remobilization phenomena within these sediments. With respect to the latter question, Pu radionuclides and ^{106}Ru have been shown to be moved in apparently soluble phases for hundreds of kilometers downstream from their points of release from Windscale (Hetherington et al., 1976a; Murray et al., 1978) and from Cap de la Hague (Deutches Hydrographisches Institut, 1980).

The transfer of reactive fallout nuclides to open-ocean sediments contrasts with the transfer to shallow coastal sediments or the transfer of fuel reprocessing reactive radionuclides in coastal areas. Early models of plutonium sinking (Noshkin and Bowen, 1973) seem to overestimate the actual rates which prevail. In the Pacific, for example, generally about one half of the water column inventory is still in the upper 1000 m (Bowen et al., 1980). Further, measurements of ^{239,240}Pu inventories in 18 cores collected in the last few years in the central North Pacific (Bowen and Livingston, 1980) show that the sediments contain an average of 0.058 mCi km^{-2} of ^{239,240}Pu; this is only approximately 2% of the water column inventories in this area (Bowen et al., 1980).

After their delivery to marine sediments, reactive radionuclides are available for studies of such processes within the sediment column as bioturbation, buildup of sediment strata, and diagenesis. Livingston and Bowen (1979) attribute the profiles of ^{239,240}Pu and ^{137}Cs measured in coastal sediments to benthic organism mixing effects; likewise Aller and Cochran (1976) and Benninger et al. (1979) considered the effects of mixing on the observed profiles of ^{234}Th and ^{210}Pb which they found in nearshore sediments. Koide et al. (1975), in contrast, argued that the profiles of Pu, ^{210}Pb, and Th which they measured in anoxic varved basin sediments on the shelf off southern California have been produced by sediment accumulation and, therefore, record the delivery of these radionuclides within the sediment. Hetherington (1976) and Aston and Stanners (1979) suggest that coastal and estuarine sediments in the Irish Sea are similarly recording the discharge of reactive nuclides from Windscale. Our present inclination is that organism mixing effects have been underestimated in the Irish Sea studies and may have exerted greater control on the observed profiles.

Although mixing in sediments is more intense and rapid in productive coastal waters (Livingston and Bowen, 1979), there is evidence of its effects in deep pelagic sediments. Bowen et al. (1976) reported ^{239,240}Pu and ^{137}Cs profiles in sediment cores from the central North Pacific which they attribute to mixing. Williams et al. (1978) reached the same conclusion from later studies of the distribution of bomb-produced ^{14}C in sediments from the same area.

3 Sampling and Measurement

Apart from some notable exceptions, marine studies of artificial radionuclides involve collection and analysis of quite large samples because of the extremely low concentrations which generally characterize sea water, sediment, or biota. One radionuclide which is an exception is tritium for which water samples of the order of 1 liter or less have been found adequate. Around points of release of radioactivity to the sea such as fuel reprocessing plants, local concentrations of the various nuclides released have often been high enough to permit sensitive measurement on samples substantially smaller than are needed for either fallout nuclide measurement or fuel reprocessing nuclide measurement at a distance from the point of discharge. As an alternative to collection of very large samples of sea water, various workers have used chemisorbers to collect radionuclides from sea water either *in situ* or aboard ship. Details of these, and of a host of other sampling considerations, and caveats, were summarized in the report of a panel convened by the International Atomic Energy Agency (IAEA) to address problems and techniques associated with the measurement of marine radioactivity (IAEA, 1970).

As in the analysis of marine materials for other trace constituents, the typical measurement process starts with the extraction or concentration of the radionuclide sought, followed by a sequence of purification steps, and ending with the quantification of the isolated radioactivity by a sensitive detection system. Typically, artificial radionuclides are present in the ocean at concentrations very much less than those of the naturally occurring radionuclides. In many cases, purification of an artificial radionuclide isolate from this latter group can be analytically more demanding than purification from other artificial radionuclides.

The type of detection device used for measurement of a particular radionuclide is determined by the principal forms of radiation it emits during its decay. Special low-level techniques have had to be developed which allow reliable detection and quantification of the minute amounts of artificial radioactivity in the oceans.

Thanks to rapid progress in electronics over the past decade or so, there has been an impressive increase in the quality and quantity of measurements of radionuclides whose decay is via emission of alpha particles. The availability of high-quality multidetector alpha-spectrometry equipment at relatively low costs had led to increased counting capacities. The increase in the amount of time available for sample measurement has allowed sensitivity limits to be substantially reduced. The detection of beta- and gamma-emitting radionuclides at ultra-low levels necessitates the use of anticoincidence detectors and circuitry to reduce as far as possible the background levels against which measurement is to be made. Detectors used for gas counting of nuclides such as ^{14}C and ^{85}Kr have been improved by substantial reduction in their volumes, leading to a corresponding decrease in level of their background counting rates (Loosli and Oeschger, 1978). Measurement of gamma-emitting radionuclides using gamma-spectrometric techniques with either NaI (Tl) or Ge (Li) detectors can sometimes be used to reduce or eliminate the need for radiochemical separations. The nuclides sought are then separated from others present by their resolution by spectrometry rather than by chemistry. This approach is usually practicable in situations where the concentrations of the radionuclides sought are relatively high—such as around nuclear fuel reprocessing plant releases (Aston and Stanners, 1979) or around the United States' Pacific nuclear weapons testing grounds

(Noshkin et al., 1975). Gamma spectrometry measurements find frequent application also in measurement of a pair of radionuclides of the same element (as ^{134}Cs, ^{137}Cs) where both are not separable by radiochemical means.

There are two methods of measurement of tritium at oceanic levels: beta counting and ^{3}He regrowth. In the beta counting technique, from several hundred milliliters to 1 liter of water are enriched (usually by electrolytic enrichment) to a few milliliters, converted to a counting gas and counted in a proportional counter for 8 hours or more (Östlund et al., 1974). The detection limit of this technique is about 0.05 to 0.1 T.U. (1 T.U. = 1 tritium per 10^{18} hydrogen atoms) and the precision is about 3%. The former is governed by a combination of contamination introduced by handling and reagents and background counting statistics. The latter is usually dominated by uncertainties in the enrichment procedure, with minor contributions from counting statistics and uncertainty in counter efficiency.

The ^{3}He regrowth technique (see Clarke et al., 1976) involves the degassing of about 50 ml of sea water, storage in a low-He-permeability vessel for an appropriate period of time (usually about a year), and mass spectrometric measurement of the regrown ^{3}He. The detection limit is comparable to the beta technique (0.05 to 0.1 T.U.) but the analytical precision is somewhat better (about 1%). The former is governed by the instrumental detection limit of ^{3}He (which is currently around 5000 to 10,000 atoms), so that the detection limit for tritium is inversely related to (water) sample size. Further, the procedure is less susceptible to contamination than the beta technique since all processing is performed *in vacuo*. The analytical precision is limited by ^{3}He ion counting statistics, instrumental stability, and uncertainty in the assumed ^{3}He/^{4}He and C(He) of the operating standard (air).

A comparison of the techniques shows the beta counting technique to have the advantage of immediacy of results and low initial capital investment, whereas the regrowth technique appears to have lower susceptibility to contamination, greater precision, and potentially lower detection limit (if sample size is increased).

4 Future Evolution of Artificial Radioactivity in the Sea

It will be obvious that ultimately a complete approach to any prediction of the evolving patterns of artificial radioactivity in the oceans is not possible unless future sources of supply of radioactivity are known. Lacking this knowledge we can only make predictions which are highly personalized. We hope, for example, the oceans will not receive any further input of significance from nuclear explosions in the atmosphere. On the other hand, it seems likely that more use will be made of the oceans in connection with radioactive waste disposal activities. We would predict that pressure will develop to increase the rate at which the dumping of low-level radioactive waste will take place. This activity is not likely to have other than a localized impact on marine distributions of artificial radioactivity and may not impact significantly on marine radioactivity studies on a global or even regional scale. At the Third International Congress on the History of Oceanography Templeton (1980) gave a review of past and present dumping practices and the ongoing efforts being made to upgrade techniques of assessment of their impact on the ocean. He also drew attention to the concept of use of the sea floor or subseabed for storage of high-level radioactive waste—a subject of considerable international study and concern. The likelihood of this eventually requires use of a crystal ball not available to the present authors and the timing of such activities, if they were to proceed, is not likely to be in the short-term future. Barring the unforeseen, presently considered releases to the oceans subsequent to disposal would be unlikely to impact upon the oceans on time scales shorter than those of global ocean mixing. A fairly safe prediction is that the oceans will receive additional input of radioactivity in consequence of low-level waste discharge from coastally located nuclear fuel reprocessing plants. On the one hand, it appears probable that existing operations will continue, but at discharge rates which may be substantially reduced through higher standards or waste treatment improvements. On the other hand, it can be conceived that the number of countries carrying out this activity is more than likely to increase, potentially expanding the geographical spread of input from these sources to the world ocean. One foreseeable possibility in this regard would be the appearance of a source of this type in the southern hemisphere along southern African, Atlantic, or Pacific coastlines. As the Southern Hemisphere oceans presently contain much less artificial radioactivity than do those in the Northern Hemisphere, such inputs would represent new and different experiments in tracer oceanography which we should be prepared to exploit, should they begin.

Subject to these uncertainties over possible future inputs, we may now focus our attention on the probable evolution of the artificial radioactivity now present in the ocean and the processes which affect its distribution.

Based on presently observed radionuclide distributions in the sea, some general trends can be envisioned. Certainly the striking pictures of the soluble nuclide distributions in the Pacific (Fig. 5) and the Atlantic (Fig. 3) give us a strong qualitative grasp of the processes at work.

Surface water concentrations of the global fallout nuclides will continue to decrease at a rate greater than their respective half-lives, due to down-mixing and dilution with relatively radionuclide-poor deeper waters. This rate of decrease will be proportionately greater in areas of high upwelling and/or weak vertical stability, and in general where water masses are formed (in near-polar regions). One's initial expectation would be for a reduction of lateral gradients of radionuclides in surface waters with time; but the apparent survival of close-in fallout patterns in the Pacific (Bowen et al., 1980) over time spans approaching a decade suggests that this may not be the case. Further, frontlike features in radionuclides such as tritium (Broecker and Östlund, 1979) are likely permanent features of transition regions between circulation systems rather than "shock waves" of tracers entering the oceans. We therefore expect distinctive spatial patterns to remain in surface waters, although the concentrations will decrease with time. The persistence (or lack thereof) of these features, plus the magnitude of the decrease in surface concentrations with time will provide valuable and quantitative constraints on regional mixing and dynamics that otherwise elude direct measurement.

Penetration of these tracers into the intermediate waters will proceed on a more or less episodic fashion. This can be readily conceded when one considers the dynamics of mode water formation and the variability in the climatic processes during formation. Time series hydrographic data from O.W.S. *Bravo* (Needler, 1979) graphically demonstrate a variability on the subdecade to decade time scales in the ventilation of the Labrador Sea. A similar modulation can be seen in the Sargasso Sea (Jenkins, 1980). What is important about this variability is that it exists on time scales comparable to the time span over which the tracers have been injected, so that we must be careful about drawing general conclusions from a statistically small data base. For example, the GEOSECS tritium section was made in autumn of 1972 in the Labrador Sea, at the end of a prolonged period of poor or nonexistent ventilation of the deep water there (Needler, 1979). The picture would likely have been rather different if the section had been taken the following year, since the winter that followed the GEOSECS expedition produced a generous ventilation of Labrador Sea Water.

The impression given by the radiocarbon and tritium data from the GEOSECS Atlantic expedition is that the deep waters are more or less compartmentalized by the tight gyre circulation. That is, the radionuclide concentrations have a characteristically frontlike distribution (Fig. 3 at about 40°N). Rather than seeing this apparent "shock wave" of tracers progressing southward in a smooth fashion, we would expect the front to remain stable in position and the concentrations of the respective "compartments" to change. It appears that the bulk of the southward transport takes place by eddy exchange across the fronts or jetlike flow (see Fig. 4, from Jenkins and Rhines, 1980). The latter process then injects the tracer into the interior by shedding and reentrainment; for example, it has been estimated on the basis of tritium measurements that the Deep Western Boundary Current in the Atlantic has 90% exchanged with the interior by the time it has reached Hatteras (Jenkins and Rhines, 1980).

The main thermocline represents a substantial barrier to the penetration of tracers into the abyss. Estimates of vertical turbulent diffusivities range from a maximum of 1 cm^2 sec^{-1} down to 0.01 cm^2 sec^{-1} (see Garrett, 1979), giving a diffusive travel time through the thermocline in excess of several decades. It appears, though, that horizontal processes are more important in ventilating the thermocline (Jenkins, 1980) and the abyss (Armi, 1978). If the isopycnal-outcrop or boundary-impingement models are more representative of reality, then the penetration of these tracers into the oceans may yield a different kind of information than the reemergence of waste radioisotopes that have been disposed of in the abyss. Monitoring of the disposal and reemergence of radionuclides from deep-sea disposal will therefore be an additional source of information about ocean transport and dynamics.

As the overall concentration of the bomb-produced radionuclides decreases, due both to decay and to dilution, it is hoped that analytical techniques will keep pace regarding detection limits and accuracy. The kind of information yielded will take on a different character as the tracer distributions approach longer time scales and larger spatial scales. Further, as the signal-to-noise ratio will inevitably decrease as time goes by for the global fallout tracers, the radio-tracers produced by nuclear fuel reprocessing and perhaps even waste disposal will take on a new importance.

The information yielded by this second set of tracers is generally more on a local scale at present, but already significant perturbations on fallout inventories are being seen on a more regional scale (Livingston et al., 1982), and it is expected that the global-scale circulation will soon be appreciably tagged. For example, future studies should take advantage of the expected emergence of soluble radionuclides discharged from Windscale (^{90}Sr and ^{137}Cs) in the Norwegian Sea Overflow and subsequently, the North Atlantic Deep Water. It is conceivable that future reprocessing of thermal oxide fuel may introduce significant amounts of tritium in the same way (Livingston et al., 1982).

Unlike other nuclear fuel reprocessing byproducts, ^{85}Kr behaves as a global tracer, since it is an unreactive and otherwise stable gas. Its atmospheric concentrations are relatively uniform so that its downward penetration will be controlled more by vertical processes initially rather than lateral atmospheric or oceanic transport. In that respect it will prove a useful tracer in future years.

It is convenient to consider the evolution in sediments viewed in the context of their proximity to the continents and to the ocean surface. In nearshore, shelf, and slope sediments, it seems probable that concentrations of reactive nuclides will increase more rapidly than in sediments of the deep ocean. This would be not just because of shorter delivery paths, but also because of the higher particle loads and higher productivity which frequently characterize water masses of the shelf and slope. Delivery to, and evolution within, deep ocean sediments are likely to evolve over much longer time scales.

Coastal upwelling may well turn out to be a more important source of supply of reactive radionuclides to the coastal zone than terrestrial runoff. The persistence of reactive nuclides in subsurface oceanic water columns, such as noted above for the transuranics, means that they are potentially available for supply to shelf areas via upwelling processes, over quite long time scales. One would expect to see a more rapid evolution of this nature along upwelling-prone western continental margins.

Evolution of artificial radionuclides contained within coastal, shelf, or slope sediments is certain to be a complex process which would be dependent on sediment characteristics. Such factors as sedimentation rate, type and concentration of sediment infauna, and prevailing redox chemistry would be some of the controls on nuclide evolution patterns. In areas with high rates of sedimentation and diminished infaunal mixing, one would expect to see continued burial of delivered radionuclides. In contrast, in areas where sedimentation rates are low and mixing active, one would expect radionuclide concentration patterns to evolve as a function of the rates and nature of organism mixing. It is likely that diagenetically induced nuclide fractionation effects will be demonstrable and is conceivable that gradients of increasing reactive metal nuclides towards the sediment-water interface may be maintained by mechanisms such as those advanced to explain similar gradients of stable metals in estuarine sediments (Elderfield and Hepworth, 1975). Clearly, such sediment transport processes as occur as a result of infrequent high-intensity meteorological phenomena, sediment slumping on continental slopes, or turbidity current events must be expected to play significant roles in processes of sediment nuclide evolution.

In contrast to the picture presented for near-continental sediments, a more slowly evolving scenario is predicted for deep pelagic sediments. As noted earlier in respect to the evolution of nonconservative radionuclides within oceanic water columns, transfer of these nuclides to deep ocean sediments more than likely will take place at rates which, though probably slower than would have been predicted a decade or so ago, are likely to be fast relative to the scale of world ocean mixing. Individual nuclide sinking rates would be expected to vary according to their differing biogeochemistries in the ocean. The importance of biogenic particles and particle fluxes in this context has been noted earlier. This would lead to predictions of more rapid increases in the amounts of reactive nuclides found in sediments underlying productive and biogenic particle-rich water columns as opposed to those to be expected within sediments underlying unproductive and biogenic particle-poor oceanic regions. Sediments which underlie areas of the bottom traversed by advected bottom water which has been relatively recently formed and labeled at the surface, may be expected to receive an input of reactive nuclides additional to that which they receive by the delivery of sinking particles from the local surface ocean. These would be from such concentrations of reactive nuclides which avoid removal by scavenging during advection to depth and which would be available for transfer to the sediments by scavenging by the local rain of sinking particles or by direct uptake by interaction at the sediment-water interface. Following the delivery of nonconservative radionuclides to deep ocean sediments, their evolution within these sediments would be expected to depend on the same kinds of processes as in coastal sediments. So sedimentation rates, bioturbational mixing effects, and diagenetic processes would be expected to affect nuclide distributions but the rates of these processes in the deep ocean would be expected to be considerably less than those which characterize sediments in shallower waters close to land.

5 New Species

When one asks whether the oceans can be expected in the future to contain new species of artificial radioactivity, we must define what we mean by new species. In this context we make the following distinctions:

1. Nuclides which are not widely present now in the oceans;
2. Nuclides which are present in the oceans but have not been studied to any extent;

3. Nuclides which are currently studied in the oceans but which may appear in greater abundance in certain ocean areas by the creation of new sources of their release;
4. The formation or identification of chemical speciation of specific radionuclides.

Actually, there is some artificiality to these distinctions, as some nuclides fall into two or more categories. The future appearance of nuclides which do not occur at present to any appreciable extent (in a quantitative or geographical sense) is likely to derive from activities related to the nuclear fuel cycle. Increased fuel reprocessing, both in amount and location, and use of the ocean in planned disposal of radioactive waste in a whole spectrum of forms and amounts, could well introduce more and different radioactivity. These could include such activation products as ^{55}Fe, ^{63}Ni, ^{60}Co; such fission products as ^{106}Ru and ^{99}Tc; and transuranics produced by high burnup of nuclear fuel such as ^{241}Pu, ^{242}Cm, and ^{244}Cm. Some examples of nuclides which are widely present in the oceans but which have not been measured extensively, if at all, are ^{60}Co, ^{237}Np, ^{99}Tc, and ^{134}Cs—the latter is only now beginning to move in measurable amounts from coastal areas receiving nuclear fuel reprocessing discharges. Sensitivity problems have limited many measurements of ^{60}Co and ^{134}Cs. Progress in decreasing sensitivity limits for these is likely to be made by using coincidence counting techniques for ^{60}Co and greatly enhanced sampling concentration techniques for ^{134}Cs. The lack of data for ^{237}Np and ^{99}Tc is probably attributable to difficulties in their radiochemistries as well as sensitivity problems.

One process which should not be overlooked when considering potential sources of new radioactivity in the oceans is that of production via radioactive decay. This is especially important for nuclide pairs or chains where the decay products have half-lives as long as, or longer than, their precursors. This process has been recognized in the past in the formation of ^{3}He from ^{3}H decay (Jenkins and Clarke, 1976) and ^{241}Am from decay of ^{241}Pu (Livingston et al., 1975). The production of ^{238}Pu from decay of ^{242}Cm may be an important new source of ^{238}Pu because of the enormous amounts of ^{242}Cm which accumulate in spent nuclear fuel. Formally, ^{234}U, the granddaughter of ^{242}Cm, may need to be remembered, but its slow rate of production and widespread natural occurrence may limit its usefulness. Likewise, the slow rate of production of ^{237}Np from ^{241}Am decay will delay studies of ^{237}Np from this source for many centuries.

The whole field of chemical speciation of artificial radionuclides is barely scratched. For nuclides with stable chemical analogs such as ^{3}H, ^{90}Sr, ^{137}Cs, and ^{65}Zn, speciation is clearly the same as that of the stable counterpart, with the exception of cases such as mineral phases where exchange between a radioelement and stable counterpart in the mineral phase is not achieved. For the elements without stable counterparts, ^{99}Tc and the transuranics, very little is known about their chemical speciation in either water or sediments. The role of carbonate in complexation of those transuranics such as Np or Pu which can form stable high-oxidation states is by no means established; yet, when its role in making uranium conservative in sea water is remembered, its importance for

Pu and Np complexation needs to be known. Will ^{99}Tc be shown to move in the oceans essentially conservatively as pertechnetate? What will be speciations associated with the small amounts of reactive nuclides such as ^{239,240}Pu or ^{106}Ru which are transported in fuel reprocessing waste stream dispersal over considerable distances and in apparently soluble phases?

6 New Experiments and Problems

One area which could be predicted to receive increased attention in the future is that of particle transport processes which move reactive nuclides vertically through the ocean. Recent developments in the design and operation of sediment traps make it likely that they will be used to provide information on the reactive nuclides transported by the large particles which they collect. This information should be more useful when collected at the same time as measurements of other chemical constituents moved by the same particle fluxes. Radionuclide fluxes determined by such means will have bearing on the understanding of other chemical fluxes and will themselves be more understandable in the context of developing knowledge of these fluxes. It should be possible to determine rates of removal of reactive nuclides from the euphotic zone in different parts of the ocean using sediment traps in the manner used by Bruland (1980) to study trace metal and natural nuclide removal. To be more interpretable, such sediment trap studies need to be compared with fluxes determined by carefully collected series of ocean floor sediment samples, collected over time series long enough to record the delivery increments during the intervals between sampling.

As indicated earlier, opportunities for new experiments made possible by radioactive waste disposal activities are likely to appear. Attention has already been drawn to the potential use of the tracers introduced to the northern European shelf as a result of nuclear fuel reprocessing waste discharge. These tracers provide tools for study of a wide range of physical, chemical, biological, and geological processes and should be exploited to the full. We are now at the point of being able to study the arrival and movement of tracers from this source in the Arctic Ocean and stand to advance our ideas of Arctic circulation significantly. In like manner, every new development involving the disposal of radioactive waste on the sea floor is potentially exploitable as a new experiment in tracer oceanography. The extent to which radionuclides after release are moved by deep ocean processes provides information, not only on the fate of the released radionuclides, but on the processes which transport them.

A more purposeful injection of radiotracers has been recently discussed (Broecker and Shepherd, unpublished report). Aside from the international policy problems associated with the intentional release of significant amounts of radionuclides into the ocean, development is required for sample acquisition and measurement technology needed for high-density sampling. In addition, because of the statistical nature of the mixing/stirring phenomena, it appears likely that multiple releases will be necessary. Consequently, feasibility experiments involving more routine (nonradioactive) dye releases on smaller scales will precede such experiments.

The accumulation of a large data base of radionuclide distributions will facilitate the evolution of a new class of time-dependent numerical tracer models. The early work of, for example, Kuo and Veronis (1970) used advection-diffusion models with steady-state tracers. The development of such models for the nonequilibrium tracers is an important challenge. In addition, the Lagrangian particle models (Holland and Rhines, 1980) may prove useful for purposeful release and deep-sea disposal problems. These models are experiments in themselves and represent an important feedback loop between theory and measurement. It is expected that they will play a more pronounced role in the design of future field programs.

7 New Measurements

Some kinds of new measurements of artificial radioactivity in the oceans have been referred to already; for example, those of such nuclides as ^{237}Np, ^{90}Tc, ^{60}Co, and ^{134}Cs. Others seem likely to be in response to radioactive waste disposal activities. The detection of radioactivity released from such kinds of deep-ocean disposal activities which have been used in the past for low-level waste involve potential releases which are likely to be hard to measure, especially in water. Studies have shown (Dyer, 1976; Bowen and Livingston, 1981) that contamination of sediments by released radioactivity is confined to sediments very near to the release points. Other work has not found elevated levels of radioactivity in sea water overlying such disposal sites (Noshkin et al., 1978). New methods will need to be developed and tested which will increase the sensitivity of measurement of released radioactivity in the water overlying such areas. Because of the need to sample extremely large volumes to maximize the amount of nuclide collected, *in situ* methods are likely to be required. The technology of battery-operated submersible pumps capable of sampling thousands of liters of water is well advanced. These can be used to sample: particles (by passing the sea water through an appropriate size filter); or soluble radionuclides (by passing the water through a chemical absorber designed to collect the nuclide sought). Such absorbers as ion-exchange resins loaded with cupric ferrocyanide have been found to concentrate cesium isotopes from sea water (Folsom and Sreekumarau, 1970) and preparations of precipitated manganese dioxide have been shown to concentrate plutonium isotopes (Wong et al., 1978). A problem common to all such methods is reliable monitoring of the volume passing through a given collector and maximization and determination of the efficiency of a collector for a given nuclide. Such techniques will not, of course, be restricted to use around deep-ocean waste disposal sites but used as appropriate to increase the range and quality of data which describe the dispersion and mixing of radionuclides from atmospheric weapons testing fallout and from nuclear fuel reprocessing waste dispersal.

Another *in situ* collecting device which is not new, but likely to see continued use, is the marine organism. The IAEA panel which considered sample collection techniques (IAEA, 1970) noted that organisms were the first, most persistent, and probably useful *in situ* collecting devices. The use made recently of the blue mussel, *Mytilus edulis,* as an *in situ* collector for determining levels of pollutants, including artificial radionuclides, in United States coastal waters points up the validity of this form of biological approach (Goldberg et al., 1978).

The extension of the detection limit of many species will no doubt improve the utility of a number of measurements and perhaps allow us to study new areas. For example, the degassing and analysis of larger water samples will allow the detection of smaller tritium concentrations in the ^{3}He regrowth technique (Clarke et al., 1976). For example, analysis of liter samples should decrease the detection limit from 0.1 T.U. down to less than 0.01 T.U. This would be particularly important in the deep North Atlantic or the shallow Antarctic where tritium levels are typically less than 1 T.U.

The development of small active volume proportional counters has allowed the measurement of such isotopes as ^{85}Kr and ^{39}Ar. This latter isotope has a half-life (269 years) very well suited for circulation and thermocline studies, in addition to being inert. It is likely that some measurement will be made, but the larger water sample size required due to the low atmospheric activity (10^{-3} dpm liter^{-1} air) will represent a hindrance to large-scale sampling. Work is progressing toward measurement of this nuclide by detection of its daughter (^{39}K) by laser-Raman spectroscopy, so that the sample size problem may be reduced.

Current analytical precision for ^{14}C is about 3 to 4 $^\circ/_{\circ\circ}$ using beta counting. Direct measurement (using high-energy mass spectrometry or accelerators) holds the promise of greater sensitivity and hence smaller samples (Bennett et al., 1978) although the precision of the technique is an order of magnitude poorer than beta counting. It is, in principle, possible that the necessary improvements in the precision can be made. This would enhance the ^{14}C measurement program greatly by removing the need for very large samples.

8 Conclusions

We have attempted here to outline the present status of the study of radioactive tracers, primarily anthropogenic, in the sea, and to suggest where the future lies in this field. In fact, our discussion of the anthropogenic radionuclides is by no means comprehensive and does not include all of the significant contributions to the field. We have, however, attempted to give a broad-brush, impressionistic picture of where we have come from and where we are going in the field. We conclude with a few generalized recommendations as to what we feel are important directions for future work.

Regarding the soluble/conservative tracers, there are five points to make. First, the general observation that initial and boundary conditions to transient tracer models are as important in determining spatial/temporal distributions as interior dynamics suggests that a good deal of work is needed on the study of the delivery and time histories of these tracers. Second, decade time-scale variability in water-mass formation and mixing requires a longer time base and greater caution before conclusions are drawn. Third, we see a general, first-order similarity between the spatial patterns of the soluble radionuclides, so that it is the second-order differences generated by input history differences which will ultimately yield information about interior transport processes. This will in the future require synoptic multitracer studies. Fourth, careful use of the differing input functions or boundary conditions, for example, between ^{3}H and ^{85}Kr or ^{3}H and ^{3}He, may provide valid tests of diagnostic models. And, fifth, as the data base increases, so do the sophistication and subtlety of the models and interpretations. Increased rapport and cooperation between analysts and theorists will be crucial in the realization of the potential of the tracers.

For both conservative and nonconservative tracers, new inputs associated with the nuclear fuel cycle operation and radioactive waste management represent new tracer experiments. In addition to continuing studies of weapons test-derived radionuclides in the global fallout experiment, such releases as those from nuclear fuel reprocessing plants have potential use ranging from local to oceanic scales. Sediment/nuclide association studies are feasible and needed in areas such as have been used for dumping low-level waste. It is worthwhile reiterating the obvious point that the information yielded through studies of the oceanic movement and behavior of man-made radiotracers provides information on the fates and behaviors to be expected from a wide range of other nonnuclear pollutants.

Acknowledgments

Support for H. D. Livingston in preparation of this manuscript was provided by the U.S. Department of Energy under contract EY-76-C-02-3563 and DE-AC02-81EV10694 and for W. J. Jenkins by NSF-OCE-78-21734 and NSF-OCE-79-19815. We acknowledge this support with thanks. This is Contribution No. 4965 from the Woods Hole Oceanographic Institution.

References

Aller, R. C., and J. K. Cochran. 1976. ^{234}Th/^{238}U disequilibrium in nearshore sediment: particle reworking and diagenetic time scales. Earth Planet. Sci. Lett. 29, 37–50.

Armi, L. 1978. Some evidence for boundary mixing in the deep ocean. J. Geophys. Res. 83, 1971–1979.

Aston, S. R., and D. A. Stanners. 1979. The determination of estuarine sedimentation rates by ^{134}Cs/^{137}Cs and other artificial radionuclide profiles. Estuarine Coastal Mar. Sci. 9, 529–541.

Bennett, C. L., R. P. Beukens, M. R. Clover, D. Elmore, H. E. Gove, L. Kilius, A. E. Litherland, and K. H. Purser, 1978. Radiocarbon dating with electrostatic accelerators: dating of milligram samples. Science 20, 345–346.

Benninger, L. K., R. C. Aller, J. K. Cochran, and K. K. Turekian. 1979. Effects of biological sediment mixing on the ^{210}Pb chronology and trace metal distribution in a Long Island Sound sediment core. Earth Planet. Sci. Lett. 43, 241-259.

Bowen, V. T., and H. D. Livingston. 1981. Radionuclide distributions in sediment cores retrieved from marine radioactive waste dump sites. In: Impacts of Radionuclide Releases into the Marine Environment. IAEA, Vienna, pp. 33-63.

Bowen, V. T., and W. Roether. 1973. Vertical distributions of Strontium-90, Cesium-137 and tritium near 45° North in the Atlantic. J. Geophys. Res. 78, 6277-6285.

Bowen, V. T., V. E. Noshkin, H. L. Volchok, H. D. Livingston, and K. M. Wong. 1974. Cesium-137 to strontium-90 ratios in the Atlantic Ocean 1966 through 1972. Limnol. Oceanog. 19, 670-681.

Bowen, V. T., H. D. Livingston, and J. C. Burke. 1976. Distributions of transuranium nuclides in sediment and biota of the North Atlantic Ocean. In: Transuranium Nuclides in the Environment. IAEA, Vienna, pp. 107-120.

Bowen, V. T., V. E. Noshkin, H. D. Livingston, and H. L. Volchok. 1980. Fallout radionuclides in the Pacific Ocean: horizontal and vertical distributions, largely from GEOSECS stations. Earth Planet. Sci. Lett. 49, 411-434.

Broecker, W. S., and H. G. Östlund. 1979. Property distributions along the $\sigma_\theta = 26.8$ isopycnal in the Atlantic Ocean. J. Geophys. Res. 84, 1145-1154.

Broecker, W. S., and T-H. Peng. 1980. The radiocarbon age of North Atlantic deep water revisited. J. Geophys. Res. Submitted.

Broecker, W. S., T-H. Peng, and M. Stuiver. 1978. An estimate of the upwelling rate in the equatorial Atlantic based on the distribution of bomb radiocarbon. J. Geophys. Res. 83, 6179-6186.

Broecker, W. S., T-H. Peng, and R. Engh. 1980. Modelling the carbon system. Radiocarbon 22(3). In press.

Bruland, K. W. 1980. Oceanographic distributions of cadmium, zinc, nickel and copper in the North Pacific. Earth Planet. Sci. Lett. 47, 176-198.

Clarke, W. B., W. J. Jenkins, and Z. Top. 1976. Determination of tritium by mass spectrometric measurement of ^{3}He. Intl. J. Appl. Rad. Isotopes 27, 515-522.

Craig, H. 1969. Abyssal carbon and radiocarbon in the Pacific. J. Geophys. Res. 74, 5491-5506.

Deutches Hydrographisches Institut. 1980. Meerwasser. In: Umweltradioaktivitat und Strahlenbelasting. Jahresbericht, 1977. Der Bundeminister des Innern, Bonn, pp. 58-63.

Dreisigacker, E., and W. Roether. 1978. Tritium and ^{90}Sr in North Atlantic surface water. Earth Planet. Sci. Lett. 38, 301-312.

Duursma, E. K., and M. G. Gross. 1971. Marine sediments and radioactivity. In: Radioactivity in the Marine Environment (A. H. Seymour, Ed.). U.S. Natl. Acad. Sci., Nat. Res. Council, Washington, D.C., pp. 147-160.

Dyer, R. 1976. Environmental surveys of two deep-sea radioactive waste disposal sites using submersibles. In: Proceedings of an International Symposium on Management of Radioactive Wastes from the Nuclear Fuel Cycle. Vol. 2. IAEA, Vienna, pp. 317-338.

Elderfield, H., and A. Hepworth. 1975. Diagenesis, metals and pollution in estuaries. Mar. Pollut. Bull. 6, 85-87.

Farges, L., F. Patti, R. Gros, and P. Bourgeon. 1974. Activité du krypton-85 dans l'air hemispheres nord et sud. J. Rad. Chem. 22, 147-155.

Folsom, T. R., and C. Sreekumarau. 1970. Some reference methods for determining radioactive and natural cesium for marine studies. In: Reference

Methods for Marine Radioactivity Studies. Tech. Rep. Ser. No. 118, IAEA, Vienna, pp. 129-186.
Garrett, C. 1979. Mixing in the ocean interior. Dynam. Atmos. Ocean 3, 239-265.
Goldberg, E. D., V. T. Bowen, J. W. Farrington, G. Harvey, J. M. Martin, P. L. Parker, R. W. Risebrough, W. Robertston, E. Schneider, and E. Gamble. 1978. The mussel watch. Environ. Conserv. 5, 101-125.
Hardy, E. P., P. W. Krey, and H. L. Volchok. 1973. Global inventory and distributions of fallout plutonium. Nature (London) 241, 444-445.
Hetherington, J. A. 1976. The behavior of plutonium radionuclides in the Irish Sea. In: Environmental Toxicity of Aquatic Radionuclides: Models and Mechanisms. Ann Arbor Science Publishers, Ann Arbor, Mich., pp. 81-106.
Hetherington. J. A., D. F. Jefferies, and M. B. Lovett. 1976a. Some investigations into the behavior of plutonium in the marine environment. In: Radiological Impacts of Releases from Nuclear Facilities into Aquatic Environments. IAEA, Vienna, pp. 193-212.
Hetherington, J. A., D. F. Jefferies, N. T. Mitchell, R. J. Pentreath, and D. S. Woodhead. 1976b. Environmental and public health consequences of the controlled disposal of transuranic elements to the marine environment. In: Transuranium Nuclides in the Environment. IAEA, Vienna, pp. 139-154.
Holland, W. R., and P. B. Rhines. 1980. An example of eddy-induced ocean circulation. J. Phys. Oceanogr. 10, 1010-1031.
IAEA (International Atomic Energy Agency). 1970. Reference Methods for Marine Radioactivity Studies. Tech. Rep. Ser. No. 118, IAEA, Vienna.
IAEA (International Atomic Energy Agency). 1976. Transuranium Nuclides in the Environment. IAEA, Vienna, IAEA-SM-199.
IAEA (International Atomic Energy Agency). 1978. The radiological basis of the IAEA revised definition and recommendations concerning high level waste unsuitable for dumping at sea. Tech. Doc. 211, IAEA, Vienna.
Jenkins, W. J. 1977. Tritium-helium dating in the Sargasso Sea: a measurement of oxygen utilization rates. Science 196, 291-292.
Jenkins, W. J. 1980. Tritium and ^{3}He in the Sargasso Sea. J. Mar. Res. 38(3), 533-569.
Jenkins, W. J., and W. B. Clarke. 1976. The distribution of ^{3}He in the western Atlantic ocean. Deep-Sea Res. 23, 481-494.
Jenkins, W. J., and P. B. Rhines. 1980. Tritium in the Deep North Atlantic Ocean. Nature 286(5776), 877-880.
Joseph, A. B., P. F. Gustafson, I. R. Russell, E. A. Schuert, H. L. Volchok, and A. Tamplin. 1971. Sources of radioactivity and their characteristics. In: Radioactivity in the Marine Environment (A. H. Seymour, Ed.). U.S. Natl. Acad. Sci., Natl. Res. Council, Washington, D.C., pp. 6-41.
Kautsky, H. 1977. Stromungen in der Nordsee. Umschau Kurzberichte 77, 672-673.
Kigoshi, K. 1962. Krypton-85 in the atmosphere. Bull. Chem. Soc. Japan 35, 1014-1016.
Koide, M., J. J. Griffin, and E. D. Goldberg. 1975. Records of plutonium fallout in marine and terrestrial samples. J. Geophys. Res. 30, 4153-4162.
Kuo, H. H., and G. Veronis. 1970. The distribution of tracers in the deep oceans of the world. Deep-Sea Res. 17, 29-46.
Labeyrie, L. D., H. D. Livingston, and V. T. Bowen. 1976. Comparison of the distributions in marine sediments of the fallout-derived nuclides ^{55}Fe and $^{239-240}$Pu: a new approach to the chemistry of environmental radionuclides. In: Transuranium Nuclides in the Environment. IAEA, Vienna, pp. 121-137.
Livingston, H. D., and V. T. Bowen. 1976. Americium in the marine environ-

ment—relationships to plutonium. In: Environmental Toxicity of Aquatic Radionuclides: Models and Mechanisms (M. W. Miller and J. N. Stannard, Eds.). Ann Arbor Science Publishers, Ann Arbor, Mich., pp. 107-130.

Livingston, H. D., and V. T. Bowen. 1977. Windscale effluent in the sediments and waters of the Minch. Nature (London) 269, 586-588.

Livingston, H. D., and V. T. Bowen. 1979. Pu and ^{137}Cs in coastal sediments. Earth Planet. Sci. Lett. 43, 29-45.

Livingston, H. D., D. L. Schneider, and V. T. Bowen. 1975. ^{241}Pu in the marine environment by a radiochemical procedure. Earth Planet. Sci. Lett. 25, 361-367.

Livingston, H. D., V. T. Bowen, and S. L. Kupferman. 1982. Radionuclides from Windscale discharges I: nonequilibrium experiments in high latitude oceanography. J. Mar. Res. 40, 253-271.

Loosli, H. H., and H. Oeschger. 1978. ^{39}Ar, ^{14}C and ^{85}Kr measurements in groundwater samples. In: Intl. Symp. Isotope Hydrology. IAEA-SM-228/50.

Luykx, F., and G. Fraser. 1978. Radioactive effluents from nuclear power stations and nuclear fuel reprocessing plants in the European Community: discharge data 1972-76 radiological aspects. Health and Safety Directorate, Comm. Europ. Comm., Rep. EVR 6088, EN, RF.

Mitchell, N. T. 1977. Radioactivity in surface and coastal waters of the British Isles, 1976. Pt. 1: The Irish Sea and its environs. Ministry Agr. Food and Fish., Fisheries Radiobiol. Lab. Lowestoft, Tech. Rep., FRL 13.

Miyake, Y. 1963. Artificial radioactivity in the sea. In: The Sea, Vol. 2. Interscience Publishers, New York, pp. 78-87.

Munk, W. 1966. Abyssal recipes. Deep-Sea Res. 13, 707-730.

Munnich, K. O., and W. Roether. 1967. Transfer of bomb C-14 and tritium from the atmosphere into the ocean: internal mixing of the ocean on the basis of tritium and C-14 profiles. In: Radioactive Dating and Methods of Low Level Counting. IAEA, Vienna, pp. 93-117.

Murray, C. N., H. Kautsky, M. Hoppenheit, and M. Damian. 1978. Actinide activities in water entering the northern North Sea. Nature (London) 276, 225-230.

Needler, G. T. 1979. Comments on high latitude processes for ocean climate modelling. Dynam. Atmos. Oceans 3, 231-237.

Noshkin, V. E., and V. T. Bowen. 1973. Concentrations and distributions of long-lived fallout radionuclides in open-ocean sediments. In: Radioactive Contamination of the Marine Environment. IAEA, Vienna, pp. 671-686.

Noshkin, V. E., K. M. Wong, R. J. Eagle, and C. Gatrousis. 1975. Transuranic and other radionuclides in Bikini Lagoon: concentration data retrieved from aged coral sections. Limnol. Oceanogr. 20, 729-742.

Noshkin, V. E., K. M. Wong, T. A. Jokela, R. J. Eagle, and J. L. Brunk. 1978. Radionuclides in the marine environment near the Farallon Islands. Lawrence Livermore Laboratory Report UCRL 52381, 1-17.

Nozaki, Y., J. Thomson, and K. K. Turekian. 1976. The distributions of ^{210}Pb and ^{210}Po in the surface waters of the Pacific Ocean. Earth Planet. Sci. Lett. 32, 304-321.

Oeschger, H., U. Sieganthaler, U. Schotterer, and A. Gugelmann. 1975. A box diffusion model to study the carbon dioxide exchange in nature. Tellers 27, 168-192.

Östlund, H. G., and M. Stuiver. 1980. GEOSECS Pacific radiocarbon. Radiocarbon 22, 25-53.

Östlund, H. G., H. G. Dorsey, and C. G. Rooth. 1974. GEOSECS North Atlantic radiocarbon and tritium results. Earth Planet Sci. Lett. 23, 69-86.

Östlund, H. G., H. G. Dorsey, R. Brescher, and W. H. Peterson. 1977. Tritium Laboratory Data Report No. 6. University of Miami, RSMAS.

Rooth, C. G., and H. G. Östlund. 1972. Penetration of tritium into the Atlantic thermocline. Deep-Sea Res. 19, 481–492.

Pannetier, R. 1970. Original use of the radioactive tracer gas krypton-85 to study meridian atmospheric flow. J. Geophys. Res. 75, 2985–2989.

Schröder, J. 1975. Krypton-85 in the ocean. Z. Naturforsch. 30a, 962–967.

Stuiver, M., and H. G. Östlund. 1980. GEOSECS Atlantic radiocarbon. Radiocarbon 22, 1–24.

Stuiver, M., and P. D. Quay. 1980. Changes in atmospheric carbon-14 attributed to a variable sun. Science 207, 11–19.

Sugihara, T. T., and V. T. Bowen. 1962. Radioactive rare earths from fallout for study of particle movement in the sea. In: Radioisotopes in the Physical Sciences and Industry. IAEA, Vienna, pp. 57–65.

Telegadas, K., and G. J. Ferber. 1975. Atmospheric concentrations and inventory of krypton-85 in 1973. Science 190, 882–883.

Templeton, W. L. 1980. Artificial radionuclides in the oceans. In: Oceanography, The Past (M. Sears and D. Merriman, Eds.). Springer-Verlag, New York, pp. 420–437.

Toonkel, L. E. 1980. Quarterly ^{90}Sr deposition at world land sites. Appendix A, in Environmental Quarterly Report EML-370, Environmental Measurements Laboratory, U.S. Department of Energy, New York, A-3.

Volchok, H. L., V. T. Bowen, T. R. Folsom, W. S. Broecker, E. A. Schuert, and G. S. Bien. 1971. Oceanic distributions of radionuclides from nuclear explosions. In: Radioactivity in the Marine Environment (A. H. Seymour, Ed.). U.S. Natl. Acad. Sci., Natl. Res. Council, Washington, D.C., pp. 42–89.

Weiss, W. M., K-H. Fischer, B. Kromer, W. Roether, H. Lehn, W. B. Clarke, and Z. Top. 1978. Gas exchange with the atmosphere and internal mixing of Lake Constance (Obersee). Verhandlungen der Gesellschaft für Ökologie, Keil, pp. 153–161.

Weiss, W. M., W. Roether, and E. Dreisigacker. 1979. Tritium in the North Atlantic Ocean—Inventory, input and transfer into deep water. In: Behavior of Tritium in the Environment. IAEA-SM-232/98, Vienna, pp. 315–336.

Williams, P. M., M. C. Stenhouse, E. M. Druffel, and M. Koide. 1978. Organic ^{14}C activity in an abyssal marine sediment. Nature (London) 276, 698–701.

Wong, K. M., C. S. Brown, and V. E. Noshkin. 1978. A rapid procedure for plutonium separation in large volumes of fresh and saline water by manganese dioxide co-precipitation. J. Radioanal. Chem. 42, 7–15.

Zander, I., and R. Araskog. 1973. Nuclear explosions 1945–1972, basic data. Res. Inst. of National Defense, Sweden FOA4, Rep. A4505-Al, 56 pp.

Fisheries and Productivity Studies

Peter A. Larkin

1 Introduction

The major features of oceanic ecosystems have now been delineated. We know a great deal about the workings of the physical factors that combine with the complexity of the shape of the oceans to determine a nonuniform distribution of flows of nutrients and mixing conditions that in turn determine rates of production of organic material. The dependent production at secondary and tertiary trophic levels is accordingly complex, both spatially and temporally, and is characterized by variability that is only crudely predictable now. Reflecting these circumstances, many species of fish show wide fluctuations in year class abundance. In fish communities as a whole there are complex interrelations as to who eats whom among species. Fisheries harvest a selective crop from these ecosystems, the impact of the harvest being evidently substantial. The need for a theory of multispecies harvesting is central to the future of fisheries management.

In the next 50 years we will fill in on this understanding, gaining the insight necessary to take full advantage of the natural potential of the oceans for food production, and to augment it by a variety of ingenious techniques for enhancing critical rates of flow. We are on the threshold of major new discoveries for the benefit of mankind. The question is no longer whether, but how soon we will be able to harness the food producing potential of three-quarters of the planet. Investment now in fundamental and applied marine research will pay handsome dividends in the next half-century.

2 Physical Oceanography and Fishing

For the purposes of fisheries scientists, we have now accumulated a substantial descriptive knowledge of the world's oceans. The physical dimensions of the oceans and the major patterns of their circulation have been described. Most particularly, we know now where the major areas of upwelling occur, understand more or less why and when they occur, and even have ideas about how they may vary in their strength and location (Hartline, 1980).

The reason for placing such emphasis on upwelling is obvious, for it is *the* essential physical process in biological production in the ocean (Ryther, 1969). Every undergraduate now knows that an area of upwelling means available nutrients, which in turn means a community of organisms arranged in a hierarchy of trophic levels culminating in fish, some sea birds, and mammals, upon which may be superimposed a terminal trophic level called "fisheries." (It is also widely appreciated that for all living things Nirvana consists of becoming part of some future upwelling of nutrients, albeit in some other ocean.)

Bearing in mind that this broad picture has been filled in with impressive detail for virtually all the world's oceans, there is much to be proud of; but most of these findings are almost totally irrelevant to the current practice of most of the world's fishermen.

Since the earliest days of fishing, fishermen have concentrated their efforts in places that produced the largest catch, provided they could get the catch home in good shape for sale. As their vessels grew in size and the technologies of preservation were improved, the range of their activities enlarged. As early as the fifteenth century, the Grand Banks were fished by Europeans who had little if any knowledge of physical oceanography, trophic levels, productivity, or ecosystems. World fish catch grew from 40 to 70 million metric tons in the 1960s, an increase largely accomplished by the activities of distant water fleets, particularly of the USSR and Japan. To some extent, these activities were assisted by systematic use of oceanographic information, but in most instances fishermen fished where they caught fish, oceanography notwithstanding. The same was even true of the well-studied and highly productive Peruvian anchoveta fishery. The birds and their guano had abundantly demonstrated that there was a large amount of fish close at hand, and indeed, the birds were the basis for early speculations of what was there to catch (Paulik, 1971). Even El Niño was a matter of anecdotal remark before oceanographic studies had documented its occurrence and speculated about its cause.

By the mid 1970s, the world fisheries catch had leveled off at something slightly less than 70 million metric tons. This probably had very little to do with oceanographic changes but rather was a reflection of the failing economy of distant-water operations coupled with the impact of the intensive fishing operations of the previous decade. As a matter of fact, since the mid 1970s there has been a continuing advance in the technologies of locating and catching fish and this, rather than that the stocks are holding their own, may be the reason for the sustained catches of over 60 million metric tons.

In brief, I do not believe that oceanography means much to fishermen—yet. Aside from the occasional use of a thermometer to find the edge of a water mass, fishing takes place where it has taken place before, relying on sonar devices to detect fish at close range, and on the experiences of boat crews to get the fish on board. So much for oceanography and fishermen; now what can be said of those who are responsible for regulation of the catch?

3 Fisheries Biology and Fishing

Fisheries managers deal primarily with the statistics of the catch which reflect, as you may imagine, a mixture of what is there to be caught, what fishermen are looking for, and what they decide to keep. Even that kind of potentially mystifying information is not universally available, and those who can interpret it are not commonly at hand. Thus, though some of the world's fisheries are well documented and intensively managed, it is no exaggeration to state that most of the world's fisheries are virtually unmanaged.

Despite major efforts of the Food and Agricultural Organization (FAO) of the United Nations since the early 1950s, the gross statistics of world fish catches are still relatively unrefined. Even assuming that such statistics are superb is of little comfort for, taken by themselves, gross statistics of catch mean little except as they can be extrapolated on the assumption that the future will be like the past. Knowledgable management requires intensive analysis of the vital statistics of individual fish populations—their growth, mortality, reproduction, and detailed information on vulnerability to capture and the effort that is expended to catch them. The fisheries managers' bibles, written by Beverton and Holt (1957), Ricker (1958), and Gulland (1969), provide an impressive array of techniques for handling various kinds of data that in many instances simply cannot be mustered.

Where there is a body of information available, it is common practice to calculate what is called a "total allowable catch," which is set species by species with the aim of achieving some objective of management. This would be fine if the various species could be likened to rows of different kinds of vegetables in a garden, from which so many radishes and so many carrots might be harvested each year leaving enough of each for seed for the next year. Unfortunately for managers of the "gardens of the deep," the radishes and carrots are not in rows but are mixed up and yanked out together. The "total allowable catch" of the various species may thus mean some inadvertent removals of other species. There is also the haunting suspicion that, since the radishes and carrots may be competing and both may be eaten by cabbages, it just isn't realistic to think you can take a full crop of each. This suspicion was abundantly explored theoretically by Volterra in the 1920s (d'Ancona, 1954). In consequence of this kind of consideration, the common practice is to allow a total catch of all species (say for 70-odd species) that is "less than" the sum of the "total allowable catch" for each of the separate species. How much "less than" is still a matter for what is euphemistically called "creative judgment," but the basic idea seems to be to preserve the character of the mix of species primarily because that seems a prudent thing to do.

What kind of a mix of species to try to achieve in the catch is thus a particularly sore point for fisheries scientists. It is known that the mix may change from time to time. The switch from demersal to pelagic species in the North Sea is a recent experience (Jones, 1981). Another is the dramatic increase in squid in the Gulf of Thailand (Pope, 1979). The east coast fisheries of North America are well documented and provide some more examples of rather curious shifts in abundance of different kinds of fishes (Sissenwine et al., 1981).

For all of these kinds of changes we suspect that the fisheries may be partly to blame and that local variations in oceanography may also be implicated, but how much of which and how it comes about remain obscure.

First, consider the effect of fisheries on species abundance. Rather evidently, fisheries are selective of the species they catch, fishermen generally concentrating on the largest individuals of the species of greatest value. It seems logical to expect that this kind of selection would change the proportions of the different species in the community, just as the grazing of sheep changes the species mix of plants in a meadow (Harper, 1961), and logging changes the species mix in a forest. Surprisingly, what I have just said almost summarizes our present understanding of the interrelations between various kinds of fish in the ocean. Appreciating that it happens is a far cry from understanding how it occurs, which is why there is much consternation at present about what we call "the multispecies problem."

The way it stands at present, the models that appear to provide the greatest verisimilitude (such as that of Anderson and Ursin, 1977), require at least as many tuning variables as there are species because we do not have such rudimentary data as the feeding preferences of the various fish species. Just imagine: after almost 100 years of fisheries biology, collecting stomach contents has come back in vogue—with some Woods Hole residents, I might say, leading the way (Edwards and Bowman, 1979).

This raises the other consideration—local variations in oceanography. To take an extreme position, it could be that physical factors primarily determine the relative abundance of species, and that the interrelations among species are just a secondary theme. From the pioneer work of Blaxter and Hempel (1963), it was evident that the combination of limited yolk reserves and the seasonal variation in timing of plankton blooms was a potential source of wide fluctuation in the survival of larval fishes, and hence of the success of year classes. Since that time, much has been done to flesh out this line of explanation, pinpointing spatial and temporal heterogeneity, "patchiness," as a prime consideration in setting the oceanographic circumstances with which the annual crop of larval fishes must cope (Steele, 1976). Lasker (1978) has been a major contributor to the understanding of the mechanics that link the success of the annual spawning deposition to events of physical oceanography, such as the time of upwelling and the depth of mixing. Thus, if one were simpleminded, it would be easy to suppose that oceanography determines the relative abundance of species, and that the fishery only comes afterward as a factor in hastening the demise of successful year classes—a sort of clean-up operation, correcting nature's mistakes.

Most of us, of course, suspect that the natural situation is that the physical factors set the stage on which the biological dramas are enacted. Thus, if herring are very abundant because of favorable oceanography, then predators, disease, or competition among the herring will reduce their abundance; and if herring aren't abundant because oceanographic conditions are poor, it may not be long before conditions change and herring will again be abundant. The multispecies problem thus has environmental and biological dimensions that are highly complex.

Despite what seems to be a floundering of fisheries science at present, I believe that we are in the middle of an exciting period of discovery. The present situation in fisheries, especially as it is concerned with oceanography, is that now we have what we think is a fairly good appreciation of the nature of the problems. What we lack is experimental demonstration that we have indeed got it right, and theoretical understanding sufficient to enable prediction and to guide management. That is what we will get in the next 50 years, and it will give us the capacity to predict what will happen well in advance; the knowledge necessary to manage what is naturally produced with prudence and efficiency; and the insight to substantially increase food production from three-quarters of the earth's surface. I believe that nature can be improved upon, and by the year 2030, we will be doing it.

4 Fisheries and Oceanography by 2030

The three prime questions to be posed for the next 50 years of fisheries and oceanography are these: Will greater knowledge of the oceans be used by fishermen in catching fish? Will there be sufficient knowledge of processes of marine biological production to enable better management of living marine resources? Will we be able to manipulate marine organisms and the oceans to enhance production of living marine resources?

The answer to the first question, I suspect, is a resounding "No." Fishermen in the year 2030 will be more sophisticated versions of the contemporary brotherhood. They will be pleased to know that science has it figured out; will be even more pleased to catch the fish; but won't need any help from us in doing so. Most of the world's fish catch is taken within 100 miles of shore. The topography is well known and more or less constant, as are the major patterns of circulation. From that point on, searching with sonar is far more efficient than detailed science. Just as the weatherman predicts thunderstorms in a region but not by city blocks, fishermen only need general forecasts of abundance and they can then find the concentrations of fish. In fact, it would not surprise me if the fishermen of that day turned off some of their superefficient sensing devices just to make a bit more of a game of it.

There may be one exception to this speculation. For those species that roam over the wide expanses of the ocean, the trick to catching them economically is in finding them efficiently. In some measure, oceanographic surveillance may be useful, identifying the seasonal vagaries of water masses, and helping to guide fishing vessels to potentially profitable areas. But even in this case I suspect that direct aerial or satellite spotting of fish will be the more common practice, as it is today. In a nutshell, if you think the fishermen of the future will rely on oceanography, I would suggest you think again.

The answer to the second question—will there be sufficient knowledge of processes of marine biological production to enable better management?—is a resounding "yes," and not just in the sense that almost anything would be an improvement on present levels of understanding. The leaders of the 1920s to the 1940s, people like Baranov, Russell, Graham, and Thompson, developed the beginning of quantitative theories for single species approaches that were subsequently fleshed out by Ricker and Beverton and Holt. By the 1960s, at the same time as these theories were being extensively applied, there were some early warnings that the theory was too simple to suit the circumstances. Since the 1950s, biological oceanographers have put fisheries into broader ecological contexts, and the more imaginative fisheries scientists have paid greater attention to what oceanographers were publishing. While this is perhaps gross simplification in the interests of brevity, I believe it captures the mood of much of the memorable work of the 1950s to the present. Such people as Cushing, Riley, Parsons, Steele, Hempel, Dickie, Longhurst, Walsh, Dugdale, Mann, and many others not only encouraged broader conceptions, but led the way in the development of more holistic models in both national and international enterprises, much assisted in many cases by their collaboration with cooperative physical oceanographers. Moreover, they were in general somewhat better equipped than their predecessors with modeling skills and were among the first to benefit from the blessings of computers.

The vanguard has been most recently composed of a new generation of theorists who come to work fully armed, even bristling, with computer techniques, mathematical skills, and especially, imagination. I have the impression that many of these people, whom I shall not name, are somewhat overspecialized and would do well to do a lot more observing and experimenting before publishing. These impressions and feelings clearly identify me as belonging to the year class of 1950, that particular group which now thinks it can see the forest for the trees, unlike those who went before or who come after. In truth, of course, this new generation of theorists is the right thing for the times, because they bring the power of analysis which will be needed to solve the scientific problems which are now exposed in skeletal form.

Take, for example, the central issues: How do marine ecosystems maintain their structure? What factors can displace them to new configurations? To what extent can they be manipulated with impunity? The simple ideas of trophic levels have now been replaced with systems concepts that include notions of such properties as diversity, stability, resilience, and multiple equilibria (see, for example, Holling, 1973). We no longer think of simple systems of differential equations that are tractable but unrealistic. Instead, we build large-scale simulations which have as components the functional processes that relate living organisms to each other and their environment; and having observed the properties of these simulated systems, we then try to compress them into a relatively small number of equations which will describe the system in terms of its abstract properties.

In effect, this is akin to experiment, for it is not always easy to manipulate the real world on the scale that is required to do what we oldtimers would call "really hard science." Unfortunately, our solar system does not have two planets earth, one of which could be used as a control.

Attempts to build experimental microcosms have not been conspicuously successful. Where they are too small, they do not mimic natural circumstances; and where they are large, they tend to be unmanageable (Menzel and Steele, 1978). It is, of course, easier to study the lower trophic levels in miniature settings because there are not the same problems of scale; and there have been many notable advances through experimentation in the understanding of chemical processes in the ocean, the autoecology of small organisms, primary productivity, grazing by zooplankton, and pathways of decomposition. Even small fish ponds have contributed important ideas about the nature of interactions among various kinds of fishes and the character of their quantitative dependence on other parts of a microcosmic system. But these aren't the ways of doing things that will pay off in a major way in the future.

The key to future understanding almost certainly lies in large-scale, multidisciplinary, multiship sampling programs to attain simultaneously the necessary geographical scope and the necessary ecosystem coverage, and in large-scale ecosystem and management experiments which have the necessary potential for extrapolation to natural conditions. On the largest scales, of course, it still won't be possible to do the hardest kind of science but, given enough opportunity to work at the practical limits, it should be possible to predict the rest. From then on, as for modern-day physics, nothing will succeed like success in prediction.

These heroic comments make it all sound easy but in fact, of course, it won't be. For one thing, it seems likely that there will be different modeling systems for different parts of the ocean for some time to come. All the models may have the same architecture in the same sense that all houses have floors, but the emphasis in each may be on different processes. Even by the year 2030, it would seem likely that Arctic, temperate, and tropical marine ecosystems will be treated more or less separately; and within each, lagoons, coastal swamps, estuaries, inshore waters, offshore waters, and mid-ocean regions will be handled as somewhat separate entities. The theorist with a universal model is likely to find that it fits everything poorly. The theorist with a particular model is likely to find it has no scientific generality. The moral for those who are not theorists is to expect the need for major efforts at large-scale and intensive investigation in many different settings.

For all these enterprises I foresee one major technical problem: in a word, systematics. For many years I have clung to the belief that by grouping organisms by what they do ecologically and by what size they are, it should be possible to deduce the characteristics of the ecosystem in which they participate. Many approaches of this kind have indeed been illuminating, that of Swingle (1950) coming to mind for its historic interest, and those of Sheldon et al. (1972) and Steele (1976) for their significance. But viewed another way, these are only properties of ecosystems which we have observed. They do not tell us which species will fit in with which others to produce what kind of consequences. For fisheries managers, this is the crucial point of interest. It thus becomes necessary to identify not just the principal actors in the current play, but all of the potential actors in all of the potential plays. Bearing in mind some recent developments in fish population genetics, there may also be the possibility that some of the actors may change their roles.

Systematics is not just an ordinary stumbling block, and I see no ordinary or quick solutions to its miseries. Methodologically speaking, except for better microscopes, greater capacities for handling information, some biochemical techniques of great promise, and more museum collections, systematics today plods along the way it always did, dependent primarily on long hours of tedium and long memories on the part of its slaves.

Despite these obstacles, it seems likely that 50 years from now we will have sufficiently fleshed out our understanding of marine ecosystems to be much better managers. I would guess, though, that we will not harvest much more than we do now where there is dependence on the natural processes as we find them today. While there is much speculation about eating phytoplankton soup and krill paste sandwiches, it seems likely to be as uneconomical and unappetizing then as now. But we *will* control harvests with great effectiveness to obtain a sustainable yield of what we wish to harvest. The invention of trawls which will exclude fish above a certain size, as well as include those above a threshold size, will be an important development in increasing yield (Reed, 1980).

The big increases in marine production will come from large-scale manipulation of the species in which man has an interest, and from manipulation of the ocean environment. The answer to my third question: will we be able to manipulate marine organisms and the oceans to enhance production of living marine resources? is "yes" and it's fun to watch the techniques developing now. Conider first the manipulation of organisms. Those that have a suitable combination of commercial value and ease of culture are already the target for substantial activity. At present, a significant portion of world oyster production is aided and abetted by man; shrimp culturing is becoming a way to make money in present-day economic settings; and seaweed culture will almost certainly supersede harvesting of "wild" seaweed. Dozens of examples all serve to underline that mariculturists are on the move, and that by the year 2030 they will be major contributors to marine production. It is variously rumored that culturing techniques already produce 10% of the world total production of sea foods, and by the year 2000 it could be 30%.

In some parts of the world this kind of enterprise is costly, largely because at present they are labor intensive. Much more attractive for the adventuresome are the potentials for sea ranching rather than sea farming. For example, for well over a century salmon culturists have dreamed of producing millions of salmon just by providing a helping hand in the early life history. Most of their early successes owed far more to inadvertence than to *savoir faire,* and there were far more failures than successes, but in recent decades there have been many major advances in the relevant sciences of fish physiology, behavior, nutrition, and disease, as well as a great variety of experience with a wide range of mechanical contraptions. Disenchantments abound, but out of the wreckage there have emerged some major successes. Japanese salmon hatcheries now assist in an annual production of chum salmon that exceeds the total annual production of all five species of salmon from the west coast of Canada. The North Pacific rim is currently overflowing with hatchery schemes (many of them are harebrained) from which there should emerge by 2030 the potential for maximizing North Pacific salmon production.

There is a big "if" to this dream, for to achieve it will require sufficient understanding of the ecology of the North Pacific that one could orchestrate the production of the various races of the various species so as to get the most each year. Elsewhere I have speculated about this possibility, and some of its other research components such as genetic manipulation to develop new races of salmon that would migrate to "desirable" parts of the ocean (Larkin, 1980). Salmon, of course, aren't the only kinds of fishes, even if they are the most important to me, so suffice it to say here that there are great potentials in ranching them at sea. What is more exciting is that similar potentials are waiting to be explored for other species. The British have already tried plaice nurseries. If one likes haddock from Georges Bank, why not do the research to back up haddock hatcheries? Or, for that matter, any other local species. The ultimate scenario is that they are caught close to maturity, brought to spawning, stripped of eggs and milt, raised in billions to an optimal size for release at carefully selected locations that guarantee both their subsequent success and their return for capture inside one's 200-mile limit when they are just the right size for the market. I'm midly baffled not to have read somewhere that someone has already tried it and failed, but is dauntlessly carrying on. I certainly hope to see the day when I read of the first successes.

5 Fisheries of the Future

With any luck, by the year 2030 potentials such as these will have been demonstrated for enough species that there will be an enormous proliferation of such activity. Sophisticated biological oceanographers from all over the world will rage at the incompetence of hatchery managers who failed to take their advice about when and where to plant the young fish, and hatchery managers will convene to discuss the impracticality of most of the suggestions they get from biological oceanographers. It will be a great time for all participants, livened by uninformed political promises of more to come and well-informed political threats about the consequences of failure to produce.

At just about that time, the physical and chemical oceanographers will be again making noises about the potentials (that they will say were first talked of "a century ago") for manipulating the physical circulation of parts of the ocean. There will be talk again of damming the Straits of Belle Isle to modify climate and to control ice, of manipulating circulation in fjords, and of resurrecting the scheme to close off the Bering Sea from the Arctic. There will be many trial efforts to induce upwelling, some of which will be said to show great promise.

Those more oriented to the market potentials will be planning the possibility of large, artificial floating islands that contain cities surrounded by thousands of mariculture ponds fed by nutrient-rich water pumped from the deeps. To the likes of these, the world population explosion will be a laugh, for they will think of themselves as the first colonizers of the empty three-quarters of the earth's surface. But I'm going beyond my topic . . . .

To return to extrapolating the here and now, I sincerely hope that, by 2030, we will have seen the limitations of what is now called "optimum sustainable yield," which is supposed to be effected through regional councils, bureaucrats, or some mix of the two.

The basic premise of these approaches is that in managing fisheries, scientific judgment should be compromised in a hierarchy of social and economic considerations. I have remarked before (Larkin, 1977) that it is a recipe for heaven or hell. Thanks to what economists call discount rates, and what sociologists refer to as adjustment problems, the cold hard core of the rationale for "optimum sustainable yield" can be nothing but shortsightedness. In its own way, it is a concept that can be more punishing to the natural world than the concept of maximum sustained yield that it replaced. Both are founded on beliefs in social efficiency, and both assume adequate scientific knowledge. In my view, it is premature to apply either of them today at the risk of the basic wherewithal for the future. Far better would be to acknowledge our present lack of understanding, to encourage the most vigorous pursuit of new knowledge, and meanwhile to desist from the temptation to push natural systems to what we presently may perceive as their limit. Eventually, by which I mean the year 2030, I hope we will harvest within the constraints of some principles of management of natural ecosystems.

Translated, this burst of rhetoric has a simple moral. The past century has seen enormous gains in our understanding of marine production. We are now aware of the major issues, scientifically speaking, and given the encouragement can promise that by the year 2030 we will have major contributions to make to mankind. Maintaining the natural capital resources that will be essential to this promise should be a first consideration. Now is a good time for conservation in the most conservative sense of the word.

A postscript about Woods Hole seems appropriate. One of the facts of life is that marine science generally has been much tied to relatively large institutions. In Canada, where I come from, virtually all of the marine work of real substance has come from the activities of the biological stations of what used to be called the Fisheries Research Board of Canada. The same type of thing is true in the United States, the United Kingdom, and many other parts of the world. Ocean science is big science. It requires many people from many disciplines working as teams on long-term national and international projects. The accomplishments of the past century have clearly shown that it is institutional assemblages such as that of the Woods Hole complex that have led the way, and it will be their fortunes that dictate the pace of advances in the future.

References

Andersen, K. P., and E. Ursin. 1977. A multispecies extension to the Beverton and Holt theory of fishing, with accounts of phosphorus circulation and primary production. Medd. Danm. Fisk-og Havunders. N.S. (7), 319–435.

Beverton, R. J. H., and S. J. Holt. 1957. On the dynamics of exploited fish populations. Fishery Invest. (London) Ser. 2(19), 533 pp.

Blaxter, J. H. S., and G. Hempel. 1963. The influence of egg size on herring larvae *(Clupea harengus* L.). J. Cons. Intl. Explor. Mer 28, 211–240.

d'Ancona, U. 1954. The Struggle for Existence. E. J. Brill, Leiden.

Edwards, R. L., and R. E. Bowman. 1979. Food consumed by continental shelf fishes. In: Predator-Prey Systems in Fisheries Management (H. Clepper, Ed.). Sport Fishing Institute, Washington, D.C., pp. 387–406.

Gulland, J. A. 1969. Manual of Methods for Fish Stock Assessment. Part 1. Fish population analysis. FAO Man. Fish. Sci. 4, 154 pp.

Harper, J. L. 1961. Approaches to the study of plant competition. Symp. Soc. Exptl. Biol. 15, 1–39.

Hartline, B. K. 1980. Coastal upwelling: physical factors feed fish. Science 208, 38–40.

Holling, C. S. 1973. Resilience and stability in ecological systems. Ann. Rev. Ecol. Syst. 4, 1–23.

Jones, R. 1982. Species interactions in the North Sea. Can. J. Fish. Aquat. Sci. In press.

Larkin, P. A. 1977. An epitaph for the concept of maximum sustained yield. Trans. Am. Fish. Soc. 106(1), 1–11.

Larkin, P. A. 1980. Pacific salmon: scenarios for the future. Donald L. McKernan Lectures, Washington Sea Grant Publ., University of Washington, Seattle, 23 pp.

Lasker, R. 1978. The relation between oceanographic conditions and larval anchovy food in the California current: identification of factors contributing to the recruitment failure. Rapp. Proc.-Verb. Reun. Cons. Intl. Explor. Mer 173, 212–230.

Menzel, D. W., and J. H. Steele. 1978. The application of plastic enclosures to the study of pelagic marine biota. Rapp. Proc.-Verb. Reun. Cons. Intl. Explor. Mer 173, 7–12.

Paulik, G. J. 1971. Anchovies, birds and fishermen in the Peru Current. In: Environment: Resources, Pollution and Society (W. W. Murdoch, Ed.). Sinauer Associates, Inc., Stamford, Conn., pp. 158–185.

Pope, J. 1979. Stock assessment in multispecies fisheries, with special reference to the trawl fishery in the Gulf of Thailand. South China Sea Fish. Dev. Coord. Prog. 19, 106 pp.

Reed, W. J. 1980. Optimum age-specific harvesting in a non-linear population model. Biometrics 36(4), 579–593.

Ricker, W. E. 1958. Handbook of computations for biological statistics of fish populations. Bull. Fish. Res. Bd. Can. 119, 300 pp.

Ryther, J. H. 1969. Photosynthesis and fish production in the sea. Science 166, 72–76.

Sheldon, R. W., A. Prakash, and W. H. Sutcliffe, Jr. 1972. The size distribution of particles in the ocean. Limnol. Oceanogr. 17, 327–340.

Sissenwine, M. P., B. E. Brown, J. E. Palmer, and R. J. Essig. 1981. An empirical examination of population interactions for the fishery resources off the northeastern United States. Can. J. Fish. Aquat. Sci. In press.

Steele, J. H. 1976. Patchiness. In: The Ecology of the Seas (D. H. Cushing and J. J. Walsh, Eds.). W. B. Saunders, Philadelphia, pp. 98–115.

Swingle, H. S. 1950. Relationships and dynamics of balanced and unbalanced fish populations. Agr. Exptl. Stn. Alabama Polytechnic Inst., Bull. 274, 74 pp.

The Impact of Oceanography on the Military and Security Uses of the Ocean

Alan Berman

1 Introduction

As oceanographically oriented scientists, we can examine how, with changing naval technology, oceanographic factors have affected the ways that nations have used the sea to suit their strategic purposes. Although the technology of naval warfare has undergone major changes through the centuries, the general objectives have been relatively invariant with time. As the technology of naval warfare has changed, so has the oceanographic support needed by naval planners and operators.

Traditional naval theories (Mahan, 1890; Richmond, 1943; Gorshkov, 1967) have interpreted the objectives of national naval power very broadly; the ability to protect resource exploitation at sea (principally fishing); the ability to insure the continuity of a nation's surface-borne commerce; the ability to deny an enemy or a hostile nation the use of the seas for commerce or for force projection; and the ability to project sea-based force upon another country in order to implement some requirement of national policy. Contemporary discussions of the military and strategic uses of the seas have tended to focus very narrowly on the issue of force projection. In particular, modern naval theorists (MccGwire, 1979; Sokolovskij, 1968) have tended to concentrate on the problems and issues related to the deployment of nuclear weapons, from aircraft carriers which are capable of launching both nuclear and conventionally armed aircraft and from nuclear submarines carrying 16 to 24 ballistic missiles with a larger total number of independently targetable nuclear warheads. While such systems represent more than a trivial extension of the technology of the 12-inch gun of the pre-World War I capital ship and the carrier-based attack aircraft of World War II, they are, in the final analysis, only an extension of the traditional precepts of the use of naval power for force projections. The new dimension inherent in the operation of these modern capital ships is that they are capable of launching weapons of such awesome destructive force that they can, by their own power, terminate a major war or force the resolution of a conflict on terms that were initially unacceptable to the nation targeted by these systems.

2 Sea Power in the Past

Traditionally, the strategic use of the seas—sea power—has depended on the effective existence of both fighting and commercial ships. The material foundation from which the British Navy evolved was the shipping and shipbuilding capabilities of Tudor England. Henry VII initiated the policy of fostering the industry of shipbuilding. He gave bounties for the construction of larger ships, which were suited for both foreign trade and combat; he stimulated the increase of ships and seamen by a Navigation Act which was more rigorous in its terms regarding imports in foreign bottoms than any of its predecessors; and took steps to expand the fishing industry. When the occasion arose, the King preferred to rely upon hiring private ships for purposes of warfare. This was possible because the differentiation between a fighting ship and a commercial vessel was slight.

Henry VIII, in contrast to his father's abstention from interference in European conflicts, had a clear understanding of the national dependence on sea power and a passion for the sea. A practical seaman by background, he improved the design of the fighting ship and made heavy artillery its prime weapon. The modest fighting fleet of around a dozen ships which he inherited from his father was expanded to over 80. The preamble to his Navigation Acts of 1540 opened with words in which the dependence of England upon the sea was emphasized (Richmond, 1943).

When Henry VIII fought France in 1512, 1522, and 1544, the British Navy enabled the English Army to reach continental theaters of battle. This was the traditional role of force projection. The navy was also the principal defense against invasion. The British Navy of that day did not become involved in an attempt to control seaborne commerce. Economic pressure did not figure in the strategy of Henry VIII because it was a form of pressure to which France was not susceptible.

Queen Elizabeth I inherited both a navy and a traditional strategy. She stimulated national shipping by a continuation of the Navigation Acts. In the first year of Elizabeth's reign, an act was passed which restricted coastal shipping to English bottoms. This act was followed by others which gave encouragement to the building of vessels suitable to foreign trade and to the development of chartered companies. These companies all had a definite relation to the strengthening of the nation's sea power because shipping was rightly recognized as an essential element in national security. The larger types of trading vessels—those over 100 tons—were potentially useful as fighting ships. In addition to other measures, the Queen in 1562 (Richmond, 1943) added a third fish-eating day—Wednesday—to the calendar with the express object of "restoring the Navy of England" so that it might, "thereby by God's grace be able to defend the Realm against all Foreign Powers." Here we see an early application of what was to become a classical example of resource exploitation as part of a nation's military and strategic use of the sea.

Formal war with Spain, as distinguished from a state of reprisals, may be said to have begun in 1585 when it became certain that the King of Spain was making preparations for the invasion of England. A counterinvasion of Spain was out of the question. However, Spain was clearly vulnerable to economic pressure. The power of the economic weapon or of the denial of the use of the seas for commerce had been recognized for over a century before Elizabeth's day. Spain's vulnerability to external pressure was due to her need of imports from overseas. About a fifth of her revenues were derived from the bullion she received from her western empire and in the commerce of her East Indian possessions. She needed money to pay her armies, to buy the materials required for building and equipping her navy, and for the administration of her scattered interests in Europe. Her shipbuilding materials (like those of England) came from the Baltic. The treasures which her *Flota* and galleons carried across the Atlantic were Spain's lifeline. The varied produce of the trade with South and Central America were brought to Havana, which was strongly fortified. From Havana, great convoys numbering as many as 70 ships were escorted across the Atlantic by well-armed ships of war. These escorts were further strengthened when they reached the danger zone in the approaches to the Azores and the Spanish ports, by powerful squadrons which went out from Cadiz and Lisbon to meet them. In modern times an identical system was used to protect shipping against surface raiders, submarines, and aircraft.

Because of the importance of this trade to Spain, its destruction became the primary object of England's strategic use of the sea. Attack was not confined to naval action. Attempts were made to cripple Spain's fighting power and to reduce or cut off her supplies of goods from neutral sources. The great object was the destruction of the *Flota* itself. The most effective steps which England could take for her own security were held to be destroying, preventing the sailing of, or, best of all, capturing the treasure fleets. English policy followed what was later to become traditional military posture aimed at the denial of an enemy of her lines of seaborne commerce.

The route of the treasure was not the only line of sea communication on which Spain depended. Her armies in the Low Countries were mainly supplied by sea across the Bay of Biscay and up the Channel. The Spanish fleet depended on the naval stores of the Baltic. Her fisheries off the Atlantic seaboard and her coastal trade were all vital elements in her national and military economy. In 1585, events clearly showed that King Philip intended to invade England. The British sent a fleet with a body of troops to the West Indies to attack the Spanish trading organization in its three nodal points of San Domingo, Cartagena, and Panama. Another squadron went to the coast of Spain to cut up the

Spanish fisheries industry, whose importance lay not only in the seamen it furnished to the Navy of Spain, but also because the salted fish it provided was needed for the victualing of the ships. The coastwise trade was also an essential factor in the distribution of goods in Spain.

Hard as these blows were, however, they were not decisive. The injuries were temporary and could be repaired. King Philip of Spain was not blind to the fact that a succession of such applications of naval power would be inimical to Spanish interests. From Spain's point of view the situation needed to be brought to a rapid end. The most direct way of doing so was to crush England once and for all. Therefore, preparations were begun in 1586 for the invasion of England. A brilliant campaign conducted by Drake on the coast of Spain in 1587 retarded the preparations. Unfortunately, Drake's advice to repeat the offensive form of defense in the spring of 1588 was rejected by the Queen, and the Spanish Armada was completed without further interruption. It sailed in the summer and met with its fate at the hands of the English Navy and the howling gales of the Channel and the North Sea.

To exploit this great victory, Drake wished to strike a decisive blow at the residue of Spanish seapower by attacking Lisbon where the bulk of the Portugese ships, the best units of the Spanish Navy, rode at anchor.

A fleet under Drake carrying an amphibious force was sent to Lisbon. Unfortunately, the army was ill equipped, insufficiently provided, and wanting in discipline. The whole operation was a failure. While the strategic plan of Drake would have been endorsed by Mahan, by Richmond, or by Gorshkov, its failure lay in execution rather than inadequate technology or inadequate understanding of the strategic use of sea power.

3 Oceanography and Sea Power

British naval actions of four centuries ago contained all of the elements of traditional military and security uses of the oceans. While oceanography did not exist as a formal profession in the 1580s, the successful execution of sea power, both military and commercial, required a detailed understanding of oceanographic factors. In Elizabethan days, winds and currents determined trade routes, and little deviation from these routes was possible. As a result blockading forces, pirates, and corsairs all knew precisely where to position themselves without the availability of radar or other modern ocean surveillance systems. The exploitation and control of fishing grounds was a central element in the strategic use of the sea. Even by modern standards, our Western European ancestors had a remarkable knowledge of the factors which affected success in the exploitation of fisheries. The same understanding of oceanographic factors which controlled the yield of the European fishing industry also allowed military planners to know when and where naval vessels should be stationed so as to deny fishing banks to an adversary in a most effective manner. In this critical phase of economic warfare, knowledge of marine biology was clearly of overwhelming significance.

It is interesting to speculate on the proposition that if King Philip's Navy had had even a rudimentary capability to forecast weather, or even synoptic weather information, the Spanish Armada might not have been destroyed, and the course of European and American history would have been altered substantially.

The Elizabethan example provided a number of classical military and strategic examples of how a state can use the oceans to achieve its objective through the diplomacy of compulsion. While this doctrine was broadly understood, it was not until 1890 that Mahan first codified the nascent doctrine of "sea power." Great Britain and the United States, the preeminent oceanic powers, came to institutionalize both the term and the instruments of sea power. One hundred years ago, sea power's strategic weapon was expressed by the image of the battle-fleet and the individual capital ships which were its component parts.

The first usage of the phrase "capital ship" began during the first Anglo-Dutch War in 1652 (Burchfield, 1972). Until the revolution in weapons technology after 1850, the term did not extend beyond a synonym for "battle ship," a ship able to lie in the line of battle. By 1889, however, the exponential pace of technology had produced a battleship—exemplified by White's *Royal Sovereign*—of vastly increased power and ocean-ranging endurance (McMurtie, 1921). Conceived almost simultaneously with the doctrine of sea power (Mahon, 1890), these ships became the distinct emblems of national power and prestige. No longer merely the premier warship, the new high-freeboard, twin-turreted ironclad became the concrete symbol of naval power. With their extended endurance, capital ships could project a nation's power thousands of miles away from the homeland. These ships could blockade ports, destroy commerce, reduce coastal fortifications to rubble, and provide cover for the debarkation of expeditionary forces. Their mere presence, or even the threat of their presence, usually led to a prompt resolution of diplomatic disputes.

The "capital ship" as it began to evolve in the 1880s was a formidable platform, whose effectiveness was dependent in a very significant way on oceanographic knowledge. The 8-inch guns of the 1880s rapidly grew to 10, 12, and ultimately 16 inches long with commensurate increases in range. To attack a port or to affect a battle ashore, these ships had to be brought as close to the beach as possible. Because these ships were the first deep draft vessels, knowledge of inshore bathymetry became extremely important. Without an excellent knowledge of water depths and inshore current systems, inshore action could become a hazardous venture indeed. Gunfire directed against shore targets could only be as accurate as the charting of the area. Hydrography and oceanography became a major and a respected peacetime occupation of many naval officers. Admiral George Dewey, the great American naval hero of the Spanish American War, spent much of his military career between 1865 and 1898 engaged in charting and surveying (Halstead, 1899).

As a vessel that was no longer driven by wind power, the capital ship of a century ago could undertake world ranging missions in support of national policy. Gunboat diplomacy was limited only by the requirements for fuel and logistics support, and by the problems of command and control. The British Navy, and to some extent the U.S. Navy, solved these problems by the creation of a system of worldwide island bases. These became the coaling and cable stations that were the *sine qua non* of sea power at the end of the nineteenth century. To make a remote island into a useful naval base, inshore waters needed to be surveyed, tidal cycles needed to be predicted, channels needed to be marked, and the marine fouling properties of the local waters needed to be understood. Even today, the basic hydrographic charts for many obscure islands contain the charming notation "from the British Admiralty survey of 1873." The need to keep all of these stations in rapid communications demanded the deployment of extensive networks of undersea cables. These requirements placed a major strain on the oceanographic data base of the nineteenth century. Finding the best route to get an undersea cable to places such as Mauritius or the Seychelles demanded major deep water ocean surveys. Military hydrography was forced to move beyond inshore surveys which were its major preoccupation prior to 1880 and to concern itself with open-ocean bathymetry. Without a cable link the battle ships of a century ago could not be directed to provide the intimidating displays of power that were the hallmarks of gunboat diplomacy of the end of the nineteenth century.

Oceanographic surveys became an important secondary mission of all major deployments of naval vessels. The earliest postulation of the existence of the Mid-Atlantic Ridge in the Southern Hemisphere appeared to have resulted from a line of soundings between Luanda, Angola, and Cape Fria, Brazil, that was made by the USS *Essex* in 1878, as part of its duties with the South Atlantic Squadron of the U.S. Navy (Schley, 1904). At times, hydrographic surveys served as a form of gunboat diplomacy. When the American schooner *General Sherman* was seized and burned by the Koreans in 1869, the U.S. Government responded by sending a squadron of five ships backed up by a German cruiser to perform "inshore hydrographic measurements" in Korean waters. When the Koreans had the poor judgment to fire their antique cannons at this peaceful oceanographic expedition, the U.S. Navy promptly landed a force of 600 men and destroyed the Korean forts (Schley, 1904). The Koreans got the message. Never again did they interfere with hydrographic surveys or merchant shipping.

The major countermeasure to capital ships evolved before the first of the modern ironclads was placed on the drawing boards. In 1869, when Robert Whitehead successfully tested his first torpedo, he created the first unguided, pilotless missile capable of destroying the capital ship at long range (Hovegaard, 1920). His creation would become in the mind of public and professional alike the capital ship's deadliest opponent.

During its first 75 years, the torpedo was inherently unguidable. It required a suitable delivery platform to bring it to the scene of the battle and to point it in the right direction. In the late 1880s this launching platform was the torpedo boat. Small, cheap, and potentially deadly, the torpedo boat exerted an almost hypnotic appeal to the world's lesser navies. Here was an equalizer. A weapon with which even the Third World states of the 1880s and 1890s could hope to whittle down the capital ships of the big industrial nations. For a second-rank naval power envious of Britain's predominant fleet, the torpedo boat seemed like a cheap and easy way to challenge British supremacy at sea.

By 1892, France had 20 torpedo boats, Russia had 152, and Germany had 143 (Ropp, 1937). At speeds approaching 30 knots this weapon system possessed the glamor and promise of progressive technology as well as the low unit cost, to suggest that here indeed was "the weapon of the future."

The almost pathological level of fear that the threat of torpedo boats instilled in the minds of captains of battleships at the end of the nineteenth century is illustrated by contemporary accounts (Azoy, 1964) of the voyage of the USS *Oregon* at the beginning of the Spanish American War. At that time the *Oregon,* one of the most powerful battleships in the United States fleet, was stationed in California. The captain was directed to redeploy his ship to Cuban waters. This required a 14,000-mile cruise around the tip of South America including a passage through the Straits of Magellan. Every cove in the Straits provided a potential hiding place for torpedo boats. For lack of positive intelligence, the *Oregon*'s captain had to assume that the threat was real. From the time he entered the Straits until he reached the Atlantic, he was on the bridge of his ship at full battle alert.

From 1886 to 1892, France employed the torpedo boat as a spear point in the radical transformation of contemporary naval doctrine. France became the first major naval power to institutionalize the anti-capital ship weapon as the centerpiece of its naval strategy, while at the same time retaining an option on the conventional capital ship force (Charmes, 1885, 1886; Reveillere, 1893).

Going beyond incorporation of new technologies, the French navy strove to create a revisionist structure of sea warfare, one in which the traditional battle fleet would be reduced to passivity. According to Theophile Aube, who became Ministre de la Marine in 1886, torpedo boats and fast cruisers would strike directly at British seaborne trade, and the "Goliaths" of the British ironclad fleet with their "feet of clay" (Charmes et al., 1900). The panic brought by this massive interdiction of sea lines of communication, he reasoned, would soon bring Britain to her knees.

So great was the immediate impact of Aube's doctrine in Great Britain that no new battleships were laid down at all in 1887 and 1888 (Marder, 1940). The apparent impotence of the ironclad in the face of a torpedo attack generated a raging debate in Parliament and throughout the naval community. Lord Northbrook, the first lord of the Admiralty, concluded that advancing technology had all but ended the utility of the capital ship (Parkes, 1956).

The doctrine of Theophile Aube was not totally incorrect. Its weakness was in the choice of the launch platform for the weapon. The fast torpedo boat of the 1880 to 1900 era was unable to operate in high sea states and had an extremely limited cruising range. Once detected, it could be destroyed by the rather moderate fire power available on the torpedo boat destroyers which began to be developed at the turn of the twentieth century. For a while, it appeared that the threat of the torpedo to the capital ship was alleviated. The destroyer appeared to be the appropriate countermeasure to the torpedo boat.

4 The Submarine Era

The situation changed radically in the early decades of this century when the torpedo became the principal weapon of the primitive submarine as it began to enter into the inventory of the world's navies. While the early diesel and battery powered submarines had numerous operational problems, they proved to be formidable weapon platforms. They were remarkably effective in their ability to attack seaborne commerce. To the degree that they could sink enemy warships or mine enemy harbors, they went a long way toward wresting military control of the sea.

The effectiveness of a submarine was of course related to the fact that it could not be seen. While primitive acoustic sound navigation and ranging systems (SONAR) were developed in the early years of this century, these devices were generally ineffective as tools for finding submarines.

Most sinkings of submarines occurred when they were trapped near the surface while engaged in the time-consuming business of charging their batteries. Often they were sunk immediately after they had attacked a surface vessel. A recently torpedoed merchantman or a "flaming datum" as it came to be called, was a sure sign that a submarine was in the area. Surviving escort vessels would seed the suspected area with depth charges. Occasionally some would sink or damage the submarine.

With the advent of the modern nuclear or even the modern schnorkel submarine, the need of a submarine to wallow on the surface while charging batteries disappeared. Moreover, the standoff range of modern torpedos became sufficiently great that an escorting surface vessel could rarely be expected to overtake a high-speed nuclear submarine even in the vicinity of a "flaming datum."

While the survival of the naval vessels on the ocean surface depends, among other things, on success in antisubmarine warfare, the survival of the submarines depends on the failure of the antisubmarine warfare efforts of an adversary. The detection and localization of submarines depends on the use of equipment, tactics, and techniques which exploit oceanographic factors.

Traditionally, submarine detection uses either acoustic or nonacoustic techniques. Acoustic techniques are subdivided into active or passive techniques. In active acoustic systems, high-powered pulses of sound are generated and the time and direction of return of reflected signals are measured. Various signal processing techniques are used to enhance the echo received from valid targets and to cancel the signals received from the ocean boundaries or other spurious sources. In passive acoustic systems hydrophones are used to detect minute signals radiated by submarines. These signals are amplified and processed to eliminate unwanted background.

The success of either passive or active acoustic systems has long been recognized to be a sensitive function of a broad spectrum of oceanographic factors. The primary enemy of the underwater acoustician is noise. Accordingly, generations of geophysicists, marine biologists, acousticians, and signal processors have studied the attributes of noises generated by marine organisms, wind and wave stresses at the ocean boundaries, microseisms, underwater volcanism, and man-made sources. The propagation of an acoustic signal through the ocean is an even more sensitive function of oceanographic factors. As a result, broad studies have been pursued which were ultimately motivated by the needs of underwater acousticians to identify the temporal and the three-dimensional variation of the temperature and salinity of the ocean. The acoustic reflection of ocean sediments has been the subject of intense worldwide study. In recent years, the realization that long-range acoustic propagation is dependent on the detailed nature of the bathymetry has led to rather extensive and detailed surveys of the ocean bottom. An interesting byproduct of these surveys has been a better understanding of the structure of mid-ocean ridges and the tectonic forces that formed them. The need to achieve greater directional sensitivity in detectors by the use of spatial arrays of hydrophones has led to a need to understand the fine-scale factors in ocean structure which affect signal coherence and array performance. This need in turn has motivated oceanographers to understand the correlation distances and other statistical properties of such variables as temperature, salinity, and currents. Very detailed studies of the upper mixed layers of the ocean are necessary to achieve a complete understanding of the performance of complex acoustic systems.

Attempts to detect submarines by nonacoustic means have led to major interactions between naval programs and the oceanographic community. The desire to detect the remnant magnetic fields associated with submarines has led to the development of ever more precise magnetometers. In turn these devices have proved to be valuable instruments for the exploration geophysicist. As magnetometers become more sensitive, they also become more sensitive to local fluctuations in the Earth's magnetic field. These fluctuations, where they have a defined spatial variation, must be measured, catalogued, and understood. Thus, there has been a very substantial symbiosis between the military requirements for magnetometry and marine geophysics.

A submerged submarine produces a number of additional effects. First, it displaces several thousand tons of water. Its propellers tend to mix waters of slightly different temperature and density. Water displaced by the passage of a submarine is no longer at a depth which is stable for water of that temperature and density. As normal hydrodynamic processes take place equilibrium is reestablished. The process of reestablishing stable density stratification normally overshoots and depending on the local Brünt-Vaissala period, internal waves may be generated. The interaction of internal waves with surface waves and the dissipation of these waves with time are areas of great interest to physical oceanographers. They are also a source of great interest to naval planners and strategists who are concerned about any observables which might be generated by a submerged submarine and thus lead to its detection.

5 Remote Sensing

While much of the modern military sponsorship of oceanographic studies has been driven by the need for information which is necessary for the successful conduct of pro- or antisubmarine warfare, other needs for military oceanography exist. The operation of modern surface forces is still limited by winds, waves, storms, fog, ice, and other oceanographic factors. Fleet operators need both short-term predictions of oceanographic conditions and long-term projections. There are limiting conditions in which even a 100,000-ton nuclear aircraft carrier cannot launch aircraft. The performance of smaller surface vessels which is dependent on their ability to launch helicopters and vertical takeoff and landing (VTOL) aircraft is even more sensitive to oceanographic factors. The planning of modern naval operations must involve oceanographic considerations. What is the probability of conditions which permit underway replenishment? What percentage of the time in the month of January at latitudes North of $60°$ will a 3500-ton frigate be able to launch an antisubmarine warfare (ASW) helicopter? What is the probability of encountering dense fog or floating sea ice? Will surf conditions permit amphibious landings? Will turbidity, inshore currents, or sharks interfere with the operations of frogmen? The list of questions of this type is very long indeed. Successful naval planning simply is not possible without the availability of vast amounts of oceanographic data.

As comprehensive as current needs for oceanographic data are, new requirements may be anticipated in the future. Even as the primitive torpedo of 80 to 100 years ago had a revolutionary impact on naval warfare, modern technology has introduced two new factors which are equally revolutionary and which will require new types of oceanographic information. The technologies associated with remote sensing and long-range missiles have had an impact on surface forces which is as severe as the impact was of the introduction of the modern torpedo-equipped submarine.

From a naval standpoint, the technology of remote sensing includes any method that permits the remote detection, identification, tracking, and targeting of surface vessels. The techniques used include aircraft- or satellite-based radar, infrared (IR), visible, and radiometric systems. These are supplemented by high-frequency (HF) direction finding and HF radar involving ionospheric and surface reflection paths. The performance of each of these types of systems is not characterized by a single measurement. All of these are affected by a variety of factors. For example, surveillance radars are limited not only by signal-to-noise ratios, but by signal-to-clutter ratios. Radar clutter depends on sea state, ocean surface wave spectrum, the grazing angle of the electromagnetic signal, its polarization, and its frequency. To understand the performance and capabilities of both their own and hostile radars, naval operators must be supplied with detailed oceanographic data. As the sea state or ocean surface wave spectra change, the ranges at which radars are clutter limited and at which they are able to detect targets will change in a nonlinear way. Such factors must always be assessed in tactical planning.

As another example, consider the issue of ocean surveillance using ionospheric and/or surface reflected HF radars. Huge arrays of antennas may be used to project HF electromagnetic energy on to the ocean surface at distances of several thousand miles. Backscattered energy may be detected which may reveal the presence of surface ships. Typically, such radars are operated in a pulsed Doppler mode to discriminate against ocean clutter. However, as we know, the ocean surface moves, so that it also produces Doppler shifts in the observed returns. Since the electromagnetic wave lengths in the HF bands are comparable in length to the predominant length of ocean swells, a resonance scattering condition generally exists. Some component of the swell results in intense signals characterized by the simultaneous up and down Doppler of advancing and receding waves. The Doppler signal of the target must be separated from the strong signal of the moving surface. Such a process is clearly difficult unless a detailed understanding of oceanographic processes is available along with data concerning ocean currents and surface wave conditions. With such oceanographic knowledge and data, HF radar becomes a powerful technique for the tracking of remote surface vessels. Without oceanographic information, it is unlikely to be an effective technique.

The dependence of all military remote sensing technologies on oceanographic factors is paralleled by the impact of similar oceanographic factors on missile guidance systems. When inertial guidance is not used, recourse must be had to radar, IR, optical, or antiradiation seekers. All of these seekers must be designed to detect targets and to discriminate against the ocean background and against enemy decoys and countermeasures. The designer of missile seekers must be able to characterize the signals received from the ocean under all ranges of sea state, wave heights, fog, rain, and ice conditions which are likely to be encountered. The implications of second and higher order effects must be understood (for instance, what does a ship's wake look like to an IR seeker?). Clearly, the performance of any missile seeker must degrade under adverse oceanographic conditions. The circumstances and extent of these degradations must be understood from both an offensive and a defensive point of view.

6 The Future Relationship

In the future, the Navy undoubtedly will require the availability of near real-time oceanographic data. Presumably, remote sensing satellites and moored buoy instrumentation will become the source of much of the required oceanographic data. Parameters measured by these sensor systems will be communicated to appropriate shore sites, where they will be processed and routinely retransmitted to naval forces afloat.

New areas of military oceanographic interest are readily foreseen. Knowledge of the locations of oceanographic fronts and mesoscale eddies will be essential. Increasing attention will be devoted to the scale and structure of microscale features of the ocean. All aspects of Arctic oceanography are likely to receive increased emphasis. Interest in the areas cited here will be driven by considerations of antisubmarine warfare and the possible routine deployment of ballistic missile submarines in the Arctic basin. A military understanding of oceanographic fronts will drive ASW tactics in such areas as broad ocean search, convoy routing, and the screening of battle groups. A detailed knowledge of the structure of ice cover, the irregularities of the floor of the ice pack, and the biological component of ambient acoustic noise in the Arctic all will affect the extent to which nations use the Arctic basin to support their strategic nuclear purposes.

While many detailed examples of the dependence of surface forces on oceanographic factors could be provided, the central role of oceanography in the operation and survival of modern naval forces is clear. Furthermore, the broad objective that nations have had in using the oceans to suit their military and strategic needs have been relatively invariant with time. There has always been a need for the equivalent of oceanographic support. As the technology of naval warfare has changed, an ever more detailed dependence on oceanographic knowledge has evolved.

References

Azoy, A. M. 1964. Signal 250! The Sea Fight Off Santiago. David McKay Company Inc., New York.
Burchfield, R. W. (Ed.). 1972. A Supplement to the Oxford English Dictionary. Oxford University Press, Oxford.
Charmes, G. 1885. *Les Torpillieurs Autonomes.* Paris.
Charmes, G. 1886. *La Reforme de la Marine.* Paris.
Charmes, G., et al. 1900. *La Guerre Avec l'Angleterre.* Paris.
Gorshkov, S. G. 1967. The Sea Power of the State. Military Publishing House, Moscow. (Naval Intelligence Support Center undated translation.)
Halstead, M. 1899. The Life and Achievements of Admiral Dewey. H. L. Barber, Chicago.
Hovegaard, W. 1920. Modern History of Warships. Conway, London.
Mahan, A. T. 1890. The Influence of Sea Power on History 1660-1783. Little, Brown, Boston.
Marder, A. J. 1940. The Anatomy of British Sea Power. Alfred A. Knopf, New York.
MccGwire, M. 1979. Naval Power and Soviet Global Strategy. International Security, Vol. 3, No. 4.
McMurtie, F. (Ed.). 1921. All the World's Fighting Ships. Sampson, Low Marston and Co., London.
Parkes, O. 1956. British Battleships, Warrior to Vanguard. Seeley Service, London.
Riveillere, R. Adm. 1893. France and Her Marine R.U.S.I. (Translated from *Marine Francaise,* Feb. 1893.)
Richmond, H. 1943. Statesman and Sea Power. Clarendon Press, Oxford.
Ropp, T. 1937. The Development of a Modern Navy: French Naval Policy 1871-1904. Dissertation, Harvard University, Cambridge, Mass.
Schley, W. S. 1904. Forty-Five Years Under the Flag. D. Appleton and Co., New York.
Sokolovskij, V. D. 1968. *Voennaya Strategijce.* Moscow. Translated by Joint Publications Research Service, U.S. Department of Commerce.

Part III
Global-Scale Oceanography

Large Scale Geochemistry

Heinrich D. Holland

1 Introduction

Globally minded geochemists tend to regard the ocean as a reservoir engaged in geologically rapid exchange with the atmosphere and the biosphere, and reacting at a more leisurely rate with the lithosphere. The aim of these geochemists is to reduce the operation of this megasystem to a set of coupled differential equations whose solutions represent the present, the past, and perhaps the future composition of the atmosphere, the biosphere, the oceans, and the crust. Needless to say, this goal is not yet in sight; but if the progress of the past 50 years is any indication, the next 50 years may bring the solution of many outstanding problems.

These problems can be divided into two groups. The first group is the problems that surround the operation of the megasystem today; work on a subset of these would normally be called "real" chemical oceanography. The second group is the problems of reconstructing the state and operation of the megasystem in the past; a subset of these comprises the field of paleooceanography.

Both groups of problems were given a considerable impetus by Barth's observation in the early 1950s (Barth, 1952) that the oceans are a reservoir very much in transit. The residence time of cations and anions in the oceans was found to be between about 10^2 and 10^8 years. The shortest of the residence times is less than the mixing time of the oceans, and even the longest residence time is barely more than 2% of the age of the Earth.

Later in the 1950s (see for instance Craig, 1957; Revelle and Suess, 1957; Bolin and Ericksson, 1959) box models were introduced, particularly for treating the distribution of carbon in the ocean-atmosphere system, and insistent questions concerning the mechanisms that control and regulate the chemistry of sea water were raised. Sillén (1961) proposed an equilibrium model which flowered in the 1960s, particularly in the hands of Garrels and Mackenzie (see for instance Mackenzie and Garrels, 1966). Although some parts of the Sillén model are still functional, other parts clearly are not. The role of seawater cycling through mid-ocean ridges has only been appreciated during the past few years (see for instance Holland et al., 1976), and the role of kinetic factors on the sulfate concentration of sea water (see for instance Berner, 1980; Lasaga and Holland, 1976), on the oxidation state of the atmosphere-ocean system (Holland, 1978, Chapter 6), and even on the deposition and resolution of carbonates (see for instance Honjo and Erez, 1978) is just now being actively explored. In formulating the differential equations which govern the operation of the megasystem we now know most of the major sources and sinks, but we are still a long way from being able to write down functional relationships between element concentrations and element fluxes that are considered believable by more than a small group of close friends.

This state of affairs describes our level of understanding of the marine geochemistry of most elements. The prescriptions for progress vary, but they generally include more precise measurements of sources and sinks, and the development of a fundamental understanding of the processes that determine the rate at which elements are added to and removed from the oceans. Progress will surely require a variety of measurements at sea and at home, as well as the execution of experiments that are geologically significant in spite of the relatively short lifespan of the experimenters.

The quest for the construction of reasonable controlling differential equations will differ in detail from element to element and from compound to compound. All I will be able to do here is to illustrate some of the problems and some of the processes that are likely to lead to solutions; I will do this by describing the attacks on three problems in global chemical oceanography.

2 The Magnesium Problem

The input rate of magnesium from rivers to the oceans is reasonably well known. In the 1960s the most important output was thought to consist of the formation of new magnesium silicates in marine sediments. However, a thorough search by Drever (1974) failed to turn up more than a rather small fraction of the required quantity of new magnesium silicates.

In the late 1970s workers discovered that magnesium was removed essentially quantitatively when sea water reacted with basaltic rocks (Hajash, 1975; Bischoff and Dickson, 1975; Mottl and Holland, 1978) at elevated temperatures (Fig. 1). The submarine hot springs in the Galapagos discovered by the *Alvin* group, the active high-temperature vents found on the East Pacific Rise at $21°N$ (Francheteau et al., 1979; Spiess et al., 1980; Hekinian et al., 1980), several new vent fields discovered more recently, the heat flow deficits found near mid-

Figure 1. Black Smokers in the East Pacific Rise. (Photo by Dudley Foster, WHOI.)

ocean ridges, and the rare gas anomalies in the water column above ridge crests (Craig et al., 1975) have all combined to suggest that the missing magnesium has disappeared into new oceanic crust, and that in its place calcium and potassium have entered the oceans (Fig. 2). These major elements are accompanied by several important trace elements, among them lithium and rubidium, iron and manganese, copper and zinc. Seawater cycling through ridges is clearly affecting the marine geochemistry not only of magnesium but of a whole variety of elements; it is almost certainly responsible for the abnormal accumulation of iron and manganese in sediments not far from the East Pacific Rise; and it may be generating copper-zinc ore bodies on the ocean floor that rival or exceed in size those which were produced by the same or similar mechanisms in the geologic past.

These discoveries have opened several new chapters in the history of ocean exploration and global marine geochemistry. It seems likely that the capabilities of manned and of unmanned deep submersibles will be improved during the next

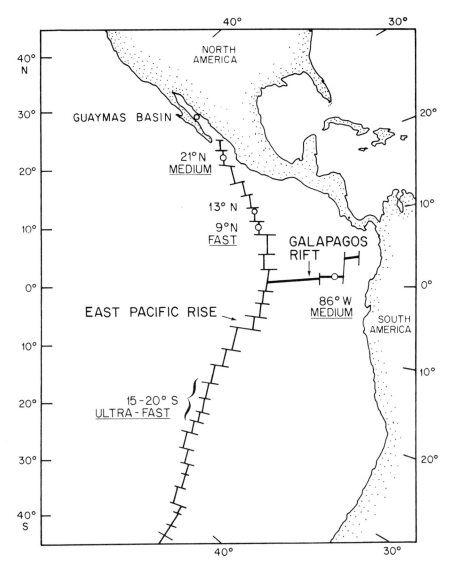

Figure 2. Locations of the active high-temperature vents found between 30°N and 20°S latitude.

two decades to take advantage of the discovery that parts of the ocean floor are dramatically heterogeneous on a very small scale, and that some of these heterogeneities have large consequences for the chemistry and mineralogy of the whole ocean.

The discoveries that have been made to date raise some fascinating questions regarding seawater cycling through mid-ocean ridges in the past. If the distribution of submarine hot springs is really as strong a function of spreading rate as seems to be required by recent discoveries, then the rate of seawater cycling may have varied significantly with time during the course of Earth history. The dolomite/limestone ratio in sedimentary rocks has varied in a curious fashion during geologic time. Archean carbonates seem to be largely limestones; Proterozoic and Paleozoic carbonates contain a large proportion of dolomites; Mesozoic and Cenozoic carbonates are again largely limestones. This pattern might be related to changes in the intensity of seawater cycling through the oceanic crust; however, the formation of dolomites is still not well understood, and other indicators, notably the lithium content of shales, do not appear to show the expected time variation. The whole question is still very much unresolved but seems to be amenable to solution on a time scale shorter than 50 years.

3 The Oxidation of the Ocean and the Atmosphere

I would now like to turn to a second favorite problem, one that has also yielded a partial and still somewhat uncertain solution. That is the problem of the controls on the oxidation state of the atmosphere and oceans, past and present. The boundary conditions of the problem are clear enough. The overall oxidation state of the crust has remained essentially constant during the past 3 billion years, and the oxygen content of the atmosphere has adjusted itself so that this has remained possible. The present, rather high oxygen pressure is required to insure that all but about 0.2% of the enormous quantity of carbon which is fixed annually by green plant photosynthesis is reoxidized to CO_2. Figure 3 shows the nature of the probable control mechanism. The oxygen production rate via photosynthesis and the burial of a small part of the organic carbon must decrease with increasing oxygen pressure. The rate of oxygen use by the oxidation of volcanic gases and of surface rocks exposed to weathering must increase with increasing oxygen pressure in the atmosphere. The sign of the net change in the oxygen content of the atmosphere is therefore positive at low oxygen pressures and negative at high oxygen pressures. It can be shown that at present the system is close to equilibrium; this is rather to be expected, because a system such as this has a stable control point: at the oxygen pressure where the rate of change of atmospheric oxygen is zero.

The mechanism that controls the oxygen content of the atmosphere therefore appears to be a classic negative feedback control system. Unfortunately, the quantitative aspects of this system are still virtually unknown. We do not know the course of either the oxygen production curve or of the oxygen use curve except in the roughest fashion. We know that the oxygen production curve is linked to the geochemistry of phosphorus and to the geochemistry of nitrogen. Of the two elements, phosphorus appears to be less accessible to organisms, and photosynthesis may rely on the availability of phosphorus rather than combined nitrogen. We know surprisingly little about the supply of dissolved phosphorus and of phosphorus bound in organic matter to the sea; we also know too

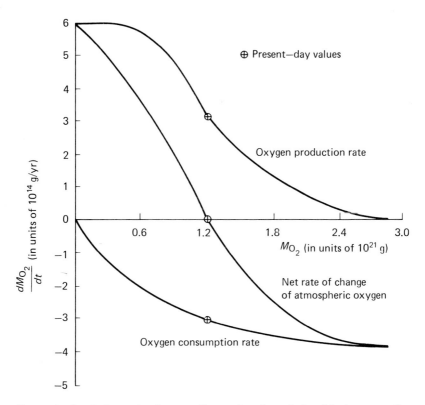

Figure 3. Semischematic diagram illustrating the relationship between the oxygen content of the atmosphere, the rate of oxygen production, the rate of oxygen consumption, and the net rate of change of atmospheric oxygen (Holland, 1978).

little about the proportion of marine phosphorus that is removed with organic matter, the proportion that is removed as a constituent of fish bones, and the proportion that is adsorbed on mineral phases. Without a clear understanding of these matters the oxygen production curve is subject to error.

The variation of the oxygen use rate with oxygen pressure is also poorly known. We know that most of the oxygen used in chemical weathering converts elemental or organic carbon to CO_2, sulfide sulfur to sulfate sulfur, and ferrous iron to ferric iron; however, we don't fully understand the kinetics of these oxidation mechanisms. To this extent we are uncertain of the response time of the control system and the sensitivity of the equilibrium oxygen pressure to a variety of tectonic, climatological, biological, oceanographic, and geographic factors.

Studies of the preservation of organic matter in a variety of oceanographic settings (Demaison and Moore, 1980) are helping to clarify some of the outstanding issues, but the effort to solve even the major outstanding problems will surely involve a good deal of research in organic geochemistry and microbiology.

When we understand more fully the chemical and bacteriological factors that control the degradation of organic matter in marine sediments, we can devise a fundamental model for the return of carbon to the oceans from anoxic marine sediments, the removal of sulfur from the oceans as a constituent of sulfide minerals, and the transfer of the many other elements across the sediment-water interface in anoxic settings. Satellites may well play a progressively greater role in monitoring biological activity on a global scale; such monitoring would yield useful data for a better understanding of patterns and rates of photosynthesis and for the analysis of man's effect on the CO_2 content of the atmosphere.

Several fairly persuasive lines of evidence suggest that the oxidation state of the ocean-atmosphere system has not changed dramatically during long periods of time (long compared to the 3 million year residence time of atmospheric oxygen with respect to loss by oxidative weathering and gain by the burial of reduced carbon, sulfur, and iron). This stability has apparently been achieved in spite of the development of widespread anoxic conditions in the lower portions of most of the world's oceans during parts of the Cretaceous period (Arthur and Schlanger, 1979) and during the curious period near the Permo-Triassic boundary when the large-scale formation of marine evaporites apparently drove the $\delta^{34}S$ value of sea water sulfate down to $+10\ ^{o}/_{oo}$ (Claypool et al., 1980) and the $\delta^{13}C$ value of marine carbonates up to $+2\ ^{o}/_{oo}$ (Veizer et al., 1980).

There is, however, persuasive evidence that the ocean-atmosphere system operated at a considerably lower oxygen pressure in the Archean and at least during the first part of the Proterozoic Era. The preservation of detrital uraninite in quartz pebble conglomerate uranium ores is difficult to understand otherwise, and the development of the huge iron ore deposits in Precambrian banded iron formations that are not related to nearby volcanism is best explained by invoking an ocean whose surface waters contained only a little oxygen and whose deeper parts were mildly anoxic. It seems likely that the increase in the oxygen content of the atmosphere toward the end of the Proterozoic Era is the result of changes brought about by biological evolution; but the validity of even this inference is not certain. Changes in tectonic processes, in the intensity of volcanism, and in paleographic settings may have been involved, as well as changes in the geochemical cycle of phosphorus.

4 Trace Elements

The third problem concerns the analysis and distribution of trace elements and compounds in sea water and in marine sediments. This area differs from the first two because it is more closely tied to the development of analytical techniques, and because it is less firmly focused on a single question. The development and application of new techniques have revolutionized our understanding of the marine chemistry of trace elements and compounds during the past 30 years. We now know a good deal about the geographic distribution and chemical behavior of elements and compounds whose presence in sea water was barely detectable in 1950 (see for instance Brewer, 1975), and it seems likely that a similar happy fate awaits many of the substances that are still beyond the reach of presently available analytical techniques. (It could, perhaps, be argued that the chemistry of an element or a compound that is present in sea water at the nano- or picomolar level is not particularly interesting. But even if this were granted, the use of these substances as tracers of important oceanographic processes and as indicators of the state of the ocean-atmosphere system would assure their study an honored place in chemical oceanography.)

The GEOSECS study of the distribution of bomb tritium and ^{14}C in the oceans has improved the estimates of the rate of downward mixing of surface waters sufficiently so that forecasts of the uptake of anthropogenic CO_2 by the oceans are becoming believable. Similar studies of the distribution of fluorochlorocarbons bids fair to supply important additional data for the rate of ocean mixing, and it seems likely that other tracers will come into their own during the next few decades. Most of these tracers will be pollutants; some will undoubtedly be harmless, but others are apt to be troublesome, and careful study of their distribution will become an end in itself rather than a means for understanding the behavior of the oceans.

Numerous mechanisms exist for the ocean to dispose of these tracers and of trace elements as a whole. Many trace metals are removed from the oceans as a constituent of carbonaceous sediments, by complicated and poorly understood processes. The overall result is a great enrichment of highly carbonaceous sediments in a suite of trace elements. The ratio of the concentration of many of these elements in carbonaceous sediments to their concentration in seawater frequently lies between 100,000 and 500,000. The degree of trace element enrichment is usually proportional to the organic carbon content of the sediments. This is shown in Figure 4(a) for molybdenum in sediments from the Black Sea. The Mo/C ratio along the lines in this figure depends, among other things, on the concentration of molybdenum in sea water. It is most interesting that the Mo/C ratio in marine sediments does not seem to have varied significantly during the Phanerozoic Era. Figure 4(b) shows this for Devonian black shales from Kentucky and West Virginia. Figure 4(c) confirms the conclusion for a range of geologic periods. Too few Precambrian black shales have been studied to show how far back in Earth history the near constancy of this slope extends. The small amount of existent data suggests that Proterozoic black shales have complements of Mo and of other trace metals that are similar to those in Phanerozoic shales.

This near constancy indicates, perhaps unsurprisingly, that the concentration of many trace metals in sea water has been roughly the same for geologically long periods of time. It reflects the near constancy of the composition of crustal rocks, the near-constancy of the proportionality of the marine trace metal supply to the rate of organic matter deposition with marine sediments, and the predominance of carbonaceous sediments in the removal of these elements from sea water. Quite likely, however, this state of affairs has not persisted throughout all of geologic time. If the oxygen content of the atmosphere was very low during the Archean Era, the release of trace metals during weathering would

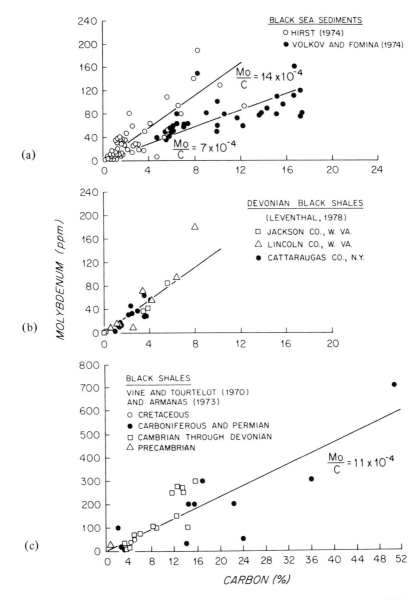

Figure 4. The correlation between the molybdenum and carbon content of (a) sediments from the Black Sea—differences between data might be related to analytical errors; (b) Devonian black shales; and (c) shales from various periods.

have been less complete, and their distribution in marine sediments would be correspondingly different. Possibly, therefore, black shales hold important clues to the evolution of the oxidation state of the atmosphere-ocean system.

Black shales are not the only sedimentary rocks whose trace element content can serve this function. Marine phosphates seem to supply similar clues. Sea water today is deficient in cesium, perhaps because Ce^{4+} is removed preferentially with manganese nodules. The cesium deficit in sea water is reflected in the rare pattern of marine phosphates (Laajoki, 1975). The patterns are strikingly similar and suggest that Ce^{4+} removal from sea water was as efficient during the deposition of these ancient sediments as it is today. It is not known where or how Ce^{4+} was removed from the oceans 2 billion years ago. The state of progress in this problem area is therefore similar to the state of progress in the other two areas which I have discussed. In all three we seem to be in full cry, pursuing an elusive quarry that has finally come into view.

I would like to close with some words from Roger Revelle's (1961) foreward to the 1959 International Oceanographic Congress: "Nothing under the sun is completely new, and each of these problems has been written and talked about for many years. But during the last few years new instruments, new techniques, and, above all, new people have combined in a vigorous approach to these old problems." It seems to me that this comment applies today as well as it did in 1959, and that there is every indication that many of the major problems in global chemical oceanography will finally be solved during the next half-century.

References

Armands, G. 1973. Geochemical studies of uranium, molybdenum, and vanadium in a Swedish alum shale. Stockholm Contrib. Geol. 27, 1–148.

Arthur, M. A., and S. O. Schlanger. 1979. Cretaceous "oceanic anoxic events" as causal factors in development of reef-reservoired giant oil fields. Bull. Am. Assoc. Petrol. Geol. 63, 870–885.

Barth, T. F. W. 1952. Theoretical Petrology. John Wiley, New York, 387 pp.

Berner, R. A. 1980. Early Diagenesis, A Theoretical Approach. Princeton University Press, Princeton, New Jersey, 241 pp.

Bischoff, J. L., and F. W. Dickson. 1975. Seawater-basalt interaction at 200°C and 500 bars: implications for origin of sea-floor heavy-metal deposits and regulation of seawater-chemistry. Earth Planet. Sci. Lett. 25, 385–397.

Bolin, B., and E. Ericksson. 1959. Changes in the carbon dioxide content of the atmosphere and sea due to fossil fuel combustion. In: Atmosphere and Sea in Motion (B. Bolin, Ed.). The Rockefeller Institute Press, New York, pp. 130–142.

Brewer, P. 1975. Minor elements in seawater. Ch. 7. In: Chemical Oceanography, Vol. 1 (J. P. Riley and G. Skirrow, Eds.). Academic Press, London, New York, San Francisco.

Claypool, G. E., W. T. Holser, I. R. Kaplan, H. Sakai, and I. Zak. 1980. The age curves of sulfur and oxygen isotopes in marine sulfate and their mutual interpretation. Chem. Geol. 28, 199–260.

Craig, H. 1957. The natural distribution of radiocarbon and the exchange time of carbon dioxide between atmosphere and sea. Tellus 9, 1–17.

Craig, H., W. B. Clarke, and M. A. Beg. 1975. Excess ^{3}He in deep water on the East Pacific Rise. Earth Planet. Sci. Lett. 26, 125–132.

Demaison, G. J., and G. T. Moore. 1980. Anoxic environments and oil source bed genesis. Bull. Am. Assoc. Petrol. Geol. 64, 1179–1209.

Drever, J. I. 1974. The magnesium problem. Ch. 10. In: The Sea, Vol. 5 (E. D. Goldberg, Ed.). Wiley Interscience, New York.

Francheteau, J., H. D. Needham, P. Choukroune, T. Juteau, M. Séguret, R. D. Ballard, P. J. Fox, W. Normark, A. Carranza, D. Cordoba, J. Guerrero, C. Rangin, H. Bougault, P. Cambon, and R. Hekinian. 1979. Massive deep-sea sulphide ore deposits discovered on the East Pacific Rise. Nature (London) 227, 523-528.

Hajash, A. 1975. Hydrothermal processes along mid-ocean ridges: an experimental investigation. Contrib. Mineral. Petrol. 53, 205-226.

Hekinian, R., M. Fevrier, J. L. Bischoff, P. Picot, and W. C. Shanks. 1980. Sulfide deposits from the East Pacific Rise near 21°N. Science 207, 1433-1444.

Hirst, D. M. 1974. Geochemistry of sediments from eleven black sea cores. In: The Black Sea—Geology, Chemistry, and Biology. Mem. 20 of the American Association of Petroleum Geologists, Washington, D.C., pp. 430-455.

Holland, H. D. 1978. The Chemistry of the Atmosphere and Oceans. Wiley Interscience, New York, 351 pp.

Holland, H. D., R. F. Quirk, and M. J. Mottl. 1976. The nonimportance of reverse weathering in the oceans. Geol. Soc. Am. Abs. Programs 8, 922.

Honjo, S., and J. Erez. 1978. Dissolution rates of calcium carbonate in the deep ocean; an in situ experiment in the North Atlantic. Earth Planet. Sci. Lett. 40, 287-300.

Laajoki, K. 1975. Rare-earth elements in Precambrian iron formations in Värylänkylä, South Puolanka Area, Finland. Bull. Geol. Soc. Finland 47, 93-107.

Lasaga, A. C., and H. D. Holland. 1976. Mathematical aspects of non-steady state diagenesis. Geochim. Cosmochim. Acta 40, 257-266.

Leventhal, J. S. 1978. Trace elements, carbon and sulfur in Devonian black shale cores from Perry County, Kentucky; Jackson and Lincoln Counties, West Virginia; and Cattaraugus County, New York. U.S. Geological Survey Open File Report 79-504.

Mackenzie, F. T., and R. M. Garrels. 1966. Chemical mass balance between rivers and oceans. Am. J. Sci. 264, 507-525.

Mottl, M. J., and H. D. Holland. 1978. Chemical exchange during hydrothermal alteration of basalt by seawater. I. Experimental results for major and minor components of seawater. Geochim. Cosmochim. Acta 42, 1103-1115.

Revelle, R. 1961. Preface. In: Oceanography (M. Sears, Ed.). Am. Assoc. of Science, Washington, D.C.

Revelle, R., and H. Suess. 1957. Carbon dioxide exchange between atmosphere and ocean and the question of an increase in atmospheric CO_2 during the past decades. Tellus 9, 18-27.

Sillén, L. G. 1961. The physical chemistry of sea water. In: Oceanography (M. Sears, Ed.). Am. Assoc. of Science, Washington, D.C., pp. 549-581.

Spiess, F., A. Carranza, D. Cordoba, C. Cox, V. M. Diaz Garcia, J. Francheteau, J. Guerrero, J. Hawkins, R. Haymon, R. Hessler, T. Juteau, M. Kastner, R. Larson, B. Luyendyk, J. D. Macdougall, S. Miller, W. Normark, J. Orcutt, and C. Rangin. 1980. East Pacific Rise: Hot Springs and geophysical experiments. Science 207, 1421-1433.

Veizer, J., W. T. Holser, and C. K. Wilgus. 1980. Correlation of $^{13}C/^{12}C$ and $^{34}S/^{32}S$ secular variations. Geochim. Cosmochim. Acta 44, 579-587.

Vine, J. D., and E. B. Tourtelot. 1970. Geochemistry of black shale deposits—A summary report. Econ. Geol. 65, 253-272.

Volkov, I. I., and L. S. Fomina. 1974. Influence of organic material and processes of sulfide formation on distribution of some trace elements in deepwater sediments of Black Sea. In: The Black Sea—Geology, Chemistry, and Biology. Mem. 20 of the American Association of Petroleum Geologists, Washington, D.C., pp. 456-476.

General Circulation of the Oceans

Pearn P. Niiler

1 Introduction

Oceanographers have three basic conceptions of the general circulation of the ocean: a three-dimensional velocity field through an appropriately small, fixed volume of space, averaged over sufficiently long time; the motion of populations of passive objects through limited areas of ocean basins; and the net effects of water motion which determine the geochemical distribution of the world's oceans. The interrelationship among these definitions can be derived theoretically and they depend upon the space and time variability of the water motion. Because we do not know enough about ocean variability, inferences from one interpretation cannot be related concisely to another. For example, on the edge of a small, stationary eddy, a long-term measurement with a fixed current meter would indicate a steady, unidirectional flow. However, objects placed into the water at the meter would not go far in this implied direction, but simply move in circles around the eddy. As another example, a geochemical constituent can spread from a source in a "tongue" or an asymmetric plume, not because there is a net water flow in the direction of the plume, but because the ocean there has a spatially nonuniform diffusivity, being very strong along the axis of the plume.

Whether the general circulation is best viewed from Eulerian (fixed sensors), Lagrangian (drifting sensors), or diffusive (no net water motion) aspect depends entirely on the space and time scale of the circulation phenomena of interest. Therefore, different mosaics of ocean circulation are difficult to compare directly because each has been made by a different artist, with very different size tiles. The unifying aspects lie in the physical explanations of the causes for the circulation phenomena as expressed in the conservation laws. To apply these laws, we need accurate descriptions of the flow rates on appropriately chosen length and time scales, and a description of both the mean and the time variable component of flow.

Before 1960 the general circulation phenomena were described by ship drift and the three-dimensional distribution of temperature, salinity, and dissolved oxygen. With these data, the surface circulation patterns can be readily identified and their plausible extensions through the main pycnocline can be derived from relative geostrophic flow. The western and eastern boundary currents and undercurrents, trade-wind currents and countercurrents, and the circumpolar circulation have been identified in every ocean. Below the main pycnocline, the horizontal distributions of the salinity and oxygen maxima and minima (and other geochemical species) are interpreted as indices of "cores" of water which sink from the surface or rise from the bottom (as in the case of resuspended constituents) and stay coherent over great subsurface distances. Therefore, relying upon the conservation of heat and salt in deep water, we conclude that bottom and deep water temperature and salinity characteristics are acquired near the ocean surface in only very few areas in polar latitudes; these properties spread equatorward, most rapidly in undercurrents on the western sides of ocean basins. The deep water property movements are thus traced between basins through a maze of sills and deep currents. Their distributions are caused both by flow and by mixing.

The circulation pictures which are based on this data set are very sensible; however, the rates of flow they imply have a high degree of arbitrariness or subjective bias. No objective method has been found to derive absolute velocity profiles with depth from ship drift and temperature, salinity, and/or geochemical tracers. Also, the horizontal extent of water mass penetration or interleaving and the implied flow characteristics from core analysis depends upon which conserved compound is used to identify the cores.

The excitement today in ocean general circulation stems from direct and accurate horizontal current measurements with current meters and remotely tracked floats. The use of radiochemical compounds as water tracers has enhanced the core method, because their source distribution is different from salinity and temperature and they augment specific flow structures which flow past their sources. Flow patterns are now measured where classical water mass analysis suggested these should exist (as in the Drake Passage) and where no direct indication was previously given (as the intense, barotropic recirculation of the Gulf Stream system). For the future, there are excellent prospects that these measurements can be used to identify the physical causes for the specific general circulation phenomena; then these can be understood better and modeled theoretically. Correct modeling is very important because the vertical component of general circulation cannot be measured by direct means and must be computed from a model. The vertical circulation of the oceans has the greatest long-term environmental consequences because it carries cold water and nutrients to the euphotic level and man-made constituents to and from the ocean surface, and its patterns strongly affect the natural geochemical equilibria of the abyss. This paper discusses the most recent, quantitative direct measurements of horizontal flow patterns in the western North Atlantic and eastern North Pacific and the associated vertical circulation implied by dynamical circulation models of the ocean.

2 The Theory of Vertical Circulation in the Ocean

Horizontal ocean circulation can be measured directly, but because the large-scale, persistent vertical circulation is weak (about 1 m to 0.01 m per day), it is very difficult to measure. Theoretical models are used to compute it; here we review the theory of how this is done for large-scale flow. Two models exist: kinematic and dynamic. Kinematic models rely upon the conservation of mass or a dissolved chemical substance. Therefore, if tongues and water masses are observed to sink or rise across an ocean basin, some portion of their apparent vertical displacement is thought to be due to a vertical movement, and some due to three-dimensional mixing. In the presence of mixing, water parcels do not follow constituent isopleths or water masses. To estimate the rate of sinking or rising, the rate of horizontal flow following a water mass, or residence times in a volume also has to be estimated or measured. Direct measurements on large horizontal scales are very rare; therefore, horizontal flow is also theoretically computed, usually by a dynamical method. Computation of residence times also requires horizontal flux estimates. While kinematical principles are sufficient for the discovery of the large-scale, three-dimensional flows, dynamical principles today are the only practical tool for computation of the flow rates, or the strength (and direction) of the general circulation.

In central ocean areas the approximate conservation equations for large-scale, steady flow are

$$\rho_0 f \times \vec{V} = -\nabla p + \partial_z \vec{\tau} + \rho_0 \vec{F} \tag{1}$$

$$0 = -\partial_z p - g\rho_0 \sigma \tag{2}$$

$$\nabla \cdot \vec{V} + \partial_z w = 0 \tag{3}$$

$$\vec{V} \cdot \nabla \Theta_m + w \partial_z \Theta_m = -\nabla \cdot \vec{q}_m \tag{4}$$

where $\vec{V}$ is the horizontal velocity, w is the vertical velocity component, p is the deviation from hydrostatic pressure, σ is the *in situ* density anomaly, and g is gravitational acceleration. In the β-plane approximation, $f = [0, 0, f_0 + \beta y]$ and x measures distance east, y north and $-z$ depth; ∇ is the horizontal gradient operator. In Equation (1), $\vec{\tau}$ is the turbulent horizontal shear stress on the water; its value on the surface, $\vec{\tau}_0$, is the wind stress. The quantity $\vec{F}$ is the force per unit mass due to horizontal turbulent transfer processes; as later used here, it is primarily due to ocean eddies. In Equation (4), Θ_m is a dissolved conservative chemical constituent as salt, or it is potential temperature, Θ, and $\vec{q}_m$ is its turbulent flux. In the approximation where Equation (4) applies to potential temperature, the *in situ* density is a function of salinity and potential temperature only. In principle, this does not alter our conclusions, but in practice the pressure effect on σ must be taken into account below 500 m.

The conservative equations are a good approximation for describing large-scale, steady circulation except in western boundary currents, such as the Gulf Stream in the equatorial undercurrents.

To see more succinctly how the circulation strength is computed and how vertical and horizontal flows are related, p is eliminated from Equations (1) through (3) and vorticity equations result:

$$f \times \partial_z \vec{V} = g \nabla \sigma + \partial_z^2 (\vec{\tau}/\rho_0) + \partial_z \vec{F} \tag{5}$$

$$f w_z = \beta V + \partial_z k \cdot \nabla \times (\vec{\tau}/\rho_0) + k \cdot \nabla \times \vec{F} \tag{6}$$

Given the knowledge of the spatial distribution of the *in situ* density, σ, and the turbulent stress convergence $\partial_z \vec{\tau}$ and $\vec{F}$, and a knowledge or measurement of $\vec{V}_0$, at some level $z = -z_0$ the entire local profile of the horizontal flow can be computed by integrating Equation (5) in the vertical direction:

$$fx \vec{V} = fx \vec{v}_0 + \int_{-z_0}^{z} [g \nabla \sigma + \partial_z^2(\vec{\tau}/\rho_0) + \partial_z \vec{F}] \, dz' \tag{7}$$

With Equation (7), we proceed to the vertical integration of Equation (6), from a level $z = -z_1$, a value for w_1 is assumed known,

$$fw = fw_1 + (z_1 + z)\beta V_0 - \beta \int_{-z_1}^{z} dz' \int_{-z_0}^{z'} [g \partial_x \sigma + \partial_z^2(\tau_x/\rho_0) + \partial_z F_1] \, dz'' - \int_{-z_1}^{z} [\partial_z k \cdot \nabla \times (\vec{\tau}/\rho_0) + k \cdot \nabla \times \vec{F}] \, dz' \tag{8}$$

Because a large data base for σ exists over the world's oceans, Equations (7) and (8) are attractive applications of the dynamic method, provided we can make strong physical or empirical arguments for neglecting the turbulent flux contributions to the vorticity components (i.e., when terms proportional to $\vec{\tau}$ and $\vec{F}$ are small). For the moment, consider a deep column of water between z_0 and z to be such that $\vec{\tau}$ and $\vec{F}$ are small, and choose $z_1 = 0$ the ocean surface where $w_1 = 0$ and $\vec{F}$ is small (in Section 3 we will discuss the empirical basis for doing this). Then Equations (7) and (8) are

$$fx \vec{V} = fx \vec{v}_0 + \int_{-z_0}^{z} g \nabla \sigma dz' \tag{9}$$

$$fw = z\beta V_0 - \beta \int_0^z dz' \int_{-z_0}^{z'} g \partial_x \sigma dz'' + fk \cdot \nabla \times (\vec{\tau}_0/f\rho_0) \tag{10}$$

Equation (9) is the familiar geostrophic relationship, and Equation (10) is the

General Circulation of the Oceans

integrated Sverdrup balance. Given the reference level velocity $\vec{v}_0$, the profile of $\vec{V}$ and W can be computed from the horizontal gradients of σ and the surface wind stress $\vec{\tau}_0$. The local, general flow pattern computed in this way conserves mass and vorticity. The practical questions are:

- At what depth z and where in x,y are $\vec{\tau}, \vec{F}$ sufficiently small so that Equations (9) and (10) are good approximations to the physical balances?
- On a scale analysis basis (Pedlosky, 1979), it is argued that at a depth of about 100 m, $\vec{\tau}$ should vanish or be very small; the last term on the right-hand side is bigger than the first two terms. Thus, the Ekman pumping, or $w_E = k \cdot \nabla \times (\vec{\tau}_0/\rho_0 f)$, is thought to be the vertical velocity at the base of a hundred-meter water column. In some areas of the ocean, this is so. In the North Atlantic trade-wind circulation system, south of 20°N all terms are the same size in the Ekman layer and a clear separation of the scale of baroclinic and Ekman flow does not exist (Behringer and Stommel, 1981). Of course, south of 10°N, this has been known for a long time. Therefore, some care has to be used in interpreting W_E as the vertical velocity at the base of a mixed layer. Here we do not make this distinction but use Equation (10) for computing the vertical velocity in areas where $\vec{\tau}$ and $\vec{F}$ are small.
- The ocean is full of time-dependent baroclinic features and any existing data set for $\nabla \sigma$ (or gradient of any dissolved constituent) is noisy. A best estimate yielding a minimum value of a measure of error can always be defined, but practically this is difficult to accomplish because over vast areas of the ocean there are not enough data to define the character of the noise (or space and time scales of eddies).

The theory of the reference level z_0 has lately received considerable attention (Schott and Stommel, 1978; Wunsch, 1978). Substitute expressions (9) and (10) and there results a set of coupled linear equations between the unknown horizontal velocity components u_0 and v_0 and $-\nabla \cdot \vec{q}_m$, evaluated as a function of depth. Therefore, if we presume that at two levels in the ocean $\nabla \cdot \vec{q}_m$ vanishes, for any single conserved quantity, and at that level the gradients $\nabla \Theta_m$, $\partial_z \Theta_m$ are well determined, we can solve the resulting linear system of equations for u_0 and v_0. In fact, any two additional physical hypotheses about our dynamical system will do (three are needed if we chose the reference level z_1, for w_1 some depth other than the surface). Schott and Stommel's (1978) in the "β-spiral method" minimizes

$$\int_{-z_0}^{-z_1} [\partial_z \nabla \cdot \vec{q}_\Theta / \partial_z \Theta]^2 \, dz$$

over a depth interval $(z_0 - z_1)$ with respect to u_0, v_0. Wunsch (1978) postulates that

$$\iint_A \nabla \cdot \vec{q}_\Theta dS dz$$

is small over three-dimensional, closed areas which are circumscribed by curves $S(x,y) = 0$ and by a vertical extent between sheets of constant Θ. He then minimizes the average reference level kinetic energy

$$\iint_A (u_0^2 + v_0^2)\, dA$$

at a particular level enclosed by the curve S. Davis (1978) suggests that because $\nabla \Theta$ is a "noisy" data set, that "geostrophic noise" should be minimized over appropriate intervals. Niiler and Reynolds (1978) minimize

$$\int_{-z_0}^{-z_1} [\nabla \cdot \vec{q}_\Theta]^2\, dz$$

with respect to u_0, v_0, and w_1. Each method yields somewhat different values for u_0 and v_0, w_1, and very different distribution of the vertical circulation. Because $w(z)$ is dependent upon $(\beta/f)zv_0$, the reference level value of v_0 very crucially affects the vertical velocity distribution with depth.

As we have seen, the problem of determining u, v, and w from existing geophysical data on the density field can be reduced to making statements about the importance of turbulent vorticity fluxes, and turbulent heat or salt fluxes. Today, we know very little about their strength over broad ocean areas from observations. Rhines and Holland (1979) point out that in two-layer, eddy-resolving numerical models of the northwestern subtropical area, where the western boundary current eddies are strong, and which bear close resemblance to the eddy energy levels measured (Schmitz and Holland, 1982), an important term in Equation (8) for w is that last term on the right-hand side. There, the approximate theoretical value for the vertical velocity across the main thermocline, when averaged over the area A_{ψ_2}, bounded by the bottom layer (of depth D_2) closed streamline curve C_{ψ_2} is

$$f\iint_{A_{\psi_2}} w\, dA \cong \iint_{-H}^{-D_2} \vec{k} \cdot \nabla x \vec{F}\, dA = \int_{-H}^{-D_2} \oint \vec{F} \cdot d\vec{s}_{C_{\psi_2}} \quad (11)$$

In their numerical model, the horizontal eddy flux divergence forces a mean vertical flux across the main thermocline. Its predicted strength is 3 to 4 m day^{-1}, which is two orders of magnitude larger than the value computed from the last term on the right-hand side of Equation (10), the Ekman pumping (Leetmaa and Bunker, 1978). Therefore, one needs not only a knowledge of the density data to compute the circulation, but also the eddy flux vorticity divergence.

In summary, vertical circulation in the ocean cannot be measured directly. It is computed from a model, based on kinematical or dynamical conservation principles. In both of these theoretical computations, we find that vertical and horizontal circulation are related; and both of these depend upon turbulent diffusion rates of vorticity and heat and salt in the ocean. We will not understand the ver-

General Circulation of the Oceans   237

tical circulation in the ocean until we understand much more completely the
turbulent transfer rates. A measurement of the horizontal mean flow may not be
sufficient for computation of the vertical flow. For a complete computation, the
deep ocean eddy field on various vertical and horizontal scales and the distribution of the horizontal, wind-induced stress in the upper ocean will also have to
be measured or modeled.

3 Direct Measurements

Over the last two decades, the technique was developed for maintaining moored
current meters in the deep ocean for 12 to 18 months (Gould et al., 1974) and
long time series of horizontal currents at fixed levels were obtained. In the North
Atlantic, these mostly were used to study ocean variability in such programs as
MODE (Mid-Ocean Dynamics Experiment) and POLYMODE, and in the Antarctic International Southern Ocean Studies (ISOS) Program they were used to
study the general circulation through the Drake Passage. Because the variability
and general circulation are very closely related dynamically and statistically, it is
not surprising that we inadvertently measured elements of the general circulation
in the North Atlantic and the variability in the Drake Passage.

In the late 1960s meteorologists began to use satellite-located balloons for the
study of tropical wind field, and this technique was readily adapted to ocean
surface drifting buoys. By 1975 the technique of long-term surface location
became affordable, and deployments of swarms of surface buoys were made in
the Caribbean (Molinari and Kirwan, 1975), the North Pacific (McNally, 1981),
the Gulf Stream (Richardson et al., 1977), and the Southern Ocean (Cresswell,
1976). About the same time, neutrally buoyant floats were developed which
could be tracked remotely over a long period of time in the North Atlantic
SOFAR (sound fixing and ranging) channel first by coastal hydrophones (Freeland et al., 1975) and more recently, by remote listening stations (Richardson
et al., 1981). Such a broad spectrum of space and time scales of motions has
been described by these long records that a general review is not possible in
terms of the framework we have adopted here for the discussion of ocean
circulation. Because we are interested in the quantitative aspects of the three-
dimensional circulation, we discuss the measurements which relate most closely
to the theoretical models by which the flow strength is computed. These are:
the vertical structure of large horizontal scale flow in the North Pacific and
North Atlantic upper layers and its relationship to the surface wind, important
for determining the vertical structure of w there; the problem of determining the
structure of the turbulent convergence of heat and salt in the subtropical Atlantic thermocline, which is needed for assessing horizontal reference level schemes
for z_0 and the vertical profile of w; and the eddy convergence of the vertical
vorticity component, $\vec{F}$, near the Gulf Stream and Kuroshio, which is the principal forcing mechanism for the vertical velocity across the main thermocline.

The atmospheric wind stress is transferred to the ocean surface, and because (we believe) vertical mixing in the water is much reduced below the mixed layer (Dillon and Caldwell, 1980), its gradient $\partial_z \vec{\tau}$ produces a force on the right-hand side of Equation (5). In the mid-latitudes this force is balanced on the left-hand side by the Coriolis acceleration, the Ekman layer balance. Theoretically, the gradient of this stress depends upon the sea state, the wind stress, the stratification, the Coriolis parameter, and the shear of the local currents on a variety of time scales. It has not been measured in the upper ocean because of the difficulty of measuring the fluctuating vertical velocity component in the presence of surface gravity waves. However, direct measurements of the low-frequency upper ocean velocity profiles with depth have been made and these have been related to the large-scale wind field.

Davis et al. (1981) describe the low-frequency ($w < f$) currents during three storms near ocean station P (50°N, 145°W) in August and September 1977. Within the mixed layer of about 30 m the currents and winds are related and in the seasonal thermocline between 30 m and 60 m and below, no relationship is found. In this late summer period, 5 m below the surface and throughout the 30 m mixed layer the currents are nearly 90° to the right of the wind; or the implied stress divergence is nearly in the direction of the surface stress (Fig. 1). These measurements imply that in the summertime stratified conditions, sinking or rising of water is forced by the horizontally nonuniform wind, and the sinking water has the properties on top of the seasonal thermocline. The stress is much reduced in the seasonal thermocline and the conservation laws which govern water motion in the main thermocline should apply here. The observed velocity profile resembles a theoretical profile in a homogeneous Ekman layer in which the stress vanishes at the base of the mixed layer, of depth smaller than homogeneous Ekman layer depth (Kundu, 1980). In the simplest model of instantaneous mixed layer, the stress is a linear function of z and the velocity field is independent of z, to the right of the wind field. The mixed layer depth varies with the storms and vagaries of summer heating.

McNally (1981) describes the wind-related motions of satellite-tracked drifters, drogued at 30 m depth between 35°N and 45°N and between 160°W and 130°W over a 9-month period. In summer, when the mixed layer is at 30 m or shallower, no discernible relationship is observed between winds and currents. However, in winter, when the mixed layer is at some distance below the drogue level, the drifters move on the average 30° to the right of the wind (Fig. 2). [Miyake (personal communication, 1978) finds that in winter the drogues at 6 m depth move 30° to the right of the monthly averaged surface wind.] Therefore, the vertical distribution of stress in the upper ocean in winter must be very different from summer. We do not have a simple theoretical model for winter conditions. These measurements imply that the vertical distribution of the mean wind drift and the mean vertical velocity in the upper ocean probably cannot be computed from a mean wind stress and mean stratification. The stress gradient, $\partial_z \vec{\tau}$, in the upper ocean is a nonlinear function of the synoptic wind and synop-

General Circulation of the Oceans

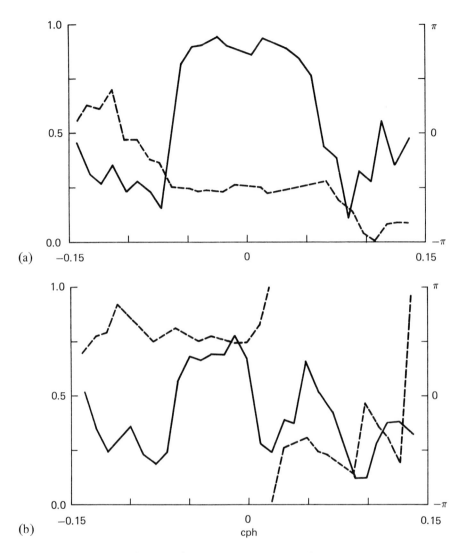

Figure 1. Rotary coherence (solid line, left ordinate) and phase (dashed line, right ordinate) of the wind stress at (a) 5 m depth and (b) 42 m depth at 50°N, 145°W, August 19 to September 5, 1977. A phase of $-\pi/2$ corresponds to flow to the right of the wind stress (from Davis et al., 1981).

tic stratification. Because there are no data at many levels over long periods, we do not know how to compute this function adequately throughout the seasons, nor how to compute the structure of three-dimensional mean circulation in the upper ocean. This is a serious inadequacy, for it has important consequences to understanding the circulation, and of course, to understanding the causes of the distribution of the upper ocean biota, chemical constituents, and air-sea exchange of heat, moisture, and gases.

Below the seasonal thermocline, and where strong eddies are not present, Equations (9) and (10) are a good model for the three-dimensional flow. If a value of the horizontal flow at a reference level $z = -z_0$ can be specified, the three-dimensional circulation can then be computed from hydrographic and wind data. Also, the profile of w can be estimated. Keffer and Niiler (1982) have done such a computation in the North Atlantic at about 28°N, 45°W (Fig. 3). Here on either side of the Mid-Atlantic Ridge, time series of currents from June 1977 to September 1979 were obtained in two arrays of current meters as part of the POLYMODE (Fu et al., 1982). The measured mean flow at 1500 m was used as the reference level velocity and the historical hydrographic data were used for computation of the baroclinic part of the flow. Figure 3 is a summary of this computation. There is the expected clockwise rotation of the horizontal

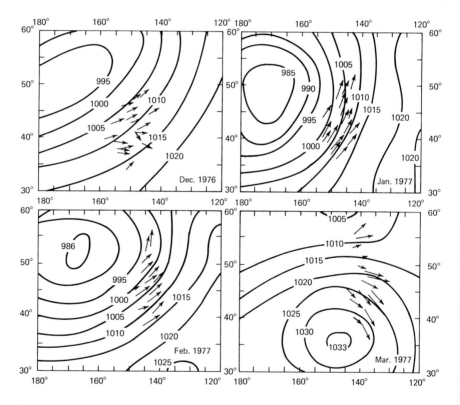

Figure 2. Monthly displacement vectors of satellite-tracked drifters, drogued to 30-m depth, superimposed on the monthly mean surface pressure charts in the northeast Pacific. The surface wind flows 30° to the left of the surface isobars (from McNally, 1981).

flow with depth; however, the surprising discovery is that the mean flow below the main thermocline is nearly 1 cm sec^{-1} to the northwest over the entire water column below 700 m. Before these direct measurements were made, such strong deep circulation in the mid-ocean was not anticipated, and even now they present a perplexing theoretical problem. This flow had not been anticipated from the analysis of the distribution of water mass properties, following tongues of North Atlantic Deep Water salinity and oxygen concentrations in the central North Atlantic as displayed by Wust (1935), and later by the GEOSECS measurements. Based on the hydrographic core method, the predominant flow below 700 m along the Atlantic Ridge should be to the south. Here the actual water motion is opposite the direction of the extended tongue. Along the Mid-Atlantic Ridge, tongues are apparently maintained by diffusion of properties against the mean flow.

The theoretical vertical velocity can also be computed from Equation (10). The Ekman divergence (the last term on the right-hand side of this equation) is estimated from Leetmaa and Bunker's (1978) analysis of the North Atlantic wind field. Water appears to sink over the entire column from the surface to the bottom (this is possible near the bottom because water flows into deeper ocean, away from the crest of the Mid-Atlantic Ridge). Drawn on Figure 4 is also the implied vertical motion, $\tilde{w}(z)$, following the isopycnals. Between 200 m and 600 m, there appears to be no flow across the main thermocline for $w = \tilde{w}$. However, below the main thermocline, the flow is not along isopycnals due to increased mixing. There is no direct evidence that an upward motion is required

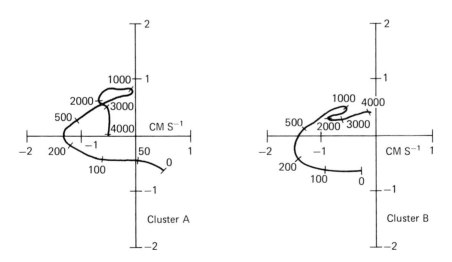

Figure 3. The hodographs of geostrophically computed horizontal currents at POLYMODE Array III, Cluster A, 28°N, 48°W, and Cluster B, 27°N, 41°W. The reference level velocity at 1500 m is obtained from direct measurements (from Keffer and Niiler, 1982).

to maintain the main thermocline against downward diffusion of heat (Munk, 1966). Keffer and Niiler (1982) also point out that in the water column between 200 and 600 m heat, salt, and potential vorticity are also conserved or uniform. We therefore discover in the subtropical mid-ocean vertical layers of very different dynamical properties. Rhines and Young (1981) suggest a theoretical explanation of this phenomena. In this portion of the main thermocline within the potential density sheets there are closed contours of potential vorticity or salinity. In such situations, horizontal diffusion of high Reynolds number very rapidly renders the interior of the closed contour devoid of potential vorticity or salinity gradients. Figure 5 is the distribution of salinity on the 26.5 potential density surface (Koblinsky et al., 1979), which at the mooring location of 28°N, 43°W is 400 m in depth. The salinity distribution is very uniform along the 26.5 σ_t surface over a large area of the western subtropical North Atlantic.

Of course, there also are direct measurements which verify an implied flow along water mass cores. Figure 6 is a mosaic of three deep flow measurements (all nearly a year long) into the deep western North Atlantic basin, all of which were currently predicted by water mass distribution: (a) displays the northward

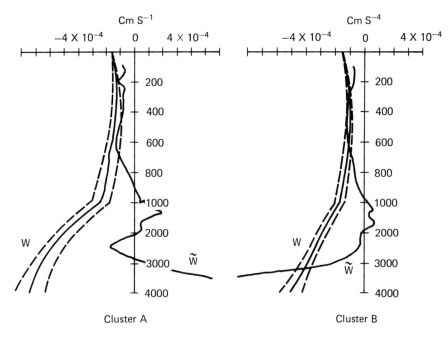

Figure 4. Theoretical vertical velocity distribution with depth w, computed from Sverdrup balance with Ekman divergence reference value at the surface, and $\tilde{w}$, computed from the conservation of potential density when $\partial_z w = \partial_z \tilde{w}$, potential vorticity is conserved. When $w = \tilde{w}$, there is no flow across isopycnals (for location see Fig. 3) (from Keffer and Niiler, 1982).

flow of the Antarctic Bottom Water on the northwestern edge of the Demerrara Abyssal Plains (Koblinsky et al., 1979); (b) displays the southward flow of the North Atlantic Deep Water in a strong western boundary undercurrent along the Blake Escarpment Rise (Jenkins and Rhines, 1980); and (c) displays the deep westward flow of the Denmark Straits Water through the Charlie Gibbs Fracture Zone (Hogg and Schmitz, 1980). It is not surprising that these bottom flows exist, but it is remarkable that they are so persistent, organized, and strong.

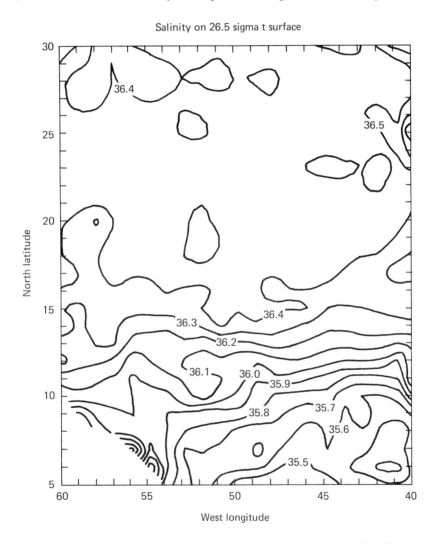

Figure 5. Subtropical Atlantic salinity distribution on the 26.5 σ_t surface (from Koblinsky et al., 1979; see Rhines and Young, 1981 for the distribution of potential vorticity on the 26.5 σ_t surface).

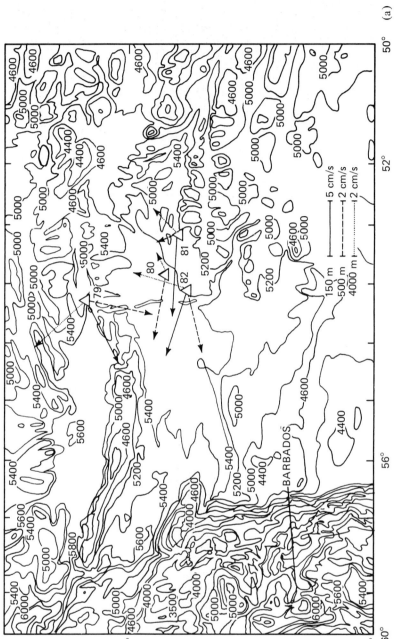

Figure 6. (a) Year-long records of velocity in the Atlantic North Equatorial Current. The 4000 m level indicates northwestward flow of Antarctic Bottom Water (from Koblinsky et al., 1979).

General Circulation of the Oceans

Direct measurements of circulation by current meters, water-mass analysis, and theory together provide powerful constraints on our interpretation of the three-dimensional thermocline circulation patterns. In the future, there are excellent prospects that an altimeter satellite system can be utilized to measure the large-scale, persistent, sea level slope across ocean basins (Stewart, 1981). This will provide another powerful constraint to the computation and interpretation of the horizontal and vertical flow pattern. Recently, autonomous listening stations for acoustic signals in the SOFAR channel have been developed (Richardson et al., 1981) and these open the possibility of tracking neutrally buoyant floats across large areas of the deep, remote oceans. Today, we do not

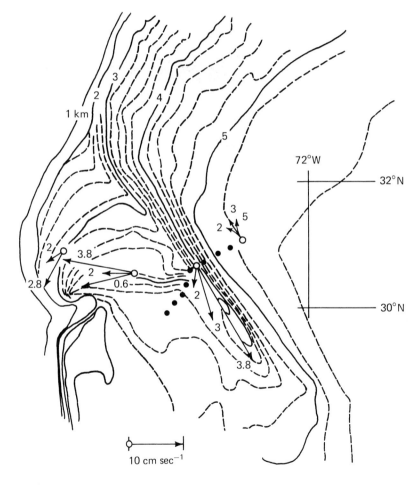

(b)

Figure 6. (b) Deep western boundary undercurrent along the Blake Escarpment.

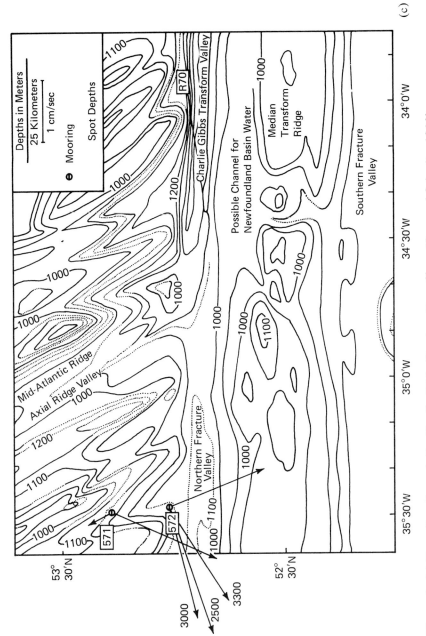

Figure 6. (c) Mean flow through the Charlie Gibbs Fracture Zone (from Hogg and Schmitz, 1980).

have a realistic, tractable model of this circulation over broad ocean areas because we do not have tractable models or measurements of turbulent transport over large areas of the open ocean.

When Swallow in 1959 (most easily accessible in Swallow, 1971) discovered intense, variable motions with neutrally buoyant floats below the main thermocline in the Sargasso Sea, Iselin (1961) suggested that the source of this variability was the Gulf Stream to the north. In a series of papers Schmitz (1976, 1977, 1978, 1980) describes the eddy kinetic energy distribution of the deep western North Atlantic as measured by moored current meters and neutrally buoyant floats. The general picture is that the eddy kinetic energy at all levels, between 70° and 50°W, increases by a hundredfold from the center of the North Atlantic Subtropical Gyre at 28°N to the Gulf Stream at 34°N and then again decreases to the north of the Stream. About the same time, Worthington (1976) pointed out that the salinity and oxygen distribution of the western North Atlantic, when taken together with the measured large easternward mass transport of the deep Gulf Stream, implied that a rather tight recirculation pattern should occur directly south and east of the Gulf Stream. Stommel et al. (1978) describe the deep dynamic topography of the recirculation area and point out that its baroclinic signature or a relative high in the 1500- to 3000-m dynamic height extends across the entire North Atlantic. Direct measurements of sufficient extent and density to adequately describe the horizontal flow of the Gulf Stream and the recirculation are available at only 55°W longitude (Schmitz, 1980). The recirculation strength is about 90×10^6 m^3 sec^{-1}, or three times the volume flux of the Florida Current. There the recirculation occurs in the area where the eddy energy gradients are large. Because such a volume of mass flow cannot theoretically be driven by the North Atlantic wind systems (Leetmaa and Bunker, 1978), its dynamical cause is most likely the eddy field of the Gulf Stream. Much more extensive direct measurements than we now have are needed of the distribution of eddy energy to directly compute the eddy momentum (and vorticity) convergence, or the "eddy force" distribution in the Gulf Stream area. But while we wait for the observational data base to increase, some physical insight into how eddies drive the ocean general circulation can be gained from numerical modeling.

Two essential properties are required for ocean eddies to drive mean horizontal circulation patterns. The first is that an energy source exist for the eddies so that they can grow quite spontaneously and coexist with the mean flow. Holland (1978) and Holland and Rhines (1980) discovered that in numerical models, western ocean basin eddies can receive their energy from both the mean kinetic energy of the western boundary currents flow (like the source of turbulence in a pipe) and the stored potential energy of the main thermocline in the western boundary area (like the source of weather systems in the atmosphere). The second property is that the eddies be reasonably confined, or that their horizontal distribution is nonuniform in the ocean. Thus, large eddy stress gradients occur, which produce a mean force on the fluid. Model-generated eddies are confined in the area where they are generated and do not fill the entire basin because a relatively strong current field is needed for their existence. Observations around the Gulf Stream area reveal that both of these criteria are met. Swirls, rings, and meanders of the Gulf Stream abound and these are very much confined to the vicinity of the Stream. Schmitz and Holland (1982) have tuned the parameters of two-layer, wind-driven numerical models (the bottom stress parameter and basin width parameter) so that the numerical model eddies have the energy levels, vertical structure, and spatial scales of the observed western North Atlantic eddies. Figure 7 is their observational data and numerical model data comparison. As in the observations, in the model a tight recirculation of mass occurs to the south and east of the western boundary current. The deep general circulation or the recirculation in the lower layers does not occur in a

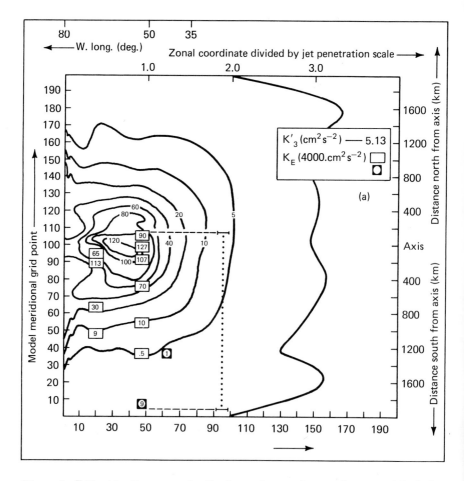

Figure 7. Eddy kinetic energy in the lower layer of a two-layer model of the western North Atlantic (contours) and direct measurements (boxed numerals).

wind-driven model in which no eddies are produced (by strongly increasing the bottom friction).

Therefore, we have a physically plausible model of the forces and strength of the horizontal deep circulation in strong eddy areas of the ocean. Figure 8 is the mean streamline pattern of the two-layer numerical models. This circulation is three-dimensional, for a vertical circulation also exists across the main thermocline. Like the horizontal flow, it is driven by western boundary current eddies. Holland and Rhines (1980) note that if the Equation (8), which is our model equation for the vertical circulation, is integrated over the area $A(\psi_2)$ enclosed by the curve $C(\psi_2)$ a lower layer stream function ψ_2, and where $Z = -H$ are the ocean flat bottom where $w_1 = 0$, and Z equals the mean depth D_2 of the lower layer, there results the approximate balance expressed in Equation (11). From it,

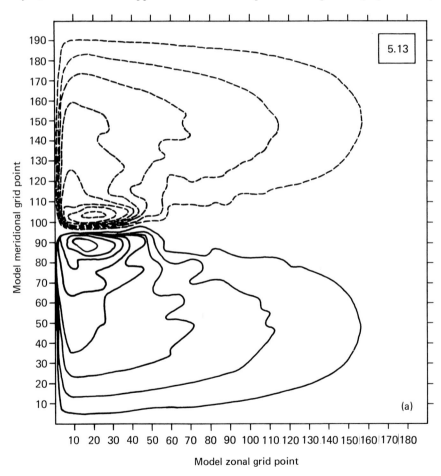

Figure 8. (a) The mean streamline pattern of upper layers of the model calibrated by the observed eddy field of Figure 7.

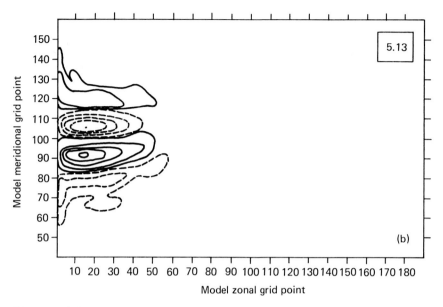

Figure 8. (b) The mean streamline pattern of lower layers of the model calibrated by the observed eddy field of Figure 7.

using numerical model eddy data, we estimate a vertical velocity of about 3 or 4 m day^{-1}. In the water mass which circulates clockwise, it is downwelling through the main thermocline and in the water mass which circulates counterclockwise, it is upwelling. Schmitz (1980) notes that the observed deep clockwise recirculation has a weak vertical shear and therefore its signature should appear in the dynamic deep relative topography. In Figure 9(a) is displayed the 1500- to 3000-db deep dynamic topography of the North Pacific (Stommel et al., 1978) and in Figure 9(b) is the 1000- to 2000-db deep dynamic topography of the North Pacific (Wyrtki, 1974). The weak geostrophic shear of both the clockwise and counterclockwise circulation patterns occurs in each ocean. These theoretically are eddy-driven flows. Their vertical motion across the main thermocline is predicted to be two orders of magnitude larger than in the open ocean, downwelling to the south of the Gulf Stream and Kuroshio and upwelling in the southern gyre. This prediction could not have been made without a combined analysis of the direct measurement of the eddy field and numerical eddy-resolving models of the general ocean circulation.

4 Conclusion

Colorful but broad-brush mosaics and a precious few direct measurements explain how oceans circulate. In this field, more direct, accurate, quantitative physical measurements and modeling of ocean turbulence or variability need to be accomplished. The most difficult step in any discipline is its elevation to a quantitative science. This step demands a strong disciplinary commitment and requires intensive training of students in the classical and narrow fields of mathematics and the physical sciences. The discipline of the physical circulation of the sea has matured and it now demands the full energies and attentions of its small band of scientists. Today in physical oceanography there are many more problems to study than there are qualified students to take up the challenge. The Woods Hole Oceanographic Institution can provide no better investment in the future ocean than by attracting the best students to Woods Hole and teaching them reading, writing, and arithmetic!

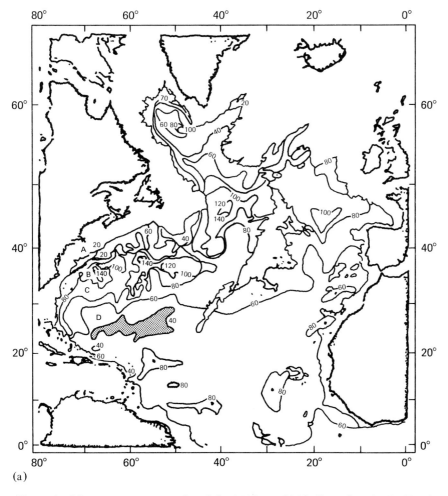

Figure 9. (a) Dynamic topography of the 1500- to 3000-db surface in the North Atlantic (from Stommel et al., 1978).

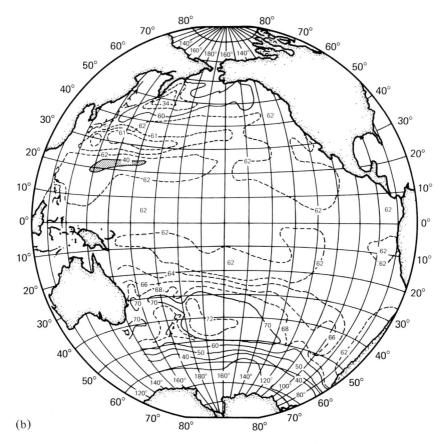

(b)

Figure 9. (b) Dynamic topography of the 1000- to 2000-db surface in the Pacific (from Wyrtki, 1974).

Acknowledgments

This work was supported by the Office of Naval Research at Oregon State University and by NOAA at the Joint Institution for Marine and Atmospheric Research.

References

Behringer, D. W., and H. Stommel. 1981. Annual heat gain of the tropical Atlantic computed from subsurface ocean data. J. Phys. Oceanogr. 11, 1393–1398.
Cresswell, G. R. 1976. A drifting buoy tracked by satellite in the Tasman Sea. Aust. J. Mar. Freshwater Res. 27, 251–262.
Davis, R. E. 1978. On estimation of velocity from hydrographic data. J. Geophys. Res. 83, 5507–5509.

Davis, R. E., R. de Szoeke, and P. P. Niiler. 1981. Variability in the Upper Ocean During MILE: Part II, Modeling the Mixed Layer Response. Deep-Sea Res. 28A, 1453–1475.

Dillon, T. M., and D. R. Caldwell. 1980. The Batchelor spectrum and dissipation in the upper ocean. J. Geophys. Res. 85, 1910–1916.

Freeland, H. J., P. B. Rhines, and H. T. Rossby. 1975. Statistical observations of the trajectories of neutrally buoyant floats in the North Atlantic. J. Mar. Res. 33, 383–404.

Fu, L. L., T. Keffer, P. P. Niiler, and C. Wunsch. 1982. Observations of mesoscale variability in the western North Atlantic: A comparative study. J. Mar. Res. In press.

Gould, W. J., W. J. Schmitz, and C. Wunsch. 1974. Preliminary field results for a mid-ocean dynamics experiment (MODE-0). Deep-Sea Res. 21, 911–931.

Hogg, N. G., and W. J. Schmitz. 1980. A dynamical interpretation of low-frequency motions near very rough topography–The Charlie Gibbs Fracture Zone. J. Mar. Res. 38, 215–248.

Holland, W. R. 1978. The role of mesoscale eddies in the general circulation of the ocean–numerical experiments using a wind-driven, quasi-geostrophic model. J. Phys. Oceanogr. 8, 363–392.

Holland, W. R., and P. B. Rhines. 1980. An example of eddy-induced ocean circulation. J. Phys. Oceanogr. 10, 1010–1031.

Iselin, C. O'D. (1961). An interpretation of the deep current measurements. Oceanus 7(3), 9.

Jenkins, W. J., and P. B. Rhines. 1980. Tritium in the deep North Atlantic. Nature (London) 286, 877–880.

Keffer, T., and P. P. Niiler. 1982. Eddy convergence of heat, salt, density and potential vorticity in the subtropical North Atlantic. Deep-Sea Res. 29, 201–216.

Koblinsky, C. J., T. Keffer, and P. P. Niiler. 1979. A compilation of observations in the Atlantic North Equatorial Current. Oregon State University, Ref. 79-12, 119 p.

Kundu, P. K. 1980. A numerical investigation of mixed layer dynamics. J. Phys. Oceanogr. 10(2), 220–236.

Leetmaa, A., and A. Bunker. 1978. Updated charts of the mean annual wind stress, convergences in the Ekman layers, and Sverdrup transports in the North Atlantic. J. Mar. Res. 36, 311–322.

McNally, G. 1981. Satellite-tracked drift buoy observations of the near surface flow in the eastern mid-latitude North Pacific. J. Geophys. Res. 86, 8022–8030.

Molinari, R., and A. D. Kirwan. 1975. Calculations of different kinematical properties from Lagrangian observations in the western Caribbean Sea. J. Phys. Oceanogr. 5, 483–495.

Munk, W. H. 1966. Abyssal recipes. Deep-Sea Res. 13, 707–730.

Niiler, P. P., and R. Reynolds. 1981. Circulation around the eastern North Pacific Subtroptical Front. Unpublished manuscript.

Pedlosky, J. 1979. Geophysical Fluid Dynamics. Springer-Verlag, New York.

Rhines, P. B., and W. R. Holland. 1979. A theoretical discussion of eddy-driven mean flows. Dynam. Atmos. Oceans 3, 289–325.

Rhines, P. B., and W. R. Young. 1981. Homogenization of potential vorticity in planetary gyres. Unpublished manuscript.

Richardson, P. L., R. E. Chaney, and L. A. Martini. 1977. Tracking of Gulf Stream ring with a free drifting surface buoy. J. Phys. Oceanogr. 7(4), 580–590.

Richardson, P. L., J. R. Price, W. B. Owens, W. J. Schmitz, H. T. Rossby, A. M. Bradley, J. R. Valdes, and D. C. Webb. 1981. North Atlantic Subtropical Gyre−SOFAR floats tracked by moored listening stations. Science 213, 435−437.

Schmitz, W. J. 1976. Eddy kinetic energy in the deep western North Atlantic. J. Geophys. Res. 81, 4981−4982.

Schmitz, W. J. 1977. On the deep general circulation in the western North Atlantic. J. Mar. Res. 35, 21−28.

Schmitz, W. J. 1978. Observations of the vertical distribution of low frequency kinetic energy in the western North Atlantic. J. Mar. Res. 36, 295−310.

Schmitz, W. J. 1980. Weakly depth dependent segments of the North Atlantic circulation. J. Mar. Res. 38, 111−133.

Schmitz, W. J., and W. R. Holland. 1982. Numerical eddy resolving general circulation experiments: preliminary comparison with observation. J. Mar. Res. 40, 75−117.

Schott, F., and H. Stommel. 1978. Beta-spirals and absolute velocities in different oceans. Deep-Sea Res. 25, 961−1010.

Stewart, R. (Ed.). 1981. Satellite Altimetric Measurements of the Ocean. Jet Propulsion Laboratory, Pasadena, California. 400-111, 3/81.

Stommel, H., P. P. Niiler, and D. Aniti. 1978. Dynamic topography and recirculation of the North Atlantic. J. Mar. Res. 36(3), 449−468.

Swallow, J. 1971. The *Aries* current measurements in the western North Atlantic. Phil. Trans. Roy. Soc. London 270 A, 451−460.

Worthington, L. V. 1976. On the North Atlantic Circulation. Johns Hopkins University Press, Baltimore, 110 p.

Wunsch, C. 1978. The North Atlantic general circulation west of 50°W determined by inverse methods. Rev. Geophys. Space Phys. 16, 583−620.

Wust, G. 1935. The stratosphere of the Atlantic Ocean. Scientific results of the German Atlantic Expedition of the Research Vessel "Meteor," 1925−1927. Vol. VI, Section I (English translation, W. Emery, Ed.). NSF, Washington, D.C.

Wyrtki, K. 1974. Dynamic topography of the Pacific Ocean and its fluctuations. University of Hawaii, HIG-74-5.

Remote Sensing of the Oceans from Space

John A. Whitehead, Jr.

1 Introduction

Oceanography from a specially designed space vehicle was born at 0112 (GMT) June 27, 1978, when SEASAT-A was launched from Vandenberg Air Force Base, California. This satellite, which contained an impressive array of instruments, performed brilliantly until its premature death due to a massive short circuit on 0200 (GMT) October 10, 1978. In its 105th day at the time of failure, SEASAT instruments, which were designed specifically to give measurements of surface winds, sea-surface topography, sea roughness, and temperature, had gathered approximately 99 days of data and relayed the data to earth.

Oceanography from space cannot yet keep up with its mature brother and sister disciplines which use such traditional oceanographic techniques and equipment as acoustics, chemical analysis, salinity and temperature determination, current meters and drifting buoys. It has only taken a few shaky steps, but we know it will differ considerably; its brothers and sisters can dive deep into the ocean depths and fathom the vertical distribution of many things, while space oceanography will glide over the surface and principally sample things very close to the surface—but at enormous speed. SEASAT-A tracked over 20 million miles of ocean surface at a speed approximately a thousand times faster than a ship underway! Despite vigorous efforts by oceanographers to get global data sets, its brothers and sisters cluster near familiar ports, in friendly and less stormy water, while space oceanography samples all regions equally (subject to orbital biases). Figure 1a shows the distribution of hydrocast data points archived by the National Ocean Data Center, while Figure 1b shows a standard 3-day coverage from SEASAT. It is expected that bureaucratic red tape in some

256   J. A. Whitehead, Jr.

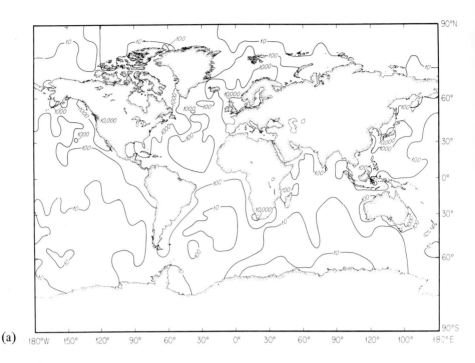

(a)

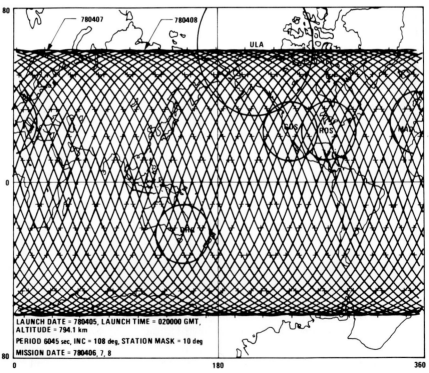

(b)

foreign waters and the increased cost of fuel will tend to force traditional oceanography to cluster even closer to home than in the past. The various disciplines use many different techniques, each with their own calibration biases and errors, while space oceanography uses the same instrument over the entire coverage zone, resulting in uniformly calibrated data sets. (However, regional biases will be generated due to regional climates which will alter the atmospheric corrections.) One is led to the conclusion that the young field can complement but not replace its brothers and sisters and one can anticipate tremendous advances over the next decades.

In this era of satellite weather photographs and LANDSAT images we forget how young satellite oceanography really is. In 1916 Helland-Hansen proposed the temperature-salinity diagram as a means for determining the density and stability in the world's oceans. Three years later, Dr. Robert Goddard published the paper, often regarded as the start of modern rocketry, entitled "A Method of Reaching Extreme Altitudes," which nominated liquid-fueled rockets as being the appropriate devices to provide lift at extremely low atmospheric densities. In 1926 Goddard launched the first liquid-fueled rocket. Meanwhile traditional oceanography was well along; Prandtl had published the first results of experiments in a new rotating laboratory and a year before Nansen had published his early description of the Nansen bottle. During World War II both oceanography and rockets advanced enormously, but the first satellite was not launched until October 4, 1957, during the International Geophysical Year which saw the generation of an impressive data set of hundreds of deep Nansen casts in the Atlantic.

Shortly thereafter there were hints that one could learn something of the oceans from space. Glimpses of the ocean were obtained from the first astronauts' photographs in the early 1960s; curious stripes in the sun glint were reported; the edge of the Gulf Stream could often be seen in both water color and a dramatic change in cloud cover. The first weather satellites, which returned visible and infrared images of the ocean, revealed similar types of features. On August 24-28, 1964, a conference was held at Woods Hole on the feasibility of "conducting oceanography explorations from aircraft, manned orbital and lunar laboratories." Ideas in the resulting volume (Ewing, 1965) are abundant and well informed and it was obvious that a community of scientists with experience in aircraft remote sensing felt that many sensors were suitable for useful space oceanography. Curiously, almost 10 years were to pass before an active SEASAT program was started; whether this delay was due to organization problems, priority of other space programs, interagency difficulties, or

---

Figure 1. (a) Density of hydrocasts per 5° square as archived by the National Ocean Data Center. This is an illustration of the uneven coverage of oceanic data as gathered by ships. There are six 5° squares in the South Pacific and one in the Southern Ocean with no casts. (b) Plot of 3 days of SEASAT ground track. Some instruments measured swaths lateral to this track that were hundreds of kilometers wide (supplied courtesy of the Jet Propulsion Laboratory).

other problems of those times, it does appear to have dampened the enthusiasm of many of the oceanographers who were enthusiastic in the mid-1960s.

Remote sensing from space was occurring in a disguised form because active meteorological satellite programs were rapidly progressing. The earliest satellites were called experimental and funded through NASA; later ones were called operational and funded through the Department of Commerce largely due to a need for timely information by the National Weather Service. Such interagency cooperation appears to greatly strengthen the acceptance of satellite programs by the government. Many of the data from the operational satellites (Table 1) are archived in the Experimental Data and Information Service (address and details in Appendix I) and are readily available to the public. The bulk of the data taken before 1979 are stored in the form of an image which is more qualitative than quantitative. All data archived from 1979 on are stored on magnetic tape.

The oceans were also observed by instruments of another satellite, GEOS-3, which was placed in orbit in 1974. Its altimeter was intended to measure the geoid—a surface of constant gravitational potential which, if the oceans and atmosphere were motionless, would coincide with the surface of the ocean. Remarkable maps of the sea surface of the entire earth have been generated from the GEOS data set; an example is shown in Figure 2. Sea surface mimics the geoid to approximately 1 meter accuracy, below which the effects of steady ocean currents must be subtracted. Geophysicists have been using the geoid generated from such data sets to deduce the density distribution inside the earth. Valuable things have been learned about the deep structure of volcanic islands, trenches, continental margins, and spreading centers.

Data from SEASAT, GEOS, and the meteorological satellites are the data sets that people presently have to work with, our real guide of present capabilities and future possibilities. To illustrate some possibilities for the next 50 years, various fundamental problems in oceanography and marine geophysics will be discussed, and an attempt will be made to assess, with trepidation, the possibility that they will be substantially affected or possibly even revolutionized by remote sensing.

Before proceeding with these examples it is useful to note that the value of remote sensing is like the difference between two very large numbers. The first number in this analogy represents the enormous advantages of remotely sensed data that we have described (routine global coverage, large data rates, and uniform calibration of sensors). However, from this must be subtracted a second number which represents all the shortcomings of the data (only the surface is sensed, atmospheric corrections may be very large, precise auxiliary data may be needed from networks of sensors at sea, orbits or gravity field must be known, and sea state corrections are poorly understood). The value of the first number represents the effort and skill of NASA scientists and engineers in constructing and placing in orbit a satellite or satellites and relaying the data to the ground. After a satellite has been constructed and placed in orbit its value is more or less fixed. The value of the second number represents the efforts of

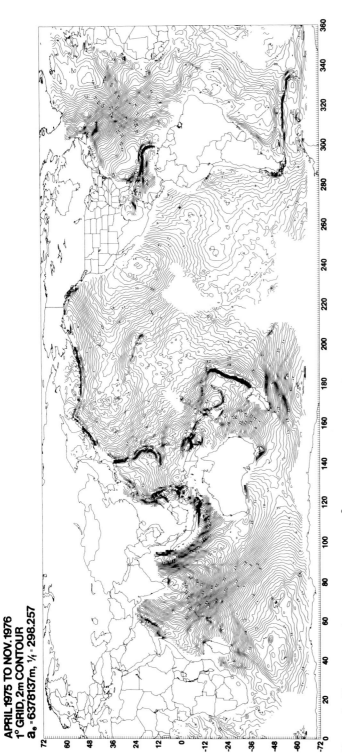

Figure 2. Mean sea-surface topography in 1° squares based upon GEOS-3 altimeter data gathered from April 1975 to November 1976 (supplied through courtesy of James Marsh of the Goddard Space Flight Center).

Table 1. Satellites and their instruments useful for oceanography

Name	Date archived	Location	Instruments
Sun–synchronous meteorological satellites			
ITOS/NOAA 1-6	10/31/70 to present	EDIS	Visible scanner Infrared scanner
TIROS-N/NOAA-A	11/6/78–11/1/80	EDIS	Very high resolution infrared ARGOS
LANDSAT 1-3	6/72 to present	EROS	Visible scanner (3 channel) Infrared scanner
NIMBUS-7	11/1/78 to present	EDIS	Visible multichannel scanner (CZCS) Scanning multichannel microwave recorder ARGOS
DMSP	2/24/73 to present	EDIS	Visible scanner Infrared scanner
NIMBUS 1-6	6/75 to present	GSFC*	Scanning multichannel microwave recorder RAMS (NIMBUS-5 only)

TIROS 1–10	4/1/60–	EDIS	Visible scanner
ESSA 1–9	11/15/72		Infrared scanner
Earth-synchronous meteorology			
SMS/GOES 1–3	11/15/74 to present	EDIS	Visible scanner
GOES-D	1980 launch	EDIS	Infrared scanner (2 channel)
			Visible scanner
			Infrared scanner (7 channel)
Ocean satellites			
GEOS-3	4/75–1/79	GSFC	Altimeter
SEASAT	7/7/78–10/10/78	JPL**	Altimeter
		to EDIS	Scanning multichannel microwave recorder
			Synthetic aperture radar
			Radar scatterometer
			Visible scanner

*GSFC stands for the Goddard Space Flight Center, Landover, Maryland.
**JPL stands for the Jet Propulsion Laboratory, Pasadena, California.

scientists and engineers in the research and academic community to convert the data to physically meaningful measurements with clearly understood error bars. As errors go down so does the value of this second number and it is reasonable to expect that this will decrease with time. Strategies to minimize this value involve cooperative research from ships, buoys, and aircraft because the instruments must be calibrated and because many of the satellite data are much more useful scientifically in conjunction with other measurements. Remotely measured surface currents, for example, would have to be used with either present or historical measurements of the density field of the ocean to determine the general circulation of the ocean.

As we discuss various scientific problems which may be greatly aided by remote sensing, it is important to remember that this is not intended to be a survey of the present state of the field. Such surveys exist both as technical reports from workshops and as studies conducted by national committees. Some material is listed at the end of the paper (Appendix II). Rather, we intend here to address the future with as realistic an appraisal as possible.

2 Large-Scale Oceanic Transport

General circulation is generally called that steady component of the motion averaged over many years on a global scale and is important in the distribution of heat, momentum, dissolved chemical species, and biological organisms. It is thus an important component in determining the climate and ecological state of the earth. The ocean circulation is driven by wind stress on the surface and by variation of heat flux into and out of the ocean in different regions. A global monitoring of the circulation and its driving forces is well beyond present oceanographic resources, but the potential for global coverage of some parameters lies in satellites.

Altimeters were flown on GEOS-3 and SEASAT. This form of measurement holds great promise, because unlike many measurements made from satellites, altimeters respond in a simple way to effects at the sea surface which represent an average current over the entire water column.

In an altimeter a short burst of electromagnetic waves is directed downward from the satellite, and the return signal is monitored by a fast-response detector. The resulting signal's delay time and shape are then used to determine the vertical distance from the wave source to the ocean surface and back to the detector. The satellite's orbit is monitored very accurately. The cycle is repeated periodically to make a map of distance between satellite and sea surface.

Altimetry data is then combined with a reference surface to determine the perturbed sea level surface from which the geostrophic velocity (that component of the velocity field in which the Coriolis force is balanced by pressure) can be calculated. The reference surface is called the geoid—a surface of constant gravitational potential. The degree to which the method will work depends upon the accuracy of the altimeter, the orbit, the geoid (for the time-independent component), and the correction for sea state, all of which must be determined to better than 5 cm to give a root mean squared accuracy of 10 cm for sea-surface pressure. How accurately can these be done at present?

The altimeter itself can measure accurately to better than 1 cm in principle, but corrections for the change of the speed of the electromagnetic waves in the atmosphere and ionosphere become rapidly more difficult below the 5-cm level. Much was learned about making these corrections from SEASAT. For instance, one channel of another instrument, the microwave radiometer, was crucial for correcting for the change in the speed of light by water vapor. SEASAT met its goal of measuring distance from ocean surface to spacecraft to an accuracy of 10 cm; an accuracy of 2 to 4 cm may be achieved in a properly designed satellite.

The orbit is determined by precision ground tracking and with the use of the equations of motion. No satellite has had tracking quality and low enough drag uncertainty to claim a 5-cm orbit. For SEASAT and GEOS-3, orbit uncertainties were eliminated by subtracting a mean (time-averaged) signal. Fortunately uncertainties due to drag on spacecraft and the unknown components of the gravity field are of long wavelength and many oceanographically important signals are unaffected. It is hoped that studies will soon be made to determine if, with an improved gravity field, a low-drag satellite, and precision tracking, the orbit can be determined to 5-cm accuracy.

An accurate geoid must be subtracted from the signal to get the steady component of the geostrophic velocity field. There is no feasible way to get a global geoid to 10 cm down to horizontal length scales of 25 km below which there is only a small component of the total energy. Long-wavelength components (approximately 200 km and above) can be obtained from a proposed gravity satellite (GRAVSAT). Also, local geoids can be generated from ship gravity data in regions of high interest, but a global program is economically unfeasible.

Finally, the correction for sea state must be made. The detailed shape of the sea surface with surface waves is not known, but it is clear that waves are asymmetrical; peaks are shaped differently from troughs. As waves get larger, the altimeter return signal gets more diffuse and the altimeter can measure wave height quite accurately. The correction of the mean sea surface due to wave height scattering is presently done empirically from data gathered in aircraft flights. It appears that this error can be reduced to a few centimeters up to gale-sized waves (less than 4 m) but the correction is still poorly understood.

The feasibility of having a dedicated altimetry program in the mid- to late-1980s has recently been studied and recommendations by the Dynamic Ocean Topography Experiment (TOPEX) science steering group have detailed recommendations concerning the studies that should be done to optimize such a mission, including tracking, shipborne research, gravity field, orbits, and coverage.

The SEASAT altimeter gave encouraging results. One example of altimetry passes over the same track during the "locked orbit" phase of SEASAT is given in Figure 3. To the left is a change in elevation of about 150 cm which corresponds to sea surface variation across the Gulf Stream. It may be possible to monitor the flux of the Gulf Stream system if this type of measurement is combined with a precise geoid and a suitable hydrographic monitoring program.

Looking at time-dependent circulation with an altimeter is particularly easy because a precise knowledge of the geoid is not required if the orbit exactly

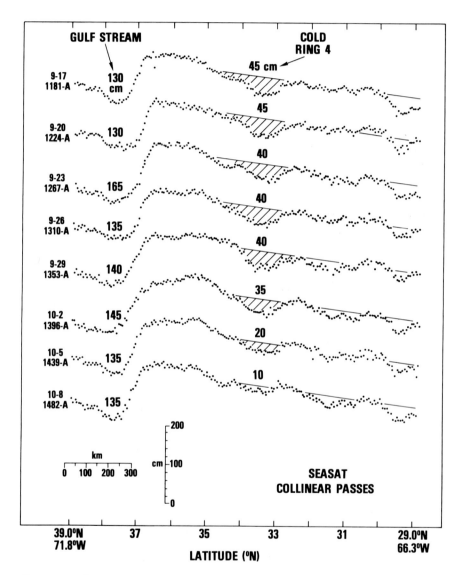

Figure 3. SEASAT altimetry (a track repeated every 3 days). Although the abscissa is labeled latitude, the actual coordinate was time, which progresses from right to left. One the left is the Gulf Stream and in the middle at approximately 33°N is a Gulf Stream ring (from Cheney and Marsh, 1980).

repeats its track over the earth. There are many important types of time-dependent flows, such as mesoscale eddies, fronts, tides, and storm surges that could be observed for such repeat orbits. To the right in Figure 3 is a bump which slowly gets smaller. It corresponds to a Gulf Stream eddy (cold core ring) which was observed by independent means to drift out from under the satellite track. Figure 4 shows an estimate of the frequency and wave number of ocean variability from the TOPEX study group report. It is for a typical ocean rather than for any real ocean based upon people's present knowledge of ocean variability. Altimetry may generate such a figure from real data.

We thus see that altimetry has, within 6 years, been demonstrated to identify and measure the sea surface to sufficient accuracy to be oceanographically useful.

Drogued drifters at the sea surface hypothetically measure the total surface velocity, not just the geostrophic component; but they are subject to windage effects whose contribution is not well understood (and may be large). In great numbers the proper design could be a potentially strong complement to the altimeters and are of major interest in their own right. Drifters can at present relay data to the ARGOS system on NIMBUS-7. By means of a timer the ARGOS system can also locate the position to the drifter to better than a kilometer. Thus, if the majority of their drag force is subsurface, or if current meters are used, their trajectories permit calculations of ocean currents. Within the last 10 years, use of buoys has become more frequent. An example of buoy trajec-

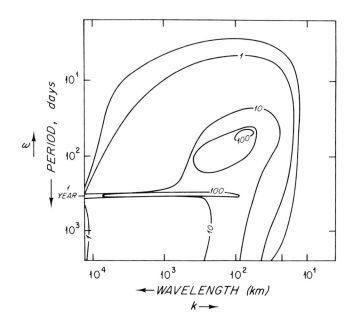

Figure 4. An estimate of the spectral composition of a "typical" deep-ocean region (from the TOPEX working group report, 1981).

tories in the North Atlantic is given in Figure 5. However, the use of drogues is as yet far from routine, and there is much not understood about buoy platform characteristics and trajectories. For studies of ocean circulation near the sea surface, drifting buoys potentially have several unique advantages compared to other techniques. They are relatively cheap (compared to, say, a moored current meter) and thus can in principle be deployed in large numbers, and they are suited to measurements of the low-frequency, vertically averaged surface layer currents (and thus are less subject to point sampling biases than are current meters).

3 Air-Sea Interaction—Temperature

Charts of the sea-surface temperature are widely used for a number of purposes. Major ocean features such as the edge of the Gulf Stream can be located. If temperature is known in conjunction with wind speed data and air temperature it is possible to estimate thermal energy exchange at the sea surface. If something is known of the vertical temperature distribution in the water it is possible to estimate ocean circulation. Temperature is a fundamental property of water masses and of interest in many problems in marine biology.

Sea-surface temperature is some suitable depth-averaged temperature of the surface "skin" of the ocean; remote sensors, if properly calibrated and corrected, measure this temperature to a known degree of accuracy. The depth of this skin is less than a millimeter for infrared radiometry and is up to a centimeter or more in microwave radiometers. The infrared radiometer is the most widely used so far but is capable of measuring only during cloud and fog-free conditions, whereas microwave radiometry has potential for all weather except rain, but with a larger footprint (25 to 150 km).

The infrared radiometer is a bolometer that detects emitted radiation from the ocean that has passed through a spectral window. The intensity of the upwelling radiance is a measure of the temperature of the ocean. It is necessary to sense in two or more channels of differing frequencies in order to eliminate the effects of atmospheric absorption and reemission. Scanning radiometers have been aboard a number of NOAA (the National Oceanic and Atmospheric Administration of the U.S. Department of Commerce) satellites and SEASAT. Data from many of the above are available from the Environmental Data Information Service of NOAA in image form, but the transformation of images to temperature data is difficult. Data gathered after January 1979 are also available on magnetic tape.

Performance is limited to clear weather and is degraded by very high humidity. The absolute precision varies with the instrument and care in the algorithm, with absolute precision demonstrated to better than approximately $\pm 1.0°C$ for cases where the data are processed carefully. The global data from operational satellites is sometimes as much as $5°C$ off, but where monthly averaging techniques have been used that utilize only high-quality data, standard devia-

Figure 5. Trajectories of surface drifting buoys in the North Atlantic for various scientific projects (compiled by Phil Richardson).

tions of 0.6°C between remotely sensed and ship-derived data have been obstained (Strong and Pritchard, 1980).

A good example of the intercomparison of a satellite temperature field with data gathered by ship is shown in Figure 6. Some details are well resolved spatially and accurately to 1°C, such as a temperature minimum on the right-hand side of the figure. However, the highest temperatures are 3° to 5°C different south of Cape Cod. Is this due to temporal or sampling errors (in this region in May the water is warming by almost $\frac{1}{2}$°C per day) or proximity to the coast?

In the scanning microwave radiometer, microwave radiation is detected by passive sensors at one or (usually) more frequencies. Knowledge of the emission properties of sea water or sea ice plus corrections for absorption through the air column is used to estimate sea-surface temperature, the effects of capillary roughness and sea foam (from which it is hoped to measure wind speed), air column water content (vapor and liquid), and sea ice extent.

Scanning microwave radiometers were aboard NIMBUS 5, NIMBUS 6, NIMBUS 7, and SEASAT. The lowest frequency channel (6.6 GHz in SEASAT) has approximately a 100-km spatial resolution and higher frequency channels have better resolution. Temperature data can be obtained for almost all weather conditions, except when rainfall exceeds some value. (The predicted value of 0.5 mm/hr appears to be an underestimate of tolerable rainfall.) Temperatures from the scanning multichannel microwave radiometer (SMMR) on SEASAT are presently being analyzed. The first analysis had a 3° to 4°C bias and a standard deviation of approximately 1.5°C when compared with models of the National Marine Fisheries Service, which are generally claimed to be accurate to approximately a degree. However, the accuracy was dramatically reduced to less than 1.5°C when a different data processing scheme was adopted a short time ago and tested against an independent data set which was gathered during the Joint Air Sea Interaction (JASIN) experiment. Data from the SEASAT SMMR are not of sufficient accuracy at present to help in estimating air-sea heat exchange, but they may be improved to that point in the future.

Drogue buoys can routinely measure temperature at a variety of depths and telemeter the data, via satellite, to land or ships. For gross ocean structure, such as mixed layer depth or structure in the upper thermocline, buoy measurements can provide extensive horizontal sampling for extended periods (more closely than shipboard or moored measurements) without the sampling problems associated with instantaneous measurements.

4 Air-Sea Interaction and Polar Oceanography—Ice

Sea ice includes a variety of different types of ice whose composition varies both physically and chemically (principally in the percentage of brine). The presence, absence, and structure of sea ice strongly affects air-sea interaction by influ-

Remote Sensing of the Oceans from Space 269

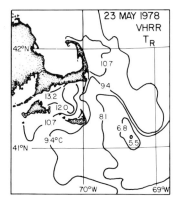
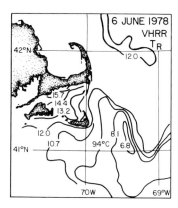
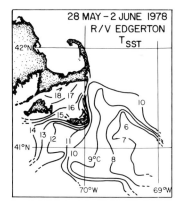

Figure 6. Intercomparison between the use of the infrared temperature sensors, the very high resolution radiometer (VHRR) aboard NOAA-5, and ship-gathered data (from Legeckis et al., 1980).

encing heat transfer, windstress into the ocean, and by forming both very salty (during freezing) and very fresh (due to melting of ice in thaws) water in polar oceans. The atmosphere is affected by sea ice through its effect on heat transfer, surface roughness, and reflected and emitted radiation.

The distribution of sea ice can be measured directly by high-resolution (1 to 4-km) visual-band images in the absence of clouds and during daylight, and by infrared sensors night or day in cloudless regions. However, the synoptic scale data needed to study many important ice processes cannot be acquired with visible and infrared (IR) wavelength sensors for two reasons: the polar regions are dark for half the year and ice regions are frequently cloudy. At the boundaries of the sea ice packs, it is cloudy most of the time. Fortunately, since sea ice has different microwave emmisivity than sea water, it is possible (for the first time) to obtain a good estimate of the ice coverage of polar seas throughout the seasons with the SMMR which can detect the microwave radiation through clouds and in the dark. This has led to a significant jump in synoptic knowledge (every 3 days, since 1974) of the distribution of sea ice in persistently cloudy areas. Scanning microwave radiometers on NIMBUS-5, NIMBUS-7, and SEASAT have allowed scientists for the first time to obtain maps of ice cover in the polar seas and even to distinguish multiyear ice from first-year ice, and ratios of the mixtures of these two ice types. Movies have even been made of the ice pack's metamorphosis over a span of more than a year.

A surprising and major discovery occurred when a large elongated ice-free region or polynya (see Fig. 7) (200 km by 1000 km) was observed in the Weddell Sea pack ice in the vicinity of 0°E. This persisted throughout the 1974 winter and again in 1975 and 1976. The ice concentration in the polynya was less than 15%. During the first winter that the pack ice was observed (1973) there was no polynya. The presence of this polynya significantly helps to resolve some debates concerning the negative heat flux necessary to cool the southern ocean and to form Antarctic Bottom Water.

5 Air-Sea Interaction–Stress

Similarly, the scatterometer measurements made on SEASAT can determine in principle the stress of the wind on the sea surface, an essential ingredient for specifying an important boundary condition of the ocean. There is inadequate experience to determine the precise accuracy or resolution of the method, but it clearly works in general. To test the scatterometer a number of field exercises were conducted while SEASAT was up. The first test was performed with data gathered in the Gulf of Alaska. A number of improvements were made to the algorithms from this data set and others and the measurements of the scatterometer were most recently tested against a withheld data set taken during JASIN.

The SEASAT-A scatterometer system (SASS) was an active microwave radar for synoptic scale global coverage of the ocean surface wind. The physical basis for this technique is the Bragg scattering of microwave radiation by wind-generated capillary waves. The intensity of the backscatter depends on the amplitude of the capillary waves which in turn depend on the wind strength. Further, the scattered waves being Doppler shifted, spatial discrimination is obtained by bandpass filtering of the return signal. The scattering process is anisotropic (i.e., it responds to preferred directions such as that imposed by the wind) making possible the determination of the wind vector.

Its footprint was comprised of two orthogonal 500-km-wide swaths which are X shaped, inclined 45° to the subsatellite track. The footprint width of approximately 1400 km is resolved into 50-km-wide sections in a region from 200 to 700 km on either side of the subsatellite track to obtain improved spatial resolution by bandpass filtering of the return signal.

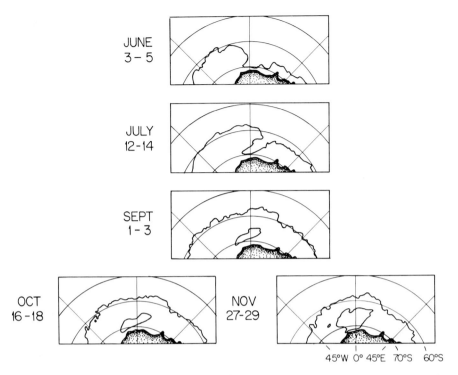

Figure 7. Edge of the Antarctic ice pack at selected times in 1975 as measured by a microwave sensor on NIMBUS-5 (contours taken from color imagery supplied by Jay Zwalley of the Goddard Space Flight Center).

The goal was to determine surface winds to an accuracy of ± 2 m s^{-1} and ± 20 degrees in direction. With less precision, the SASS was expected to measure winds out to 950 km from the spacecraft ground track. The range of winds to be measured was from 4 m s^{-1} to 26 m s^{-1}. Comparison with field data from JASIN revealed that the above accuracies were achieved.

Scatterometer data can be expected to improve as the calibration of the signal with respect to ground truth is updated. A basic understanding of the nature of the roughness of the ocean will help this process.

6 Sea State

Sea state is influenced most heavily by local wind and less by distant winds (that produce swell) and stability of the atmosphere. As such it is an indicator of a variety of meteorological phenomena. It can be measured relatively accurately by the altimeter. An example under light winds is shown in Figure 8 where the altimeter measurement of wave height is compared with a pitch-roll buoy. In this particular case, Webb (1979) located an error in one of the altimeter data processing routines and thus produced more accurate results.

7 Biomass, Chlorophyll, and Colorimetry

Ocean color (i.e., the spectrum of upwelled radiance just beneath the surface) depends strongly on the content of phytoplankton—microscopic plant organisms which form the first link in the oceanic food chain. These phytoplankton contain chlorophyll a (the dominant photosynthetic pigment) which absorbs light strongly in the blue and red region, making waters appear green in comparison to the deep blue of the pure state. Thus, the spectral band in the blue (approxi-

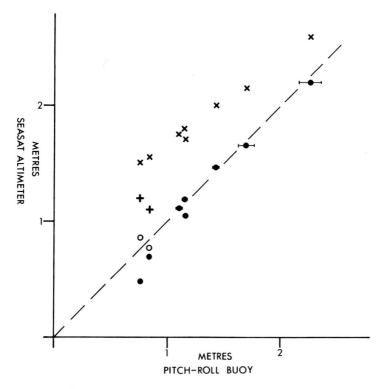

Figure 8. A comparison of the altimeter and pitch-roll buoy estimates of wave height under light wind conditions. The crosses are data with an early algorithm and the circles are data with a revised algorithm. The error bars refer to the pitch-roll estimates only. The rms error on the uncorrected altimeter values is of order ±5 cm. For these passes the buoy was between 10 and 140 km from the nearest subsatellite point (courtesy of D. J. Webb).

mately 443 nm) and the red (approximately 670 nm) are strongly affected by the phytoplankton content of water and are employed to provide a measure of the chlorophyll a content from NIMBUS-7 coastal zone color scanner (CZCS) data. Two other bands, near the so-called high point (wavelength of least sensitivity to phytoplankton content), are employed to provide a normalizing radiance value. It is thus easy to distinguish between water masses of greatly different phytoplankton content, and more importantly, between water masses of low but different concentrations. The true ocean color signal must be recovered from the measured spectrum after corrections have been applied for sun glint and atmospheric scattering. The successful development of accurate correctional algorithms will ultimately determine the limits on detection threshold and resolution in phytoplankton mass concentration. Vigorous attempts are being made to improve these algorithms. Hovis et al. (1980) and Gordon et al. (1980) describe a study in the Gulf of Mexico where spatial structure in phytoplankton concentration was discernible, after corrections, at concentrations below 0.1 mg m^{-3} (see Fig. 9). The upper limit may well be approximately 100 mg m^{-3}, although this will require use of other bands on the CZCS. Significant advances may yet be made in developing more sophisticated algorithms for atmospheric and sea-surface glitter corrections. These will largely determine the lower limits on detectability. A 12-channel color scanner is also being tested. The possibility thus exists that a global inventory of oceanic biomass can be made through the use of colorimetry, along with local studies to resolve questions of vertical patchiness (which is pronounced in some regions and not in others).

8 The Gravity Field

The Earth's gravity field, measured by geodesists in order to determine the shape of the planet, provides Earth scientists with one of their more useful geophysical observations. Anomalies in the gravity field are indicative of an uneven density distribution within the Earth. These provide insight into the mantle structure and in the mechanisms causing plate tectonics.

With the advent of satellite altimetry, developed during the GEOS-3 and SEASAT missions, it has become possible to determine the topography of the sea surface. This determines the marine geoid with an accuracy of about 1 m, although if the geostrophic current were measured by other means, the geoid could be measured to 10 cm or better.

Gravity at sea can be determined in three ways. The long wavelengths are best determined by satellite orbit analysis, as the satellite responds to long length scale geoid fluctuations. Conventional shipborne gravimeter measurements are quite accurate on very short wavelengths but can be obtained only regionally; to obtain a geoid it is necessary to integrate the data, and the constant of integration remains arbitrary. Satellite altimeters measure the marine geoid directly on all wavelengths but include error signals due to the large-scale ocean circulation

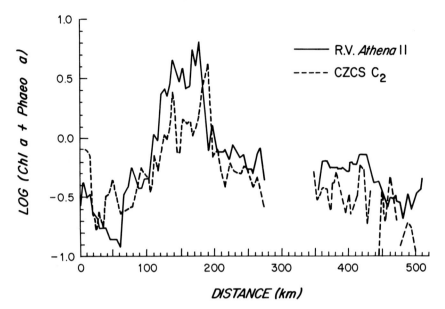

Figure 9. Values of chlorophyll a plus fragmentary pigment from chlorophyll a as measured by Gordon et al. (1980) by the coastal zone color scanner. Also shown are values measured on board a ship. This is one of the first intercomparisons yet made (copyright 1980 by the American Association for the Advancement of Science).

discussed above, tides, and other errors such as orbital inaccuracies. A number of marine geoids based both upon altimetry and orbital analysis have already been published.

The most satisfactory combination of gravity-field measurements required to study geological features on the ocean floor would therefore appear to be surface-ship measurements (for adequate resolution of short wavelengths) and gravity anomalies recovered from altimetry data (for adequate resolution of long wavelengths) plus corrections for ocean currents. The main problem with this is that although surface-ship measurements complement the coverage of altimeters in some parts of the oceans (North Atlantic, North Indian, North and Southwest Pacific Oceans) there are a number of regions where oceanographic measurements are sparse (Fig. 1). These are mainly confined to the southern oceans, south of latitude 30°S.

9 The Value of a Picture

Obviously a picture is as valuable as the idea it generates. There are now movies and pictures of the growth and shrinkage of the polar ice fields for the years 1974–1977. There are maps of ocean color during a red tide off Florida. One can follow the birth and death of a Gulf Stream or Kuroshio eddy, or watch any of a thousand important fronts advance and retreat through the years.

Figure 10 shows a high-quality infrared image of the Gulf Stream region from Cape Hatteras to east of Georges Bank. Three eddies are apparent north of the Stream. It is clear that an observer could get valuable information about wavelength, propagation speed, and growth rate from such a sequence of high-quality pictures. A glance at a bathymetric chart will enable one to verify that the distinct jump in temperature between the shelf waters and the deep oceanic waters occurs at the continental shelf break. One can also see patches of cold water which have been swept into warmer water and vice versa. Possibly, if the relation between surface temperature and near surface temperature is clarified, one can estimate the eddy mixing across the Gulf Stream from imagery.

Figure 11 is a photograph of a jet produced in a fluid mechanics laboratory. The apparatus was mounted on a turntable to simulate rotation of the earth. The jet, comprised of light fluid over heavy, also possesses large eddies being shed off

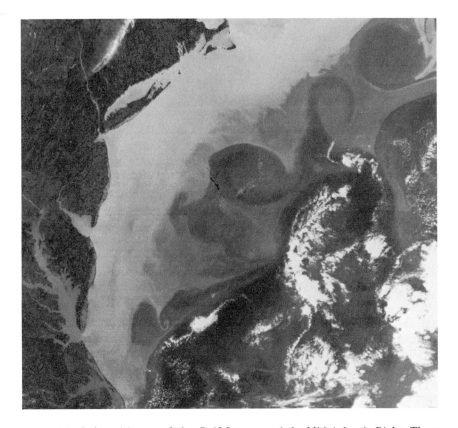

Figure 10. Infrared image of the Gulf Stream and the Mid-Atlantic Bight. Three Gulf Stream rings are visible north of the Stream (image supplied by R. Legeckis of the National Earth Satellite Service).

the side of the jet. Comparison of behavior of such models with imagery may significantly help us to understand ocean dynamics.

The use of imagery to analyze waves and eddies is now becoming routine. One oceanographic project is using imagery to locate warm core rings for the purposes of optimizing the biological sampling. Fisheries here and abroad use imagery to guide the fleet. But there is even more in the future of imagery.

An instrument has been flown on SEASAT that generates an image of incredible resolution. It is called the synthetic aperture radar (SAR) and it produced a rich image of the sea surface with a resolution of 25 m over swath widths of 100 km so that waves, swell, slicks, fronts, icebergs, and other patches which correspond to interesting ocean features (internal waves, Langmuir cells, patches of seaweed) are visible. It is an L-band (1.275 GHz) radar which is directed to one side of a spacecraft. Reflections which are scattered from small-scale (on the order of centimeters) roughness are received. The distance from the spacecraft for each echo source is calculated by recording return time to the echo. Position with regard to the spacecraft is determined by recording frequency. Since each oceanic point is sampled many times as the wide beam (hundreds of kilometers) sweeps outward, this system is equivalent to a radar with a gigantic (hundreds of kilometers) antenna. Obviously data rates are huge and processing is complex—requirements were high enough to limit performance on SEASAT to only those times when the satellite was within line of sight of receiving stations at: Rosman, North Carolina; Goldstone, California; Fairbanks, Alaska; Shoecove, Canada; and Oakhanger, England. The image is received under all weather conditions and it appears likely that the image spectrum bears some relationship to the ocean wave spectrum, in that it has a peak at approximately the same wavelength as the ocean waves. It is not clear, however, that the intensity of the peak is related to the height of the waves. The SAR remains a very powerful, but poorly understood instrument. It produces a huge amount of data and the proper data processing must be used (possibly on board the spacecraft). This cannot be done until we understand what the SAR exactly measures; which means that vigorous ground truth measurements must be conducted in conjunction with aircraft-borne SARs.

Even a picture whose exact nature is unfamiliar has value, however. Figure 12 is a SAR image for a region off Baja California, where internal wave packets can be easily seen. These packets are like those seen in the sun glint by early astronauts. Scientists have linked the packets to waves radiating off bottom topography synchronously with the tides, yet it is not known in detail how these long waves modulate the small-scale capillary waves in the ocean surface to generate such images.

10 Conclusion

We have seen how instruments to measure something in the oceans from a spaceborne platform have been highly developed in recent years. Research involving them will rarely be simple: often sophisticated temperature or speed corrections

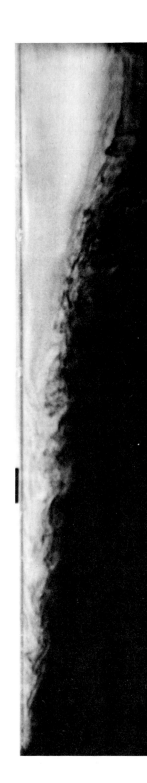

Figure 11. A turbulent jet in a laboratory experiment. The light colored fluid is fresh water which is floating out over denser clear fluid. It leans on the right-hand wall because the tank is rotating on a turntable. Comparison of such eddies under controlled conditions with remotely observed oceanic jets can lead to a clearer idea of the jet's mechanics.

Figure 12. Image of the sea surface off the coast of Baja California, showing packets of internal waves which are presumably scattered off the bottom from tidal currents (photograph supplied by the Jet Propulsion Laboratory).

must be made; sometimes the position of the sensor must be known to centimeter accuracy; the radiation that is being measured must have traveled through the atmosphere and been affected. In many cases atmospheric corrections must be made. In some there must be careful corrections for antenna characteristics.

Foam and air bubbles greatly alter reflective properties for some wavelengths. A full description of the waves on the surface is not yet in existence even under ideal conditions. Ice has varying reflectivity and emissivity. Radiation may come from the upper few micrometers (as thermal infrared) or from deeper in ice or the water. In some regions seaweed windrows or rip currents may have subtle but important effects on radiative or reflective properties.

Nor will research involving remote-sensing instruments be single component. Remote sensing will be too valuable in telling oceanographers where to go and what to measure. Although remote-sensing data can stand on their own right, an effective remote-sensing program will be coordinated with other research programs in order to gain vertical resolution. This is turn means that future research programs can be expected in which remote sensing is a full partner. These may address very different problems than have been addressed in the past. Facilities for conveniently receiving data, in some cases at sea, must be developed and made available to the research community. Possibly new centers will be needed to accomplish this. Many groups are wrestling with the difficult task of planning for these.

Many other exciting possibilities exist. Salinity could be measured from space possibly to one part per thousand but only with a "footprint" of hundreds of kilometers. Normal modes for gravity waves in the ocean have had their frequencies and structure calculated and there are now a few observations of some. Possibly they will be seen in altimetry. Possibly deep ocean and earth tides will be completely mapped soon. Possibly the response of sea surface to a cataclysmic event such as an earthquake, giant turbidity current, or volcanic explosion will be observed. Unbelievably, most of the things discussed in this paper were only imaginative possibilities in the mid 1970s. Bafflingly, most of the development has had little support from or participation by oceanographers although in the past two years interest has risen. To members of that small group who have endured workshops, advisory meetings, panels, committees, and even peril to their professional reputations—this article is dedicated. It would be surprising if remote sensing led to no major scientific discovery.

References

Cheney, R. E., and J. G. Marsh. 1981. SEASAT altimeter observations of dynamic topography in the Gulf Stream region. J. Geophys. Res. 86, 473–483.
Ewing, G. C. (Ed.). 1965. Oceanography from space. Ref. No. 65-10. Woods Hole Oceanographic Institution.
Gordon, H. R., D. K. Clark, J. L. Mueller, and W. M. Hovis. 1980. Phytoplankton pigments from the Nimbus-7 coastal zone color scanner: Comparisons with surface measurements. Science 210, 63–66.
Hovis, W. A., D. K. Clark, F. Anderson, R. W. Austin, W. H. Wilson, E. T. Baker, D. Ball, H. R. Gordon, J. L. Mueller, S. Z. El-Sayed, B. Sturm, R. C. Wrigley, and C. S. Yentsch. 1980. Nimbus-7 coastal zone scanner: System description and initial imagery. Science 210, 60–63.
Legeckis, R., E. Leggs, and R. Limeburner. 1980. Comparison of polar and geostationary satellite infrared observations of sea surface temperatures in the Gulf of Maine. Rem. Sens. Environ. 9, 339–350.
Strong, A. E., and J. A. Pritchard. 1980. Regular monthly mean temperatures of Earth's oceans from satellites. Bull. Am. Meteorol. Soc. 61, 553–559.
Webb, D. J. 1979. A comparison of satellite and sea surface measurements of significant wave height. Institute of Oceanographic Sciences Internal Report No. 86. Unpublished manuscript.

Appendix I: Archives of Remote-Sensing Data

The Environmental Data and Information Services (EDIS) of the Department of Commerce issues periodic Satellite Data Users bulletins which can be obtained at the below address. Imagery is archived as a photographic negative for data prior to 1979 and as a negative and digitally for data taken in 1979 and thereafter.

National Oceanic and Atmospheric Administration
Environmental Data and Information Service
National Climatic Center
Satellite Data Services Division
Room 100, World Weather Building
Washington, DC 20233
(301) 763-8111 or FTS 763-8111

The facility listed below can be contacted for information about the EROS Program and assistance in the interpretation and applications of the ERTS and LANDSAT imagery.

USGS/EROS Data Center
Sioux Falls, South Dakota 57198
Contact: Applications Branch
Phone: (605) 594-6511, Ext. 111

Appendix II: Additional Sources of Information

Collected papers
  Papers from OCEANS '76, JPL Publication PD 622-17.
  Boundary-Layer Meteorol. 13 (1978).
  Boundary-Layer Meteorol. 18 (1, 2, and 3) (1980).
  Science 204, 1405-1424 (1979).
  IEEE Journal of Ocean Engineering OE-5 (2), 71-181 (1980).
  Proceedings of the COSPAR/SCOR/IUCRM Symposium "Oceanography from space" (J. W. F. Green, Ed.). Marine Science Vol. 13, Plenum Press, New York (1980), 978 pp.

Survey papers
  Apel, J. R. 1976. Ocean science from space. EOS. Trans. Am. Geophys. Union 57, 612-624.
  Apel, J. R. 1977. Past, present and future capabilities of satellites relative to the needs of ocean science, Bruun. Mem. Lect. Intergov. Oceanogr. Tech. Ser. 19, 7-39, Paris, UNESCO.
  Sherman, John W., III. 1978. An overview of remote sensing oceanography in the United States. Proc. Intl. Symp. Remote Sensing of the Environment (12) held at Ann Arbor, Mich. on April 20-26, 1978.
  Huang, Norden E. 1979. New developments in satellite oceanography and current measurements. Rev. Geophys. Space Phys. 17, 1558-1568.

Apel, J. R. 1980. Satellite sensing of ocean surface dynamics. Ann. Rev. Earth Planet. Sci. 8, 303-342.

Wunsch, C., and E. M. Gaposhkin. 1980. On using satellite altimeters to determine the general circulation of the ocean with application to geoid improvements. Rev. Geophys. Space Phys. 18, 725-745.

Studies on oceans which include remote sensing issues

Committee on Geodesy, National Research Council. 1978. Geodesy: Trends and Prospects, National Academy of Sciences, Washington, D.C., 86 pp.

Committee on Geodesy, National Research Council. 1979. Applications of a Dedicated Gravity Satellite Mission, National Academy of Sciences, Washington, D.C., 53 pp.

European Space Agency. 1978. Space Oceanography, Navigation, and Geodynamics. ESA Scientific and Technical Publications Branch, ESTEC, Noordwijk, The Netherlands, 350 pp.

NASA. 1980. NASA Oceanic Processes Program Status Report. Fiscal Year 1980. Technical Memorandum 80233, Scientific and Technical Information Office, Washington, D.C.

NASA/NOSS. 1981. Science working group report. Needs, opportunities and strategies for a long-term oceanic sciences satellite program. National Center for Atmospheric Research Technical Note NCAR/TN185, PAR, 72 pp.

Jet Propulsion Laboratory. 1980. Guidelines for the Air-Sea Interaction Special Study: An element of the NASA Climate Research Program. JPL Publication 80-8, 1980.

Jet Propulsion Laboratory TOPEX Science Working Group. 1981. Satellite altimetric measurements of the ocean. March 1981, 78 pp.

U.S. North Atlantic Regional Workshop (Proceedings of). 1979. University of Rhode Island, 250 pp.

United States Distant-Water Oceanography in the New Ocean Regime

Edward J. Miles

1 Introduction

The concept of the exclusive economic zone is perhaps the most significant outcome of the Third United Nations Conference on the Law of the Sea (UNCLOS III). This concept enshrines a revolution in the law of the sea which significantly expands the geographical and functional scope of coastal state rights in the ocean. This revolution had begun as early as 1929 during the preparations for the League of Nations Conference on the Codification of the Law of the Sea and it is driven primarily by two dynamics: the patterns of marine technological advance, its effects on patterns of ocean use and the distribution of income and other values to be derived from such use; and people's expectations about patterns of marine technological advance and the link between these expectations and jurisdictional claims subsequently espoused.

The concept of the exclusive economic zone has fundamentally altered the balance between the exclusive interests of coastal states and the inclusive interests of the international community in the world ocean. The basis of the new ocean regime is, in fact, an uneasy compromise between the desire of coastal states to extend their control over resources and other activities occurring in the oceans adjacent to their coasts and the interests of states with significant distant-water operating capabilities and/or minimal resources or minimal coastlines to curtail the extent of exclusive control by the coastal state.

The emergence of this concept has already been analyzed in detail elsewhere (see, for instance, Johnston and Gold, 1973; United Nations, 1973; Aguilar, 1973; Miles, 1976, 1977); this particular outcome of UNCLOS III was shaped primarily by certain changes in the structure of the international system at the time and the dynamics of conference diplomacy. The coastal/noncoastal conflict had been exacerbated as a result of changing patterns of ocean use, and it became entwined in the global shift from a world dominated by an east/west confrontation to one dominated by a north/south confrontation. This latter characteristic meant that all issues related to control over resources, technology, and knowledge production were rising to the top of agendas pursued by an increasingly large number of national governments.

The combined effect of these two developments fed the fires of demands for spatial control by the coastal state, in the form of a single zone of 200 miles, demands for control over the acquisition of information within that zone, and demands for vesting in the coastal state property rights over all resources in the zone. This issue in the conference raised six major policy questions:

What is the extent of coastal state control over living resources in the area claimed?
What is the outer limit of the continental shelf?
What is the nature and extent of coastal state control over marine scientific research?
What is the nature and extent of coastal state control over ship-generated marine pollution?
What is the relationship between coastal state control and the traditional freedoms of the high seas?
How are the residual rights in the economic zone to be characterized? Put another way, what is the status of the zone in relation to the interests of both the coastal state and the international community as a whole?

These questions in the conference had to be answered within a primarily political "matrix" defined by the rules of procedure, the structure and size of coalitions, and the tradeoffs brought to the situation by the respective players.

In this context, the issue of marine scientific research turned out to be much more important than ever before for two reasons: since control over information was a central problem in the eyes of the Group of 77 (a group of developing nations), scientific capabilities translated into use capabilities; therefore, access to the zone for the conduct of marine scientific research was seen as a commodity to be traded. Also, that group of states (the Territorialist Group) which sought to define the zone as a territorial sea wanted the scope of exclusive coastal state control extended over all activities in the zone; science, therefore, could not be an exception.

On the other hand, the group of states fighting to protect science in the conference was much smaller than it could have been. It included primarily the United States, USSR (up to 1976), Federal Republic of Germany, and the Netherlands, with occasional muted support from Japan. The potential membership of such a group included the United Kingdom, France, Norway, Denmark, Canada, and Australia, but these players chose not to enter the fray for two basic reasons: only the United States and the USSR maintain global ocean research programs; marine scientists of other countries, of course, have global interests in the oceans but their programs are more locally or regionally based. The consequences of increasing restrictions on marine scientific research within 200 miles of the coast, in a global sense, posed a major operational problem primarily for the superpowers. Also, the fundamental tradeoff proposed at UNCLOS III was extending control over coastal state resources in return for protecting the navigational (including research) rights of others. Such countries as the United Kingdom, Canada, Australia, and Norway, which normally could be counted on to fight to protect research, chose instead to extend control over resources. As a result, they kept their own scientists on a very tight rein at the conference—not difficult, since the scientists in question were government employees.

In the final analysis, therefore, the major policy questions posed by the concept of the exclusive economic zone were answered in favor of the coastal state, as shown in Table 1. With respect to control over resources and marine scientific research, there are few constraints on coastal state discretion. With respect to resources and the basic issue of whether or not to grant consent to the conducting of marine scientific research within its economic zone, the decision of the coastal state is nonreviewable by any third party. Only on the issue of control over ship-generated pollution is coastal state discretion significantly constrained.

2 Conditions and Procedures Affecting Scientific Research

As the debate on marine scientific research developed between 1974 and 1980 in Committee III of UNCLOS III, several clues to the concerns of coastal states became apparent (Knauss, 1973, 1974; Franssen, 1974; Wooster, 1976). The dominant attitude of those arguing for exclusive coastal state jurisdiction over scientific research in the economic zone was the view that marine scientific research had so far served the purposes primarily of advanced industrial countries. It should now be made to serve the interests of developing countries as well, and access, therefore, should be regarded as a commodity to be traded. Developing countries should seek the right to participate or be represented in all states of the research from planning to evaluation of results, sharing of samples, access to all information gained, publication of results, and assistance in the interpretation of results. A more extreme view stipulated no publication without the prior consent of the coastal state.

A second attitude was more covert but seemed to be very similar to themes of the debates on the New International Economic Order within the United Nations Draft Convention on the Law of the Sea and the UN General Assembly. This view was that the primary objective of developing countries should be to narrow, not widen, the gap between developed and developing countries. The latter, therefore, should simply insist on the right to deny consent (and, therefore, access) to advanced industrial countries if all else failed. The third view, held by the Territorialist Group, was that there should be no significant constraints on the authority of the coastal state in the economic zone. The fourth

Table 1. The nature of the exclusive economic zone as defined by the Draft Convention on the Law of the Sea

Activity	Coastal state authority
Fisheries	Coastal state establishes allowable catch and decides whether surplus exists. Determines own harvesting capacity. Decisions nonreviewable. Exclusive authority to manage stocks falling within own zone.
Continental shelf	Coastal state sovereignty over continental margin to 350 miles from territorial sea baselines or to 100 nautical miles from the 2500-meter isobath. Where the outer edge of the continental margin extends less than 200 miles from the territorial sea baselines, the coastal state exercises sovereignty up to 200 miles. Small revenue-sharing obligation on production of nonliving resources outside 200 miles.
Scientific research	Coastal state consent plus obligations required for all research. Authority not as all encompassing as for living and nonliving resources but constraints on coastal state discretion not major. Basic decision on whether or not to grant consent nonreviewable.
Other economic activities	Exclusive coastal state authority *re* power generation and other economic activities in the zone.
Artificial islands and installations	Exclusive coastal state authority to construct and to authorize and regulate the construction, operation, and use of such islands and installations.
Residual rights/ status	Economic zone *sui generis*. Rights and jurisdictions of coastal state and rights and freedoms of other states governed by *relevant provisions of this convention*. [Italics added.] By implication, rules of innocent passage do not apply in the economic zone.

view was very widely shared, that marine scientific research, as conducted by the superpowers, was very often a cover for military activities of a wide variety and that the coastal state had a right to be protected against these.

The system that emerged in the Draft Convention is without a doubt largely in favor of the coastal state and significantly burdensome to the researching state, though the latter retains some small benefits and protections within 200 miles. Article 246 provides that coastal states have the right to "regulate, authorize and conduct" marine scientific research in their exclusive economic zone and on their continental shelf. Furthermore, marine scientific research undertaken in those two areas "shall be conducted with the consent of the coastal state." In connection with this, Article 296(2) specifically excepts these decisions or those suspending or terminating a research project in accordance with Article 253 from the applicability of the Settlement of Disputes provisions. It should be noted also that while a researching state can invoke a conciliation procedure if it alleges that the coastal state is not exercising its rights under Articles 246 and 253 in a manner compatible with the Convention, the conciliation commission "shall not call in question the exercise by the coastal state of its discretion to withhold consent in accordance with paragraph 5 of Article 246."

Two benefits given to the researching state are that coastal states shall "in normal circumstances" grant their consent and that "normal circumstances" may exist in spite of the absence of diplomatic relations between the two parties. However, Paragraph 5 of Article 246 specifies that the coastal state has the right of discretionary denial in four situations. These occur if the project:

- Is of direct significance for the exploration and exploitation of natural resources, whether living or nonliving;
- Involves drilling into the continental shelf, the use of explosives or the introduction of harmful substances into the marine environment;
- Involves the construction, operation or use of artificial islands, installations and structures referred to in Articles 60 and 80;
- Contains information communicated pursuant to Article 248 regarding the nature and objectives of the project which is inaccurate or if the researching state has outstanding obligations to the coastal state from a prior research project.

One significant benefit to the researching state provided by the ICNT concerns research on the continental shelf beyond 200 miles. Article 246(6) stipulates that the right of discretionary denial affecting research on the continental shelf beyond 200 miles may not apply outside areas publicly designated by the coastal state as being "areas in which exploitation or detailed exploratory operations are occurring or will occur within a reasonable period of time."

Articles 248 and 249 specify the set of obligations operative on the researching state and to which penalties for nonfulfillment may be attached. No less than 6 months (180 days) in advance of the expected starting date of the project, the researching state must provide the coastal state with a full description of:

- The nature and objectives of the research project;
- The method and means to be used, including name, tonnage, type, and class of vessels and a description of scientific equipment;
- The precise geographical areas in which the activities are to be conducted;
- The expected date of first appearance and final departure of the research vessels, or deployment of the equipment and its removal, as appropriate;
- The name of the sponsoring institution, its director, and the person in charge of the research project;
- The extent to which it is considered that the coastal state should be able to participate or to be represented in the research project.

These requirements both extend the leadtime for detailed planning of research projects and raise the possibility of increasing the costs of research to the researching state. The types of obligations which may increase costs are more specifically described in Article 249 and relate to participation of representatives of the coastal state in the research project. This may involve both travel and *per diem* costs during the planning phase and loss of ship space to the researching state by the need to accommodate representatives of the coastal state. This may occasionally require a greater amount of ship operating time in a particular area than would normally be the case. Increased costs may also arise with respect to duplicating data and samples for the coastal state and with the need to assist the coastal state in their assessment or interpretation. The full list of additional obligations is reproduced below.

1) States and competent international organizations when undertaking marine scientific research in the exclusive economic zone or on the continental shelf of a coastal State shall comply with the following conditions:
(a) Ensure the rights of the coastal State, if it so desires, to participate or be represented in the research project, especially on board research vessels and other craft or scientific research installations, when practicable, without payment of any remuneration to the scientists of the coastal State and without obligation to contribute towards the costs of the research project;
(b) Provide the coastal State, at its request, with preliminary reports, as soon as practicable, and with the final results and conclusions after the completion of the research;
(c) Undertake to provide access for the coastal State, at its request, to all data and samples derived from the research project and likewise to furnish it with data which may be copied and samples which may be divided without detriment to their scientific value;
(d) If requested, provide the coastal State with an assessment of such data, samples, and research results or provide assistance in their assessment or interpretation;
(e) Ensure, subject to paragraph 2, that the research results are made internationally available through appropriate national or international channels, as soon as feasible;
(f) Inform the coastal State immediately of any major change in the research program;
(g) Unless otherwise agreed, remove the scientific research installations or equipment once the research is completed.

Paragraph 2 of Article 249 is particularly important to researching states since it deals with the problem of controls on publication. The formulation is as follows:

This article is without prejudice to the conditions established by the laws and regulations of the coastal State for the exercise of its discretion to grant or without consent pursuant to Article 246, paragraph 5, including requiring prior agreement for making internationally available the research results of a project of direct significance for the exploration and exploitation of natural resources.

This means that as a prior condition to granting consent, the coastal state may require restraints on publication for research of direct significance for the exploration and exploitation of natural resources. Presumably, it is the coastal state that makes such a determination. Furthermore, the cross reference in Article 249(1)(e) shown above limits the obligation to make research results internationally available to possible coastal state restrictions as specified in Article 249(2). [It should be noted that another interpretation of the effect of Article 249(2) exists, to wit: Article 249(2) serves to limit the exercise of coastal state authority as provided for in Article 246(1). Discretionary denial of consent by a coastal state, therefore, is permitted only on the conditions specified in Article 246(5). Whether this interpretation is persuasive remains to be seen.] This is a matter of grave concern for university-based oceanographers in the United States, since most universities and granting agencies require open publication of research results.

Article 250 stipulates that all communications between the coastal and researching states concerning research projects "shall be made through appropriate official channels unless otherwise agreed." This means that the request for consent process has been formalized and that, in most cases, the agency to which the application must be addressed is the Ministry of Foreign Affairs of the coastal state in question via the U.S. Department of State. However, the words "unless otherwise agreed" leave the door open to less formal arrangements and we shall return to this in the next section.

The last two substantive provisions with which I shall deal are those dealing with the conditions for inferring implied consent (Article 252) and the conditions under which research projects can be suspended or terminated. The implied consent formulation is a benefit for the researching state. It reads:

States or competent international organizations may proceed with a research project upon the expiry of six months from the date upon which the information required pursuant to Article 248 was provided to the coastal State unless within four months of the receipt of the communication containing such information the coastal State has informed the State or organization conducting the research that:

(a) it has withheld its consent under the provisions of Article 246; or
(b) the information given by the State or competent international organization in question regarding the nature or objectives of the research project does not conform to the manifestly evident facts; or

(c) it requires supplementary information relevant to conditions and the information provided for under Articles 248 and 249; or
(d) outstanding obligations exist with respect to a previous research project carried out by that State or organization, with regard to conditions established in Article 249.

However, it is necessary to point out that paragraph (c) provides the coastal state with a considerable capacity for delay in the event that, for other reasons, it does not wish to deny consent outright. Moreover, paragraph (d) makes each researching institution his brother's keeper, since outstanding obligations from one institution can be the cause of denying consent and suspending or terminating the research projects of others. The point here is that it is the researching *state* which undertakes the obligation, and clearances for single institutions through formal channels in each case commit the whole state.

With respect to suspension of marine scientific research activities (Article 253), the coastal state may exercise its right if the research activities are not being conducted in accordance with the information provided under Article 248 upon which consent was based, or if the research state fails to comply with the obligations specified in Article 249. The coastal state may terminate a research project if noncompliance with Article 248 amounts to a major change in the research project or the research activities, or if situations leading to suspension have not been rectified within a reasonable period of time. On the other hand, Article 253(5) also specifies that suspension shall be lifted and marine scientific research activities allowed to continue once the researching state or competent international organization has complied with the conditions required under Articles 248 and 249.

While these are not all the articles concerned with marine scientific research in the Draft Convention, they are the most restrictive, and therefore among the most important for researching states. The fact that marine scientific research remains relatively unregulated in the water column beyond 200 miles and in the international seabed area is also important. Changes in the world ocean regime, however, have come about through unilateral actions of coastal states as well as through decisions of UNCLOS III, and these must be assessed as well.

3 Marine Scientific Research Within Zones of Natural Jurisdiction

Two studies have so far attempted to analyze trends in national legislation affecting the conduct of marine scientific research outside of the territorial sea. Wesley Scholz (1979) analyzed the national legislation of 58 coastal states which have claimed jurisdiction beyond 12 miles. Of these, 22 states have made explicit claims to regulate the conduct of scientific research. The Ocean Policy Committee, National Research Council (1982) analyzed the national legislation of 79 states. Of these, 37 states explicitly claimed jurisdiction over scientific research; nine states claimed a 200-mile territorial sea which would encompass scientific research; 16 states exercised jurisdiction over activities related to living or nonliving resources within 200 miles; seven states did not include scientific research within their zones of extended jurisdiction; and one state explicitly excluded it. Both Scholz and Katsouros (of the Ocean Policy Committee) analyze the terms and conditions governing the conduct of marine scientific research established in the national regulations of a few states. Scholz had available the regulations established by Argentina, Brazil, Columbia, Ecuador, India, Mexico, Portugal, Senegal and Trinidad-Tobago. Katsouros analyzed the national legislation of Burma, India, Pakistan, and Sri Lanka and the procedures of 28 countries as described by U.S. Embassy representatives in response to a State Department query. There is a remarkable amount of variation between these sets of national regulations but most of them fall within the policies established by the Draft Convention. In only three cases did regulations go beyond the Draft Convention; India, for instance, has excluded a whole class of physical oceanographic experiments which appear to be related to the development of antisubmarine warfare capabilities, while the government of Trinidad-Tobago has claimed ownership of all data and samples.

The conclusion is clear. Whether or not UNCLOS III produces a treaty which enters into force, the majority of coastal states of the world have already enacted legislation which is consistent with the policies established by the Draft Convention as it relates to the conduct of marine scientific research in the economic zone. In some cases, a few states have even gone beyond the requirements of the Draft Convention. As far as the future ocean regime is concerned, it is the Draft Convention which should shape the expectations of United States' distant-water oceanographers as they seek to respond and adapt to the demands of coastal states.

4 General Effects on the Conduct of United States Marine Scientific Research

The major effects with which United States distant-water oceanographers will have to contend are increases in administrative and operational costs, increased bureaucratic formalization and complexity, differences in research priorities between the researching institutions and coastal states, and the need to provide technical assistance to developing coastal states. These effects may be felt most seriously by geological/geophysical oceanographers (given their interests in continental margins) and biological oceanographers (given the importance of neritic zones for most biota). To a lesser extent, physical oceanographers interested in equatorial and Southern Ocean circulation will also be adversely affected. Chemical oceanographers may be the least affected of all (Ocean Policy Committee, 1982).

Increased costs will arise with respect to expanded planning time for research, increased travel costs for representation of coastal states in the planning of research projects, loss of ship space to representatives of coastal states, increased port calls, extension of ship operating time, sharing data and samples, and providing assistance to the coastal state in the interpretation of research results. These issues require more detailed treatment. It is not possible at this time to provide an overall estimate of the scale of increased costs which is likely as a result of the new ocean regime. Perhaps each member of the Distant-Water Operating Group of the University National Oceanographic Laboratory System (UNOLS) should be asked to prepare cost estimates according to two or three common scenarios prepared by UNOLS. What is clear, however, is that these increased costs come at a time of stringent inflationary effects, decreasing funding in real terms, and an arguably more disadvantageous approach to funding large-scale field ocean studies in the United States adopted by the National Science Board of the National Science Foundation (NSF).

In a study of trends in academic fleet support prepared for the Interagency Committee of Marine Science and Engineering in November 1974, the Chairperson of the study, Mary Johrde of NSF, made the following projection:

> . . . NSF support will show some percentage decline by 1976 in spite of increasing dollar outlays. Level dollar support of ONR for the years 1973–1976 will result in a continued percentage decline through 1976. Hence, unless support increases sufficiently from all other Federal and non-Federal sources, a deficit of more than $1 million is probable in 1976. Since the projected totals for both 1975 and 1976 are less than optimum budgets for the fleet, there is little likelihood that any part of the anticipated deficit can be made up through additional fleet economies.

By the end of 1974, it was clear that the effects of inflation (rising fuel, food, and labor costs primarily) portended a situation of level funding for several years, with a decreasing ability to support ship time. This loss of ship time appeared initially to vary considerably with each local situation. Ships were also losing crews and operating on a short-handed basis, and yard costs for normal overhaul were increasing appreciably. The solution to these problems was increased interinstitutional coordination of ship scheduling via the UNOLS mechanism. The major question underlying these trends, which has still not been answered, is to determine the most desirable fleet structure for United States oceanography, given advances in marine science, increases in inflation, and changes in the world ocean regime affecting the conduct of marine scientific research.

Wooster (1979) pointed out that the shortfall in ship support had increased to $4.2 million in 1979. Given available funds, research institutions were laying up ships. He says, "The number of layups has grown steadily to the point where 20 percent of the fleet, including more than half of the major vessels, is scheduled to be laid up in 1979."

Wooster argues that "much of the shortfall is related to two NSF practices: (1) increasing ship operating funds at a rate conspicuously less than inflation, and (2) diminishing the percentage of research funds allocated to ship operation support."

The final internal funding difficulty which United States distant-water oceanographers must now face is generated by a decision of the National Science Board of NSF at its meeting in April 1980. They recommended level funding of large-scale coordinated research projects (such as the International Decade of Ocean Exploration (IDOE) Program, characteristically heavily dependent on large ship operations) and a change in the budget structure of the Ocean Sciences Division so that large and small projects must now compete with each other for funding. In response to these recommendations, the then Director of the Division of Ocean Sciences, Dr. Dirk Frankenberg (April 23, 1980 memo) made, *inter alia*, the following observations:

> These recommendations will take effect at the same time that at least $3.7 [million] will have to be transferred from research to ship operations to cover increases in fuel costs; the cumulative effect of these two changes will be to reduce NSF ocean science research support by at least 14 percent between 1981 and 1982—such a reduction would precipitate a major crisis in U.S. Ocean Science.

The major adverse effects for the conduct of United States distant-water oceanography in the immediate future do not stem from any external changes in the world ocean regime and the conduct of coastal states. They stem from internal policies adopted deliberately by NSF as a partial response to its perceived financial constraints and as a shift in priorities away from large-scale field programs in oceanography. The change in the world ocean regime for marine science is therefore important in two respects. The increased costs generated by the change in regime are significant primarily because they fall mainly into the category of ship operating costs, which are already significantly underfunded for the United States fleet and may become more so if the policy of the National Science Board prevails. Also, the change in regime is important because increased control over research by the coastal state may lead to a decline in certain types of studies as a result of difficulties of getting clearance. This latter problem itself has two aspects: denials of clearance, trends in which are being researched; and important research not being proposed because it is assumed that approval will be virtually impossible. The latter phenomenon will be very difficult to monitor.

5 Bureaucratic Formalization and Complexity

There are several significant effects of working under increased bureaucratic formalization and complexity. In the first place, the leadtime for planning projects may increase by as much as 100%. This will occur as a result of the length of time required for submission of clearance requests through two governmental systems plus the time required to allow participation in planning by scientists from the coastal state in question.

There will probably be an extended transitional period before a Law of the Sea Treaty enters into force and is ratified by the U.S. Senate. The State Department takes the view that the United States must be consistent with its own claimed jurisdiction when submitting clearance requests to coastal states. United States distant-water oceanographers have frequently found themselves in the Alice in Wonderland position of the United States' not recognizing the claim of extended jurisdiction by a particular coastal state prior to the entry into force of a comprehensive Law of the Sea Treaty. The most effective way out of this impasse is for United States marine scientists to get the U.S. Congress to claim jurisdiction over the conduct of marine scientific research within 200 miles of United States coasts. This could be done either through an amendment of the Fisheries Conservation and Management Act (FCMA) or through the passage of a simple rider on a bill which declares: "The United States exercises jurisdiction over the conduct of marine scientific research within 200 miles of the baselines used to measure the extent of the territorial sea."

Another problem generated by increased formalization is that particular clearance requests from United States institutions can be held hostage both to other issues between a coastal state and the U.S. Government and within coastal states in the bureaucratic conflict that often exists between a ministry of foreign affairs, a ministry of agriculture and fisheries, and a department of the navy. Evidence of these difficulties can often be seen in such countries as Mexico, Peru, Brazil, and India.

Considerable slippage can occur between the information supplied by the researching institution to a ministry of foreign affairs and the information that is summarized by that ministry and passed on to others in the same coastal state. Distortions that occur at this level can generate considerable difficulties, given the increased importance of marine scientific research as a political issue and the fears of particular coastal states concerning the use of such research as a cover for military activities.

Increased formalization implies also increased administrative responsibilities for monitoring compliance with obligations by individual institutions and collectively, since all would be equally affected by the noncompliance of one. There is another incentive to monitoring compliance which lies in the fact that United States scientists will wish to avoid a reputation that stringent coastal state regulation is the only means of getting them to fulfill their obligation. The performance reputation is a commodity that should be carefully husbanded.

One possibility for minimizing the scope of increased formalization and reinforcing the significance of face-to-face contacts among scientists is for United States distant-water oceanographers to build among themselves an organization (including a legal arm) for negotiating relatively informal arrangements with different coastal states. There are two requisites for successful arrangements: the United States organization must recognize the jurisdiction of the coastal state over the conduct of marine scientific research, and clear benefits must accrue to the coastal state from its participation in such an arrangement. On the other hand, care must be taken to establish fair and reasonable conditions, since the next researching institution seeking access to the same area will be required to meet these conditions as a minimum.

It is possible, of course, for individual research institutions to have very effective working relationships of this kind with coastal states; for example, the recent agreement between the Lamont-Doherty Geological Observatory of Columbia University and the People's Republic of China. On the other hand, a real need exists for a collective organization through which umbrella arrangements can be developed, which permit the exchange of interests and activities. This would make it possible for the coastal state to avoid insisting on benefits to be disbursed on a per-vessel, per-trip, per-institution basis. More importantly, such an organization would respond primarily to scientific interests and permit some insulation from infection by external political concerns.

To develop and manage such an organization or subsection of an organization will not be easy, requiring adequate funding, imaginative leadership, and a pooling of expertise so that the whole is definitely greater than the sum of its parts. It may be more appropriate to try to develop this capability in the Joint Oceanographic Institutions, Inc. (JOI) rather than UNOLS, because JOI, Inc. already has a legal personality and a more fully developed management infrastructure.

6 Differences in Research Priorities and the Problem of Technical Assistance

There are very often very different substantive research interests between United States distant-water oceanographers and coastal states. Ironically, the group with whom the coastal state would most likely have the greatest affinity would be United States coastal oceanographers and those working primarily on applied problems. But these are not the people who find it necessary to seek coastal state consent for the conduct of research. In regions where the geographical context is sufficiently important for scientific reasons and where demand for access is high and steady among United States distant-water oceanographers, an umbrella arrangement, managed on the United States side by a functionally expanded JOI, Inc. can try to bridge the gap in interests by acting as a broker, if new program funds are available for this purpose. On the other hand, where we are faced with essentially "one-shot deals" between individual research institutions and a coastal state, such a bridging of interests is not often possible, unless the research project has links to resources of interest to the coastal state.

The developing coastal states most likely to benefit from opportunities provided under the ICNT are those with strong preexisting national commitments to developing marine scientific research and where some infrastructure already exists. Where these conditions are absent, it will not be possible for the coastal state to build any coherent program out of the widely disparate contributions of different research institutions. The only significant potential short-term benefits for these coastal states lie in seeking information and advice about their resource endowments. However, if the cost of providing this is high and the scientific importance of the area is low to medium, the research institution would probably prefer not to go there.

Where the area is sufficiently important for long-term technical assistance arrangements to be mutually beneficial, special care must be taken in designing the program.

The way this question is approached sometimes gives rise to discussions which assume that a choice must be made between developing a basic marine science capability in a recipient country or focusing exclusively on applied research and technology transfer. This kind of distinction is not fruitful since obviously, if transfer programs are to be successful, the recipient country must increase its capacity to judge which technologies should be transferred given local conditions, how they should be adapted, and what development priorities ought to be placed upon the type of marine resources available. One of the major problems here is getting a sufficiently high priority accorded to marine science by the government of the recipient country. Ayala-Castanares (1973) has effectively summarized the difficulties:

> Due to an improper conception of the problem by many governments, most of the decisions concerning marine science capability and marine resources exploitation are made by persons without proper knowledge of the problems. As a consequence, (1) marine sciences have not been considered with due priority; (2) poor planning (in educational, basic and applied research as well as technological development) and lack of continuity in programs occurs; (3) a notable shortage of well-trained personnel at the scientific level and of qualified technicians is present everywhere; (4) minimum efforts have been made for proper exploration and resource evaluation, and the necessary scientific background has been neglected, particularly in fisheries, without realizing that to create a scientific and technological capability is a long-term process which requires patience; and (5) a very small amount of funds has been invested.

It appears that low priorities accorded to marine science by governments of developing countries are to be explained primarily by two factors: lack of detailed, specialized knowledge about the adjacent ocean environment; and the low benefit/cost ratios characteristic of marine science transfer programs in the short run and the perceived greater productivity of scarce capital applied to alternate activities. It is for the latter reason that we recommend that training programs, under which we subsume marine science activities, always be linked to marine technology transfer programs, and that the latter in turn always be closely integrated into the educational structure of the recipient country. Assuming that sufficient resources exist in the adjacent marine environment to attract national governmental attention, the required technology transfer programs can be used as a lever for securing more resources allocated to basic marine science activities, and by tying these transfer programs to the local educational structure university education in the sciences will be stimulated, particularly marine science.

The approach to training programs advocated here is quite different from the education and training programs in marine science administered at least in the past by the UNESCO Office of Oceanography. We are not convinced of the utility of general fellowship programs without clear and specific domestic program links. We are equally unconvinced of the utility of constructing shopping lists of short-term courses thought, *in abstracto,* to be useful (IOC, 1971, 1972). Since we think that training and education in the marine sciences should always be linked to specific programs of technology transfer, it should be carefully differentiated according to the phased requirements of the transfer process and its objectives should be defined in terms of building a viable scientific base in a recipient country, increasing the management capacity of the recipient relative to the utilization of a variety of technologies, and increasing the skilled manpower of the recipient in terms of the operating and maintenance requirements of the technologies to be transferred.

The question of how this could be done has been addressed in detail by Liston (1979) and his recommendations are instructive. The basic principle is that "one must integrate training programs into the normal educational process of a country, so that they acquire a permanence and a clientele beyond the immediate group that short-term projects are often directed to." The initial international contact ought to be the result of people in the countries to be linked getting to know each other rather than the outcome of deliberate discussions by governmental delegations. Representatives of each institution should visit and assess each other and agree jointly on a plan to be followed. Students and faculty will have to be exchanged; this implies the development of some fluency in each other's language. The program should reflect the needs and fit the pattern of education of the recipient country. This implies that from the very beginning the assisting unit should set a realistic departure date and that "the program should be seen to be as much a product of the developing department's efforts as of the assisting department's" (Liston, 1974).

The job to be tackled by the two units linked in this fashion will usually involve staff training, curriculum development, creation or extension of a research infrastructure, and the initiation of research projects. Simultaneously, "there is also a need, expressed directly or indirectly, to assist government departments or industry with their technical problems in the interim period while the sister institution is itself developing" (Liston, 1974).

By adopting this approach, we don't claim that the gestation period for growth in recipient countries will be significantly reduced. This kind of activity remains essentially long term, difficult, and costly. This approach also has its costs since it requires the recipient to manage simultaneously a technology transfer program and the development of a basic capability in the education of indigenous marine scientists.

It should be expected (Vannucci, 1973) that the overhead costs of such an educational program will be high, since active indigenous researchers will be products primarily of the second generation. The first generation of graduates most likely will move relatively quickly into administrative and advisory positions, but this is by no means to be accounted a loss, since it is crucial that people with specialized knowledge be the ones to make continuing decisions on governmental priorities.

In additional to the kind of training program recommended here, certain other necessities will be required (Ross and Smith, 1977; Wooster, 1973, 1975). These will include:

Research funds and equipment to be used; ranging from vessels of different size and capacity to facilities for repair and calibration of instruments to funds for spare parts and travel of technicians to be trained in maintenance requirements;

The organism and participation of "interdisciplinary teams of scientists to assist developing countries upon request in planning and evaluation of programs; selection, procurement, use and maintenance of equipment; interpretation and application of data, etc.";

Travel funds and fellowships for advanced training of specialists outside of the country;

Funds for library resources and support of graduate students.

7 Conclusions

Four sets of major dynamics appear to shape the future of marine science in the United States. These are:

1. The disciplinary and interdisciplinary state of the art and the major theoretical problems which are to be solved;
2. The high costs of large-scale field programs, especially the cost of operating large research vessels;
3. Criteria for funding being applied by NSF and other federal supporters of oceanographic research. These may skew the direction of choices to be made if it is easier to raise funds for either laboratory research or applied research than for large-scale, theoretically derived, problem-oriented field investigations;
4. The change in the world ocean regime as it affects the conduct of marine scientific research.

With respect to the last set of dynamics, the United States has long maintained a large and varied international oceanographic research program at several levels and some adaptation will be required to respond to new conditions induced by the change in regime.

At the global level, the principal mechanism through which the United States has pursued cooperative marine scientific research projects has been the Intergovernmental Oceanographic Commission (IOC). But in the future, this mechanism will be useful for United States' purposes only when at least one of the following conditions is present (Ocean Policy Committee, 1982):

Where the scientific program to be investigated is clearly a global problem and cannot usefully be investigated on a lesser scale. In this situation, it is fairly easy to perceive long-term benefits to many different coastal and noncoastal states in the world, so the utilities in favor of cooperation tend to be high.

Where political problems make access to particular marine regions difficult and the informal umbrella mode of coordination infeasible. In these contexts, it is assumed that the work will be on smaller problems and that alternative regional umbrellas are either not available or are inappropriate. It is also assumed that the benefits which may accrue to a developing coastal state may be different from but complementary to those of advanced distant-water researching states.

Where internal funding procedures in a major maritime country allow new funds to be committed more easily if the IOC is seen to be the sponsor.

At the regional level, in a very few cases, a new kind of informal umbrella or formal intergovernmental arrangement may evolve, but only if certain conditions are present. As indicated previously, the specific geographical area must be very important scientifically and demand for access must be high and continuous among United States oceanographers. These arrangements will be difficult to negotiate and expensive to maintain. They will be long-term in nature and benefits to the coastal states in the region must be carefully thought out and clearly specified.

The *ad hoc* multilateral/informal umbrella-type arrangement is still the preferred approach to business because the choice of area is contingent upon its scientific importance, and face-to-face communication among scientists is primary, informal, and effective. Also the research agenda is insulated from external political conflicts, and governments see potential benefits sufficiently important to commit funds to support scientist-to-scientist contact, planning, logistic requirements, and execution of research. The change in the ocean regime, however, makes it necessary to meet the specific program interests of developing coastal states where areas controlled by these states are important to the project. Funds for assistance to developing coastal states may therefore represent additional requirements for projects of this kind. It should be no surprise if coastal state interests and programs are quite different from the objectives of the initial project.

At the bilateral level, a wide range of alternatives is available and informal arrangements are preferred. Formal bilaterals should be used as sparingly as the potential regional approach described above and for the same reasons; the area must be important scientifically and demand for access must be high and continuous. Additionally, government-to-government arrangements should be sought only if existing political problems are so difficult that research is held hostage to them. Alternatively, if political problems are not major but it is not possible to get the necessary funding to respond to the interests and priorities of the coastal state without a governmental commitment, then a formal bilateral should be sought. Formal bilaterals which are driven primarily by other foreign policy interests of the U.S. Government are not often the most efficient means of getting research done which scientists think important.

References

Aguilar, A. 1973. The patrimonial sea. In: The Law of the Sea: Needs and Interests of Developing Countries (L. Alexander, Ed.). Law of the Sea Institute, University of Rhode Island, Kingston, R.I.

Ayala-Castañares, A. 1973. The enhancement of marine science capabilities: Future directions. In: Report of the Marine Science Workshop held by the Johns Hopkins University, Bologna, Italy, 1973.

Franssen, H. T. 1974. Understanding the ocean science debate. Ocean Devel. Intl. Law J. 2, 187–202.

IOC (Intergovernmental Oceanographic Commission). 1971. Twelfth meeting of the bureau with the consultative council: Draft report of the joint session of the IOC Working Groups on Training and Education in Marine Science and on Mutual Assistance. Doc. No. IOC/B-86/Add. 1 (INF).

IOC (International Oceanographic Commission). 1972. First session of the Working Group on Training, Education, and Mutual Assistance. Selected Training Projects in Marine Science and Related Subjects, Doc. No. IOC/TEMA-1/6.

Johnston, D., and E. Gold. 1973. The economic zone in the Law of the Sea: Survey, analysis and appraisal of current trends. Occasional Paper No. 17, Law of the Sea Institute, University of Rhode Island, Kingston, R.I.

Knauss, J. 1973. Developing the freedom of research issue of the Law of the Sea Conference. Ocean Devel. Intl. Law J. 1, 93–120.

Knauss, J. 1974. Marine science and the 1974 Law of the Sea Conference. Science 184, 1335–1341.

Liston, J. 1974. Operating difficulties of overseas programs and some suggested solutions. In: U.S. Marine Scientific Research Assistance to Foreign States: Proceedings. National Academy of Sciences, Washington, D.C.

Miles, E. 1976. The dynamics of global ocean politics. In: Marine Policy and the Coastal Community (D. Johnston, Ed.). Croom and Helm, Ltd., London.

Miles, E. (Ed.). 1977. Restructuring ocean regimes: implications of the Third U.N. Conference on the Law of the Sea. Intl. Organiz. 31, 2.

Ocean Policy Committee, Commission on International Relations, National Research Council. 1982. *United States Interests and Needs in the Coordination of International Oceanographic Research.* National Academy Press, Washington, D.C., pp. 87–89.

Ross, D., and L. Smith. 1977. Training and technical assistance in marine sciences: a viable transfer product. In: International Transfer of Marine Technology: a three-volume study (J. Kildow, Ed.). MIT Sea Grant Program, Rept. No. 77-20.

Scholz, W. S. 1979. Oceanic research: international law and a survey of national legislation. Law and Marine Affairs Program, School of Law, University of Washington. Unpublished manuscript.

United Nations. 1973. Report of the African States Regional Seminar on the Law of the Sea, Yaounde. U.N. Document No. A/AC 137/79.

Vannucci, M. 1973. Developing capabilities in oceanography in developing countries. AAAS/CONAYCT Meetings, Mexico City. Unpublished manuscript.

Wooster, W. S. 1973. Marine Science and the Developing Countries. In: Report of the Marine Science Workshop. Johns Hopkins University, School of Advanced International Studies.

Wooster, W. S. 1975. Problem of Marine Science Research Assistance to Foreign States. In: Proceedings, U.S. Marine Scientific Research Assistance to Foreign States. Ocean Policy Committee, Ocean Affairs Board, National Academy of Sciences, Washington, D.C.

Wooster, W. S. 1976. The decline of marine scientific research. Marine Technology Society J. 11(2), 23-36.

Wooster, W. S. 1979. Some comments on the funding of the academic research fleet. Unpublished manuscript.

Part IV
The Human Scale

Changing Global Biogeochemistry

Bert Bolin

1 Introduction

The complex geochemical interactions on earth have been changing throughout the history of our planet and are changing today. These changes are, however, generally slow and unnoticeable by the individual; basically, we experience our global environment as steady. We know the features of this quasisteady state only in their broad outline; particularly we have incomplete knowledge about the key physical, chemical, and biological processes that maintain it.

Today man is interfering with this global system. The rate of fossil carbon burning is about 10% of the rate of net carbon assimilation by terrestrial plants. The nitrogen fixed by combustion and in manufacturing fertilizers is almost half of what plants produce naturally. Man is indeed conducting a global geochemical experiment and we do not know what the long-term impacts may be. This problem has not been considered adequately; this is understandable since the ultimate consequences are poorly understood. If changes of the biogeochemical cycles occur, however, they may well be long lasting, hundreds of years or more. To understand the principal problems of this kind is a difficult and time-consuming task. The marine sciences will play a central role in the interdisciplinary efforts that are required. Rather than discuss the general role of the oceans in biogeochemical cycles, we shall consider some specific problems encountered in developing a consistent picture of the carbon cycle. In doing so we shall consider a number of processes and interactions that are of relevance when concerned with other constituents in sea water. In this way we hope some principal features of ocean biogeochemistry will be outlined and some specific problems dealt with.

2 Modeling Biogeochemical Cycles

We need to know the main reservoirs in nature that interact on the time scale with which we are concerned (less than about 10,000 years) and the magnitude of the fluxes between them. Figure 1 provides such an overview of the carbon cycle. Many of the estimates that go into such a flow chart are quite crude, and in order to get an internally consistent picture a steady state is often assumed. Since the time scale involved in establishing a steady state for the carbon cycle is several hundred years, this implies an assumption of undisturbed conditions for at least about 1000 years. (This is a questionable assumption in view of the natural climatic variations, such as the Little Ice Age, that have occurred during the last millenium. Also, man has been interacting with the terrestrial biosphere significantly on a global scale for several hundred years in a manner about which we do not know much quantitatively.) We shall see later that such assumptions about a steady state may have led to incorrect conclusions about the capacity of the world oceans to serve as a sink for fossil fuel carbon dioxide. The changes of climate since the last glaciation may well imply that at no time during this period have the general circulation of the oceans and the distributions of chemical constituents in the sea been in a steady state. To find out whether this was so or not remains a major problem of global biogeochemistry.

A description of the carbon cycle as shown in Figure 1 is, of course, a very crude picture of reality. It summarizes in a consistent manner the sizes of the reservoirs and fluxes that are involved, but such an overview cannot serve as more than a first crude description of the dynamics of the system. To be able to understand in what way man may be changing the natural cycles we need to describe quantitatively the key physical, chemical, and biological *processes* that govern these cycles. This in turn requires the study of small-scale (small in both space and time) processes, which however, cannot be included in all their detail in any global model. The problem is very similar to the one that meteorologists and climatologists are confronted with in developing climatic models. Undoubtedly we need similar, more detailed, models in dealing with the biogeochemical cycles, and an appropriate model of the world oceans is fundamental in order to resolve with some confidence the central problems in which we are interested.

Still, simple models for the global cycles are important to investigate fundamental mechanisms for interplay and to keep the central questions in the focus of our attention. Also, models for selected regions are needed to work out schemes for the parameterization of microscale or mesoscale processes and to incorporate them properly into global models. As will be apparent later, it may be very misleading to introduce the mathematical descriptions of processes at the microscale directly into a global model.

Even though simple biogeochemical models may be useful as tools in advancing our understanding, they should be used with care when trying to predict future likely changes. Rather than a single, most likely, scenario based on

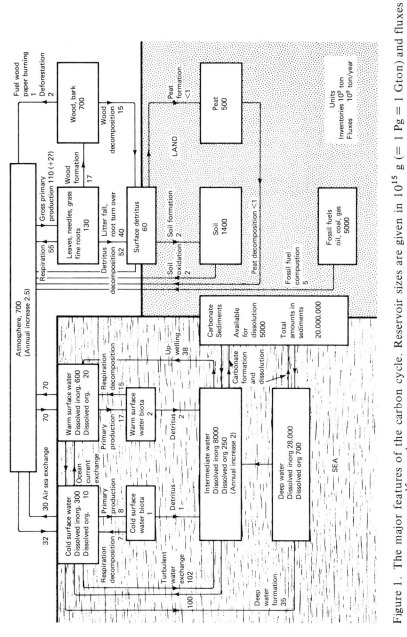

Figure 1. The major features of the carbon cycle. Reservoir sizes are given in 10^{15} g ($= 1$ Pg $= 1$ Gton) and fluxes between reservoirs in 10^{15} g yr^{-1}. For a discussion of the uncertainties in these estimates reference is made to Bolin et al. (1979).

such a simple model, a range of uncertainty should be given. More advanced models which can be validated with real data can gradually narrow the degree of uncertainty in predicting future changes.

Study of the interaction between the carbon and the fundamental nutrient cycles is necessary for a proper understanding of the global ecosystem. Its development over millions of years has established an intricate interdependence of these cycles, the understanding of which is fundamental to our understanding of the ecosystem's response to external impacts as now caused by man. The stability of the global ecosystem is governed by the mutual interplay between these cycles.

3 Terrestrial Processes

The carbon cycle is of great concern because the amount of carbon dioxide in the atmosphere has risen about 21 ppm (from 315 to 336 ppm) during the years 1958 to 1979 and by at least twice that amount since the latter part of last century (Bacastow and Keeling, 1981). Manabe and Stouffer (1980) have concluded, from extensive computations with large climate models, that a doubling of the amount of CO_2 in the atmosphere might lead to an increase of the average global temperature by 2 or 3°C. Even though the inertia of the oceans might considerably delay such a climatic change the problem obviously deserves our urgent attention. A summary of our present view of the global carbon cycle has been given by Bolin et al. (1979) and the models developed for predicting the likely future changes are summarized by Bolin et al. (1981b). In this section, we shall consider a few key problems dealing with the terrestrial part of the system.

Numerous experiments show that plants grow more rapidly in a CO_2-enriched atmosphere, if nutrients, water, or radiation are not limiting factors for such enhanced growth. This is particularly true for C-3 plants, while C-4 plants seem to be closer to their optimum rate of growth already at an atmospheric CO_2 concentration of about 300 ppm. Such a response is primarily associated with the fact that an increased transfer of CO_2 through the stomata can take place, while the counterflux of water vapor remains the same. The water economy of the plants is improved.

In trying to include a response of the terrestrial biota in carbon cycle models it has been common to assume that the net primary production is proportional to some power, β, of the atmospheric concentration and also to the amount of photosynthesizing matter (Keeling, 1973). An increase of the net primary production would then gradually become balanced by decomposition of the increased detrital matter stored. Such an approach is inadequate if dealing with the global problem. The global ecosystem consists of a mosaic of subsystems, that can be classified in a set of biomes (Whittaker and Likens, 1975). Their biomass is primarily determined by climate (temperature, precipitation, solar radiation, etc.) and by soil characteristics, primarily the availability of nutrients. The natural ecosystems have developed toward a climax state, which usually is close to the maximum values of biomass and net primary production that can be maintained under given climatic conditions with the plant species of which the given biome is composed. Even though an increasing atmospheric CO_2 concentration might possibly increase the rate of photosynthesis, it might mean merely an increase of the turnover rate of living matter, not necessarily an increase of the biomass of the biome.

From the point of global modeling the essential question therefore is in which way the climax state of the subsystems may be modified due to a changing amount of atmospheric CO_2. Since more CO_2 in the atmosphere improves the water economy of the plant communities, we might obtain an approximate idea of a quantitative relationship of this kind by studying how the biomasses of different biomes vary with precipitation under otherwise similar environmental conditions. In view of the marked dependence of the biomass of many major terrestrial biomes on availability of water, it seems quite plausible that the increase of CO_2 in the atmosphere that may have begun more than 200 years ago (see below) has led to an increase of carbon storage in parts of the terrestrial biosphere not yet exploited by man.

The brief considerations above show the danger of attempting to apply a relationship applicable at the microscale to a global model that describes the gross features of the ecosystem. The proper parameterization with regard to the ecological relations that exist at subsystem level must not be disregarded.

During the last few years there have been many attempts to estimate the magnitude of the changes of the terrestrial ecosystems due to man's impact (Hampicke, 1979). More recently Moore et al. (1981) have begun an inventory of available data on forestry and agriculture and developed a model in which the processes of regrowth and gradual change of the soils due to increased oxidation of organic matter are considered. The world has been divided into a number of geographical regions to permit the proper use of national statistics, and in each of these the major biomes have been dealt with separately. Response functions of the terrestrial biota and the soils to man's interference have been assigned to each of these. Using available information on deforestation, agricultural expansion, and these response functions, Moore et al. (1981) have estimated the annual net emission from the terrestrial biota to the atmosphere for the period 1860 to 1975 (see Fig. 2). The possible increased assimilation in natural forests as discussed in the previous section was, however, not accounted for in these integrations. A rather rapid increase of net emissions of carbon dioxide to the atmosphere seems to have occurred during the latter part of the last century and a slower one during the twentieth century. These assessments have later been extended back to the eighteenth century as more data have become available (Moore, personal communication). New integrations show a more gradual increase in the late nineteenth century, which is a result of starting the integrations earlier.

Figure 2 also shows the emissions of fossil fuel carbon by combustion (Rotty,

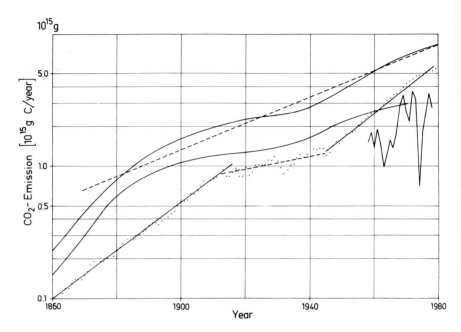

Figure 2. Rate of transfer of carbon to the atmosphere due to fossil fuel combustion (solid-dashed line and dots, according to Rotty, 1981), due to deforestation and expansion of agriculture (lower full line, according to Moore et al., 1981), and the total emissions (upper full line) during the years 1860 to 1980. The straight dashed line corresponds to an annual increase of emissions by 2.3%.

1981). They were negligible before 1860 but have risen quickly and are today about 5.2×10^{15} g yr^{-1}. The solid curve in Figure 2 gives the total emissions of carbon dioxide to the atmosphere as a function of time. The best fit of this curve by an exponential function yields an annual increase of 2.2%, compared to about 4% for the increase of fossil carbon emissions during the last 30 years. This difference in rates of emissions is important when we are concerned with the response of the oceans to an increasing amount of CO_2 in the atmosphere.

According to Moore et al. (1981) the total emissions due to changing land use can be estimated as about 150×10^{15} g C for the period 1860 to 1970 (or approximately 170×10^{15} g C if extending the period to 1980). Their later analyses (Moore, personal communication) yielded a smaller value for this period of time, but emissions before 1860 bring the total man-induced emissions of terrestrial carbon to a value of about 200×10^{15} g. The possibly increased storage of carbon in the unexploited forests due to larger atmospheric CO_2 concentration might mean a smaller net transfer from the terrestrial biosphere to the atmosphere, but we do not know how much less. An overestimate may also be due to the assumption of complete combustion during burning in the process of deforestation, which is probably not correct (Seiler and Crutzen, 1980).

4 The Role of the Oceans in the Carbon Cycle

The distribution of ^{14}C in the oceans has been the main data base for determination of the rate of ocean circulation when designing a global model for the carbon cycle. This rate is obviously of fundamental importance for determining the response of the oceans to an increasing amount of carbon dioxide in the atmosphere. The very simple models used so far (Bacastow et al., 1981) are, however, inadequate for a more precise assessment. The extensive GEOSECS observations of ^{14}C and a large number of other chemical constituents in the oceans provide an excellent basis for a validation of the adequacy of more advanced models.

Figures 3 and 4 show north-south Δ^{14}C distributions in the Atlantic and Pacific Oceans (Stuiver and Östlund, 1980; Östlund and Stuiver, 1980). Similar cross sections are available for temperature, salinity, and a larger number of other elements (also for the Indian Ocean). Of particular interest in the present context are of course those of dissolved inorganic carbon (DIC), carbonate alkalinity, oxygen, and phosphorus.

We should recall that measurements before significant amounts of bomb-produced ^{14}C had been transferred into the oceans yielded values of about $\Delta^{14}C = -40\ ^{\circ}/_{\circ\circ}$ for most surface waters except in Antarctic regions (Broecker et al., 1960). An obvious and significant increase has taken place. Östlund et al. (1976) and Stuiver (1980) have studied the penetration of bomb-produced ^{14}C into the oceans, also making use of tritium which has been almost exclusively injected by nuclear testing. They show that all surface waters, the top layers of the intermediate waters, and significant volumes of water in regions of deep water formation in the North Atlantic have been affected. The penetration into the Antarctic Ocean is more difficult to assess. The surface layers are clearly affected, and in view of the rather effective vertical exchange (see below) a significant amount has probably also been transferred into the deep waters. If the changes since before nuclear testing can be determined adequately, important information regarding the circulation and transfer rates in the top layers of the oceans can be deduced.

The distributions of Δ^{14}C in the intermediate and deep layer of the oceans (Figs. 3 and 4) obviously reflect many features of the general circulation of the oceans. The southward flowing Deep Atlantic Water is clearly recognizable between the "older" Atlantic Intermediate and Bottom Waters penetrating northward from the Antarctic. Northward flux into the Pacific Ocean (and similarly into the Indian Ocean) occurs in the bottom layers with return flow in the Pacific Deep Waters above, which is also qualitatively consistent with the observed Δ^{14}C distribution. In view of the many elements for which there are accurate measurements, it is tempting to revive Riley's (1951) idea, to deduce the ocean circulation by the simultaneous use of a number of tracers, which was tried in a very simplified manner by Keeling and Bolin (1968). There are, however, many obstacles to such an approach. The development of a dynamic theory of the world ocean circulation as initiated by Stommel and Arons (1960) and

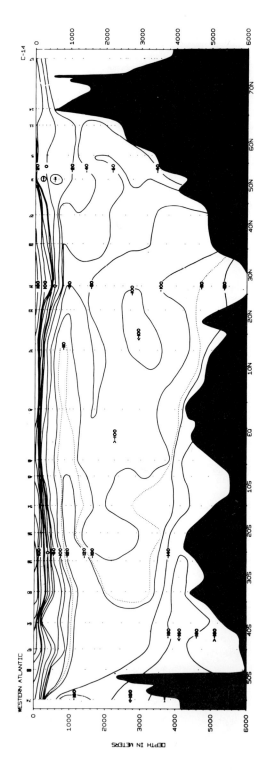

Figure 3. $\Delta^{14}C$ distribution (°/oo) in the Atlantic Ocean as obtained from the GEOSECS project (from Stuiver and Östlund, 1980).

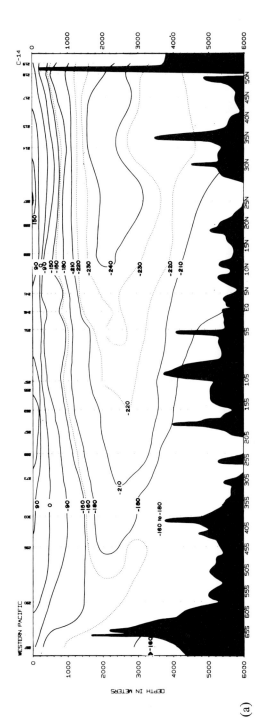

Figure 4. (a) $\Delta^{14}C$ distribution (‰) in the Western Pacific Ocean as obtained from the GEOSECS project (from Östlund and Stuiver, 1980).

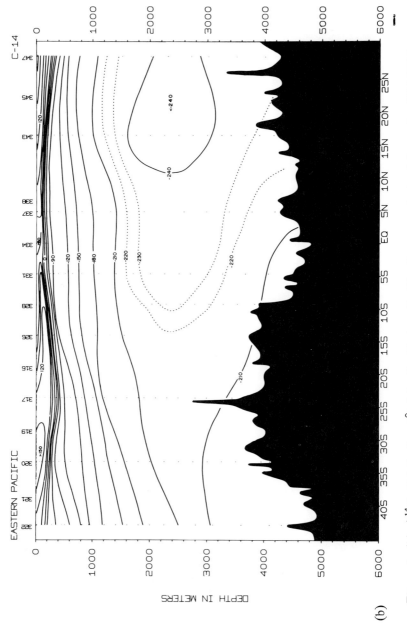

Figure 4. (b) $\Delta^{14}C$ distribution (‰) in the Eastern Pacific Ocean as obtained from the GEOSECS project (from Östlund and Stuiver, 1980).

lately generalized by Veronis (1978) clearly shows the fundamental role of boundary currents for the transfer of both heat and matter. This general result has been confirmed by many observational studies. Also the GEOSECS data show similar features (Fig. 5), where the penetration of the Antarctic Bottom Water into the Pacific Ocean along the western boundary is clearly revealed by the distribution of $\Delta^{14}C$. To account for the complex oceanic flow field and deal with vertical motions and turbulent fluxes properly requires the use of oceanic general circulation models (Bryan, 1975).

Even if we ultimately need to resort to such complex models of the ocean circulation to deal with the transfer of matter in the sea, many particular problems may be studied with simple models. Any more elaborate study should be preceded by the use of simplified models to formulate properly the more complex experiments. In trying to assess the role of the oceans as a sink for excess atmospheric carbon dioxide, all details of the ocean circulation may not be essential. A more reliable answer than provided by simple two-box or box-diffusion models (Bacastow et al., 1981) can be obtained by approximating the gross features of the ocean circulation by a somewhat more detailed but still

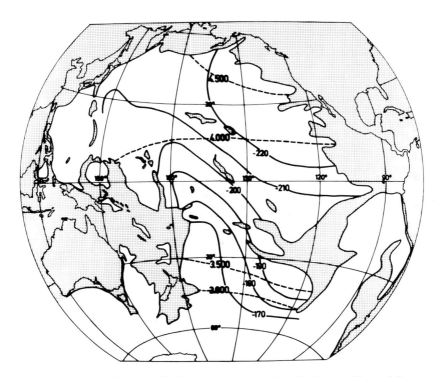

Figure 5. Depth of the $\sigma_\theta = 27.81$ surface in the Pacific Ocean (dashed lines; unit m) as an approximate boundary between the Pacific Bottom Water (below) and the Pacific Deep Water (above) and isolines for the mean $\Delta^{14}C$ value for the Pacific Bottom Water (solid lines; ‰).

coarse representation with rather few degrees of freedom. Before proceeding to the design of a global ocean model we shall, however, analyze some specific processes that we need to account for in such a model development.

Peng et al. (1979) have determined the rate of gaseous exchange between the atmosphere and the sea by measuring the rate of radon evasion (Fig. 6). Their results imply a mean exchange rate for carbon dioxide of about 20 mol m^{-2} yr^{-1}, which value agrees well with the one that can be deduced by assuming a steady state and equating the flux of ^{14}C into the sea with the decay of ^{14}C in sea water and ocean sediments. This value corresponds to a residence time for carbon in the atmosphere of approximately 7 years. A layer of about 70 m of ocean water (the average depth of the surface mixed layer) contains about the same amount of carbon (dissolved inorganic) as the atmosphere above it and the mean residence time for carbon in such a layer before transfer to the atmosphere also is about 7 years. Since the transfer of ^{14}C from the atmosphere to the sea is a first-order exchange process, the characteristic adjustment time between

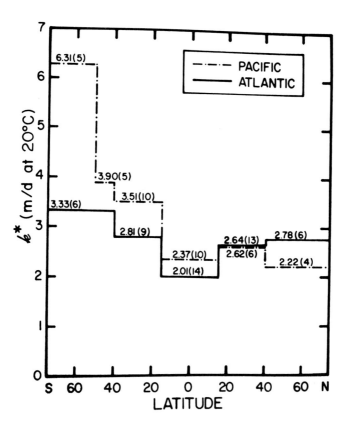

Figure 6. Rate of exchange of ^{222}Rn between the atmosphere and the sea expressed as the amount of Rn at saturation in a water column (depth in m) released per day (from Peng et al., 1979).

the two reservoirs for a disturbance of the ^{14}C amount in any one of these two reservoirs, therefore, also is about 7 years.

In the case of CO_2 transfer as a response to a disturbance of the equilibrium between the CO_2 partial pressures in the gas phase and sea water, the situation is different because of the buffering of the carbonate system of the sea. This means that the change of the partial pressure that is required in order to maintain equilibrium is ξ times larger than the change of the amount of dissolved inorganic carbon in the water. The buffer factor or Revelle factor, ξ, is about 10 at $+27°C$ and 14 at $0°C$ for present amounts of inorganic carbon in the sea. Therefore, only a small amount of excess CO_2 in the atmosphere need be transferred into the oceans to establish a new equilibrium. The total capacity of the oceans as a sink for excess atmospheric carbon dioxide is thereby reduced by a factor which is equal to ξ, while the equilibration between the CO_2 concentration in the atmosphere and the dissolved inorganic carbon in the sea is ξ times faster than that of ^{14}C—not 7 years, but rather three-quarters of a year. As will be seen, this difference is important for the interpretation of the ^{14}C data from the Antarctic Ocean. Peng et al. (1979) also found that the air-sea exchange is more rapid in polar regions than at lower latitudes (Fig. 6), particularly around Antarctica.

Peterson and Rooth (1976) have studied the water exchange in the Greenland Sea and the Norwegian Sea using ^{14}C, tritium (produced by nuclear tests) and other chemical tracers. They concluded that the time scale for deep convective mixing in the Greenland Sea is around 30 years, while vertical exchange in the Norwegian Sea is small and the exchange time of deep water between them is around 100 years. The renewal time for water in the Arctic Ocean however, is not known.

The overflow of Arctic Deep Water into the Atlantic Deep Water has also been studied using ^{14}C and tritium (Östlund et al., 1976). The rate obtained, 0.4×10^{15} to 0.8×10^{15} m^3 yr^{-1}, agrees quite well with more direct estimates by Worthington (1976). It is important to emphasize here that the penetration of Deep Arctic Water into the Atlantic is not primarily compensated for by surface water inflow into the North Polar Sea, since the nutrient balance of the Arctic waters would then not be maintained (Broecker, 1979). There must also be a return flow of Atlantic Deep Water into the Arctic Basin.

Broecker (1979) has reanalyzed the rate of renewal of deep water in the western basin of the Atlantic. Even though the quantitative treatments of the mixing process of southward flowing Deep Atlantic Water and northward flowing Atlantic Bottom Water needs further elaboration, it seems clear that the renewal time of water relative to the supplying water bodies in the north and south is merely about 100 years. In this analysis, use was made of ^{14}C, SiO_2, "NO," alkalinity, and oxygen, but the vertical diffusion from the Atlantic Intermediate Water above was disregarded.

As will be seen later, the exchange processes in Antarctic waters are crucial for a proper understanding of the overall CO_2 and ^{14}C exchange between the atmosphere and the sea. The three major oceans are interconnected by the

Antarctic Circumpolar Current, which is primarily maintained by wind stress exerted by the strong atmospheric westerlies between 40°S and 65°S. Due to this stress a northward Ekman drift is created in the ocean surface layer. To the south of the latitude of maximum wind, at about 50°S, an area of divergence and upwelling is established. This is further enhanced by a southward Ekman drift toward the Antarctic Continent caused by the easterly currents close to the continent.

Gordon (1971) has estimated the upwelling as about 1.7×10^{15} m^3 yr^{-1} to the south of latitude 53°S. The buoy measurements of sea-level pressure, and thus the geostrophic wind at these latitudes during the First Global Atmospheric Research Program (GARP) Global Experiment (FGGE) reveal, however, that previous estimates of surface winds on the average may have been 30% too low. Since the wind stress is dependent on the square of the wind, the stress creating the Antarctic Circumpolar Current may have been underestimated by a factor of two, and accordingly also the associated Ekman drift and the rate of upwelling.

Close to the Antarctic Continent, particularly in the Weddell Sea and the Ross Sea, Antarctic Bottom Water is formed during the fall and early winter months, when freezing creates exceptionally cold and saline water. Gordon (1971) has estimated the rate of water sinking to a depth below 100 m during the fall and winter as 35×10^6 m^3 sec^{-1} (the annual rate is about 0.5×10^{15} m^3 yr^{-1}). The details of the bottom water formation in the Weddell Sea have been studied by Weiss et al. (1979) using chemical tracers. Of particular importance here is their finding that sinking water is formed from upwelling deep water coming from the north, and that the time this water spends in the surface mixed layer is too short for equilibration of the ^{14}C content of the water with that of the atmosphere. The fact that these waters are ice covered during part of the year also contributes to their finding that an equilibration is not established. Assume that upwelling and deep water formation occurs in an area of 20×10^6 km^2 (Gordon, 1971), and that the upwelling water has a Δ^{14}C value of 160 °/oo, while that of the atmosphere is 0 (pre-nuclear-testing conditions). Figure 7 then shows the Δ^{14}C difference between upwelling and downwelling water as a function of air-sea CO$_2$ exchange rate and rate of upwelling. A rate of air-sea exchange larger than 20 mol m^2 yr^{-1} is likely in cases of strong winds, while an ice cover may imply considerably lower values. The diagram shows clearly that an equilibration with the Δ^{14}C values in the atmosphere is not achieved.

The surface water flowing northward from the Antarctic divergence zone primarily sinks in the frontal zone at about 45°S and forms the Antarctic Intermediate Water which occupies the region between the surface mixed layer and the Atlantic and Pacific Deep Water found below about 1200 m. This process of water formation is obvious from the salinity distribution. The rate of water intrusion into this layer and northward motion is elucidated by the distribution of ^{14}C. Along the level of minimum salinity the bomb-produced ^{14}C has clearly affected the distribution northward in the Atlantic to at least 35°S. The minimum Δ^{14}C values are found at 1500- to 1000-m depth in the South Atlantic

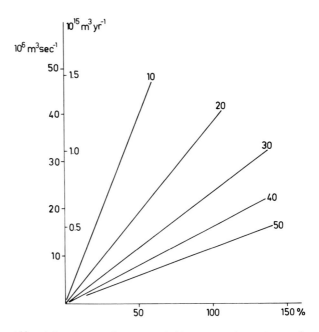

Figure 7. Difference in $\Delta^{14}C$ (%) of upwelling and sinking water in an area of 20×10^6 km² around the Antarctic Continent as a function of rate of water overturning (in 10^6 m³ sec⁻¹) and rate of gross air-sea CO_2 transfer.

(25° to 45°S) at about 750 m at the equator due to upwelling where $\Delta^{14}C = -115$ ‰. There is also, most likely, some vertical mixing across isopycnal surfaces as emphasized by Broecker and Östlund (1978). As a matter of fact, it is difficult to resolve the relative importance of vertical and quasihorizontal (along isopycnal surfaces) transfer even if also considering other tracers (tritium, SiO_2, "NO"). It seems plausible that the prebomb $\Delta^{14}C$ value for Antarctic Surface Water from which the Antarctic Intermediate Water partly originates was about -110 ‰, perhaps even somewhat lower.

In reviewing these regional studies, one is struck by the absence of almost any attempt to tie them together into an internally consistent pattern of global exchange. Obviously, requirements for overall internal consistency of distributions of tracers and transfer patterns will yield further insight into the relative importance of various mechanisms at work. For example, the transfer of ^{14}C from the ocean surface to the deep waters in the Arctic Basin must be consistent with the transfer across the Greenland-Iceland-Faero sills into the Atlantic (Broecker, 1979). Similarly the large ^{14}C flow from the atmosphere into the Antarctic Surface Water calls for adequate transfer mechanisms into the Deep Antarctic Waters and further into the interior of the major ocean basins.

As has been emphasized before, such studies ultimately will require elaborate models to provide more precise answers. A preliminary attempt using a rather crude model is still of interest, particularly with regard to the role of the oceans as a sink for fossil fuel carbon. This integral feature of the oceans seems not to be dependent on the precise magnitude of the more specific assumptions that we shall make, lacking a dynamical quantitative synthesis of the ocean circulation. A full account of such an analysis is given elsewhere (Bolin et al., 1983).

The world oceans are divided into 13 major reservoirs on the basis of classical definitions of water masses in the oceans, as used in the previous discussion. Since ^{14}C data for the Indian Ocean were not available at the time this work was done, the Pacific and Indian Oceans were combined into one major ocean, which seems acceptable in a first exploratory study (Fig. 8). Δ^{14}C values have been assigned to each reservoir in an attempt to depict the approximate steady state that presumably existed before injections of bomb-produced ^{14}C. We have further assumed that the air-sea exchange per unit area is 50% higher at high latitudes than in middle and low latitudes. On this basis, using the Δ^{14}C values assigned to the surface reservoirs and considering the area they cover, we compute the relative magnitude of the net ^{14}C transfer into the world ocean basins as shown in Table 1.

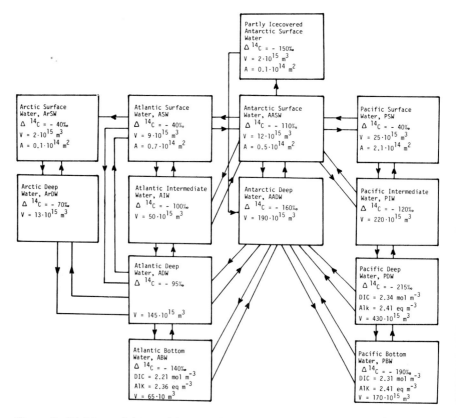

Figure 8. Division of the world oceans into major water masses; the Δ^{14}C values; volumes (V) and surface areas (A) assigned to these and the modes of water exchange between them.

Table 1. Percentage distribution of prebomb ^{14}C flux into world oceans

	Area (%)	Flux (%)
Pacific Ocean	47	31
Atlantic Ocean	20	13
Indian Ocean	15	10
Antarctic Ocean	15	42
North Polar Sea	3	4

We note that about 40% of this transfer is into the Antarctic Surface Water, even though the area is only about 15% of the total ocean area. Obviously, this value is rather uncertain, as long as the degree to which the air-sea exchange in reality is enhanced in polar regions is not well known. If we accept this value it also follows, from an estimate of the ^{14}C balance of each individual ocean, that a transfer of ^{14}C from the Antarctic Ocean in all three major oceans must occur in order to maintain a steady state, unless the air-sea $\Delta^{14}C$ difference in north Polar regions and the North Atlantic were larger than assumed here or the air-sea exchange rate significantly more than 50% enhanced as compared to low latitudes. The data presented by Peng et al. (1979) do not support such an assumption. It is possible to find a circulation pattern between the Antarctic Sea and the Atlantic Ocean, involving the northward flowing Atlantic Intermediate Water, whereby the $\Delta^{14}C$ balance of the Atlantic is maintained (Bolin et al., 1983). The result is a ventilation time of the Atlantic Deep Water flowing southwards which is of the order of 150 years, somewhat more than the 100 years obtained by Broecker (1979). There is an intense vertical exchange in the Antarctic Ocean, which corresponds to a vertical diffusivity of about 8 cm^2 sec^{-1}, which is almost 10 times the value for vertical turbulent exchange in the thermocline region (Munk, 1966). It is clear, however, that these results are dependent on the assumption of the latitude distribution of air-sea exchange.

The transfer of ^{14}C within the oceans is due both to water motions and settling of detrital matter. The contribution by the latter is, however, only minor (Keeling, 1973; Björkström, 1979). In this first computation we neglect it, when deducing the ^{14}C balance. (It is, of course, not permissible to do so when later concerned with the exchange of dissolved inorganic carbon, oxygen, or the alkalinity balance). On the basis of the $\Delta^{14}C$ distribution as given in Figure 8, assuming ^{14}C balance, water continuity, and the general features of the ocean circulation as outlined previously in this section, it is possible to deduce an internally consistent set of advective and turbulent rates of water exchanges. As a matter of fact, the problem is not fully determined and a range of uncertainty remains which has not yet been fully explored. The constraints on the solution are rather restrictive. Further details will be found in Bolin et al. (1983).

The change of the $\Delta^{14}C$ distribution in the sea since 1956 due to nuclear testing provides an additional set of data which can be used for testing the results deduced above. Figure 9 shows the $\Delta^{14}C$ values in the atmosphere in the

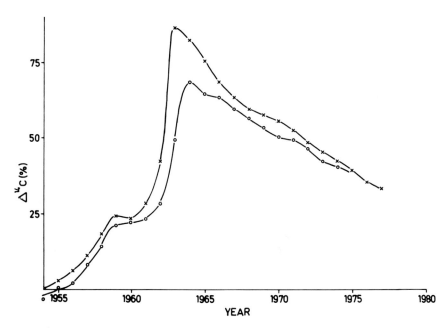

Figure 9. The ^{14}C concentration in the atmosphere (Δ^{14}C, in %). Crosses give values from the Northern Hemisphere and the circles from the Southern Hemisphere (after Nydal et al., 1978).

Northern and Southern Hemisphere since 1956. Almost all emissions to the atmosphere due to nuclear bomb testing occurred before 1962, and since then the ^{14}C concentrations in the atmosphere have steadily declined. Using the air-sea exchange rates as described above and the exchange rates within the oceans as deduced, we can compute the ^{14}C flux to the oceans as a result of the enhanced atmospheric concentrations, and also obtain the distribution among the ocean reservoirs. A comparison of such computations with the observed changes since 1963 as given by Stuiver (1980) provides an independent validation of the transfer rates as deduced on the basis of a steady state before 1956, when testing began. The data are, however, too limited to determine any changes that may have occurred in the deeper reservoirs. As a matter of fact, the GEOSECS data set is the only one that gives a reasonably reliable distribution below the surface layers and the upper part of the thermocline region. It is obviously very desirable to repeat the GEOSECS measurements sometime during the 1980s.

A comparison of the computed transfer of ^{14}C into the oceans and the observed changes in the atmosphere since 1963, with due regard to cosmic ray production and decay, permits us to estimate the net transfer into the terrestrial biota, humus, and the soils. Even though the results will be rather uncertain, it should yield some information on the rate of photosynthesis on land, which value may be compared with those estimated from observations (Whittaker and Likens, 1975).

Experiments with the model as outlined above have not yet been completed. We shall, however, present some preliminary results obtained with regard to the role of the oceans as a sink for excess carbon dioxide in the atmosphere. For an increase of the emission to the atmosphere by 2% per year, which is the approximate average for the last 100 to 150 years, we find that the oceans may take up 35 ± 5%. If we assume that the net transfer from the terrestrial ecosystems to the atmosphere has been 125 ± 50 Pg (1 Pg = 1 Gton), thus accounting for the likelihood that these still may have been overestimated (see above), the total emissions to the atmosphere during this time have been 270 ± 60 Pg, of which 180 ± 60 Pg would have stayed in the atmosphere. This corresponds to an increase of the atmospheric concentration of CO_2 by 85 ± 30 ppm (i.e., the undisturbed value would have been 255 ± 30 ppm). Before accepting such a low value, one should also determine how well an emission scenario as given in Figure 2 (or modified as indicated above) and the present ocean model can reproduce the increase in the atmosphere as observed during the last 23 years (Bolin et al., 1981b). We note finally that Brewer (1978) has deduced that the amount of dissolved inorganic carbon in the intermediate water of the Atlantic Ocean at the time of formation (about 100 years ago) may have been about 250 ppm. The uncertainty in this computation is, however, considerable.

There is one further important result that emerges from the previous discussion. The more rapid ventilation of the Atlantic Ocean and the Antarctic Deep Water indicated in these computations means that these reservoirs already may have been significantly affected by the net transfer of CO_2 to the sea. An average increase of the total amount of inorganic carbon of about 0.5% may already have occurred. This in turn means that some disturbance of the quasiequilibrium between ocean water and the sediments has occurred. How this influences the storage capacity of these reservoirs depends on how rapidly this disequilibrium is restored, about which our knowledge is limited. It should, however, be possible to assess this affect approximately using the present ocean model.

5 Conclusion

The preliminary revision of the carbon cycle, as described in the previous sections, is the outcome of a reanalysis of the role of the ocean circulation in determining the capacity of the oceans as a sink for excess atmospheric carbon dioxide. Many possible refinements of such computations should increase the reliability of the results. More detailed numerical ocean circulation models should be employed. Equally important is the use of other tracers to check the internal consistency of the computations. A preliminary calculation of the flux of particulate biogenic matter by using the GEOSECS data for dissolved inorganic carbon and alkalinity yields the result that the flux of organic carbon from the surface layers to the deep sea is about 4×10^{15} g yr^{-1} and that of carbonate carbon about 1×10^{15} g yr^{-1}. These results should be checked for consistency with the phosphate and oxygen distributions. Radium-226 provides another interesting possibility, since the source region is at the sea floor and since enrichment of ^{226}Ra relative to Ca occurs in the assimilation into biogenic matter. ^{226}Ra should also provide information on exchange rates of soluble constituents between the sediments and the sea water. This is particularly important when attempting a more careful analysis of $CaCO_3$ from ocean sediments. The close interdependence of the distribution of chemical constituents in the sea and the ocean circulation is obvious. Since direct measurements of water motions are difficult, and their representativeness is unclear, the integrated effects of the water motions as displayed by the GEOSECS chemical observations represent important data for verification of dynamical models for the ocean circulation. As has repeatedly been stressed, the ^{14}C data clearly indicate a more rapid turnover of the oceans than has previously been thought to be the case. The characteristic response time of the oceans to a disturbance in the heat flux pattern across the air-sea interphase is crucial for the study of climatic change. Most climatic models have so far dealt with the role of the oceans inadequately in this regard. The proper interpretation of the GEOSECS data may well also lead to some revision of our present understanding of the characteristics of man-induced climatic changes.

The emission of carbon dioxide to the atmosphere is but one of the disturbances of the natural biogeochemical cycles that man is initiating. To get a clearer picture of what future changes may occur, we should consider similar rather drastic scenarios, as has been envisaged for the use of fossil carbon. The total amount of sulfur emitted to the atmosphere in the process of burning fossil fuels is so far less than 1% of the amounts present in the fossil fuel reservoirs that may possibly be exploited during the next few centuries. We should also consider the likely consequences of nitrogen fixation, due to combustion and the manufacturing of fertilizers, that is maintained during a century at an annual rate equal to the rate of natural fixation. This is about twice the present rate. More difficult, but probably equally important, is to detect the pathways on a global scale of the toxic substances (such as heavy metals) that man is producing at an increasing rate. Undoubtedly the challenge for those engaged in the study of global biogeochemical cycles is far greater than can be met with present human and material resources in the field. The next 50 years will most likely mean a major step forward in our understanding of the biogeochemistry of the sea. We badly need such knowledge.

References

Bacastow, R., and C. D. Keeling. 1981. Atmospheric CO_2 Concentrations. In: Carbon Cycle Modelling (B. Bolin, Ed.). SCOPE 16. John Wiley & Sons, Chichester, pp. 103-112.

Bacastow, R., and A. Björkström. 1981. Comparison of models for the carbon cycle. In: Carbon Cycle Modelling (B. Bolin, Ed.). SCOPE 16. John Wiley & Sons, Chichester, pp. 29-80.

Björkström, A. 1979. A model of CO_2 interaction between atmosphere, oceans and land biota. In: The Global Carbon Cycle (B. Bolin, E. T. Degens, S.

Kempe, and P. Ketner, Eds.). SCOPE 13. John Wiley & Sons, Chichester, pp. 403-457.
Bolin, B., E. T. Degens, P. Duvigneaud, and S. Kempe. 1979. The global biogeochemical carbon cycle. In: The Global Carbon Cycle (B. Bolin, E. T. Degens, S. Kempe, and P. Ketner, Eds.). SCOPE 13. John Wiley & Sons, Chichester, pp. 1-56.
Bolin, B., C. D. Keeling, R. Bacastow, A. Björkström, and U. Siegenthaler. 1981a. Carbon Cycle Modelling (B. Bolin, Ed.). SCOPE 16. John Wiley & Sons, Chichester, pp. 1-28.
Bolin, B., A. Björkström, K. Holmén, and B. Moore. 1983. The use of tracers for oceanic studies and particularly for determining the role of the sea in the global carbon cycle. Tellus. To be published.
Brewer, P. G. 1978. Direct observations of the oceanic CO_2 increase. Geophys. Res. Lett. 5, 997-1000.
Broecker, W. S. 1979. A revised estimate for the radiocarbon age of North Atlantic Deep Water. J. Geophys. Res. 84, 3218-3226.
Broecker, W. S., R. Gerard, M. Ewing, and B. C. Heezen. 1960. Natural radiocarbon in the Atlantic Ocean. J. Geophys. Res. 65, 2903-2931.
Broecker, W. S., and H. G. Östlund. 1978. Property distributions along the $\sigma_\theta = 268$ isopycnal in the Atlantic Ocean. J. Geophys. Res. 83, 1145-1154.
Bryan, K. 1975. Three-dimensional numerical models of the ocean circulation. In: Numerical Models of the Ocean Circulation. Natl. Acad. Sci., Washington, D.C., pp. 94-106.
Gordon, A. L. 1971. Oceanography of Antarctic waters. Antarctic Research Series, Vol. 15. Am. Geophys. Union, National Academy of Sciences, Washington, D.C., pp. 169-203.
Hampicke, U. 1979. Net transfer of carbon between the land biota and the atmosphere induced by man. In: The Global Carbon Cycle (B. Bolin, E. T. Degens, S. Kempe, and P. Ketner, Eds.). SCOPE 13. John Wiley & Sons, Chichester, pp. 219-236.
Keeling, C. D. 1973. The carbon dioxide cycle: Reservoir models to depict the exchange of atmospheric carbon dioxide with the oceans and land plants. In: Chemistry of the Lower Atmosphere (S. I. Rasool, Ed.). Plenum Press, New York, pp. 251-329.
Keeling, C. D., and B. Bolin. 1968. The simultaneous use of chemical tracers in oceanic studies. II. A three-reservoir model of the North and South Pacific Oceans. Tellus 20, 17-54.
Manabe, S., and R. J. Stouffer. 1980. Sensitivity of a global climate model to an increase of CO_2 concentration in the atmosphere. J. Geophys. Res. 85, 5529-5554.
Moore, B., R. D. Boone, J. E. Hobbie, R. A. Houghton, J. M. Melillo, B. J. Peterson, G. R. Shaver, C. J. Vörösmarty, and G. M. Woodwell. 1981. A simple model for analysis of the role of terrestrial ecosystems in the global carbon budget. In: Carbon Cycle Modelling (B. Bolin, Ed.). SCOPE 16. John Wiley & Sons, Chichester, pp. 365-386.
Munk, W. 1966. Abyssal recipies. Deep Sea Res. 13, 707-730.
Nydal, R., K. Lövseth, and S. Gullicksen. 1978. A survey of radio-carbon variations in nature since the Test Ban Treaty. In: Proc. 9th Intl. Radiocarbon Conf., University of California, 1976.
Östlund, H. G., H. G. Dorsey, and R. Brescher. 1976. GEOSECS North Atlantic radiocarbon and tritium results. Earth Planet. Sci. Lett. 23, 69-86.
Östlund, H. G., and M. Stuiver. 1980. GEOSECS Pacific radiocarbon. Radiocarbon 22, 25.
Peng, T.-H., W. S. Broecker, G. G. Mathieu, and Y.-H. Li. 1979. Radon evasion

rates in the Atlantic and Pacific Oceans as determined during the GEOSECS Program. J. Geophys. Res. 84, 2471-2486.

Peterson, W. H., and C. G. Rooth. 1976. Formation and exchange of deep water in the Greenland and Norwegian Seas. Deep-Sea Res. 23, 273-284.

Riley, G. A. 1951. Oxygen, phosphorus and nitrate in the Atlantic Ocean. Bull. Bingham Oceanogr. Coll. 13, 1-126.

Rotty, R. 1981. Data for global CO_2 production from fossil fuels and cement. In: Carbon Cycle Modelling (B. Bolin, Ed.). SCOPE 16. John Wiley & Sons, Chichester, pp. 121-126.

Seiler, W., and P. Crutzen. 1980. Estimates of gross and net fluxes of carbon between the biosphere and the atmosphere from biomass burning. Climate Change 2, 207-248.

Stommel, H., and A. B. Arons. 1960. On the abyssal circulation of the world ocean: I. Stationary planetary flow patterns on a sphere. Deep-Sea Res. 6, 140-154.

Stuiver, M. 1980. ^{14}C distribution in the Atlantic Ocean. J. Geophys. Res. 85, 2711-2717.

Stuiver, M., and H. G. Östlund. 1980. GEOSECS Atlantic radiocarbon. Radiocarbon 22, 1-24.

Veronis, G. 1978. Model of world ocean circulation: III. Thermally and wind driven. J. Mar. Res. 36, 1-44.

Weiss, R. F., H. G. Östlund, and H. Craig. 1979. Geochemical studies of the Weddell Sea. Deep-Sea Res. 26, 1093-1120.

Whittaker, R. L., and G. E. Likens. 1975. The biosphere and man. In: Primary Production of the Biosphere, Vol. 14 (H. Lieth and R. H. Whittaker, Eds.). Springer-Verlag, Berlin, pp. 305-328.

Worthington, L. V. 1976. On the North Atlantic circulation. Johns Hopkins Oceanographic Studies, No. 6. Johns Hopkins University Press, Baltimore.

Innovative Ocean Energy Systems: Prospects and Problems

Abrahim Lavi

1 Introduction

The ocean can serve as a source for minerals and scarce chemicals needed by developing industrial societies. Under its floor lie vast deposits of coal, natural gas, and oil; on its floor rest metals in great abundance. Its deep water layers are so rich in nutrients that some believe they can supply all of mankind's needs for protein. Today, as energy and mineral resources become increasingly scarce, the ocean offers new possibilities to meet the needs of a civilized world. Universally available, ocean resources can satisfy the energy needs of industrial nations for centuries to come. Technology to harvest ocean energy exists. The impact of ocean energy can be significant if the economic viability and environmental acceptability of innovative ocean systems can be ascertained.

The commonly discussed techniques for electric power generation are based on tidal current, surface and deep ocean current, wave motion, salinity gradient, and ocean thermal gradient. Except for tidal power, no ocean energy system has been commercially developed, so, I will focus on the prospects and problems of the other concepts, giving particular emphasis to the ocean thermal energy conversion technology, which is close to commercial readiness.

Tidal power is well understood. An installation in France has operated successfully for years. The economic uncertainties are minimal, the environmental consequences can be made tolerable, and the sites for substantial power production are finite. In North America, major prospective tidal sites are located at the Bay of Fundy/Passamaquoddy, Cook Inlet, British Columbia, Ungava Bay, and Frobisher Bay. Also, there are numerous sites throughout the world suitable for major installations, 500 MW or larger (Charlier, 1977). The maximum extractable output from all of these sites combined ranges between 5 GW and 25 GW, depending on the estimater.

Ocean currents provide an obvious but small regional resource. Wave energy provides another resource. The worldwide potential of wave energy is significant. With very moderate investment, wave energy has been exploited for navigational buoy systems. The salinity gradient between fresh water and sea water can, in principle, supply substantial quantities of electric power. Natural sites are mouths of rivers where they discharge into the ocean. Ocean thermal energy conversion (OTEC), just emerging from the research and development stage, capitalizes on the temperature difference between surface and deep ocean waters. OTEC's proponents make serious claims for its high potential in terms of cost, readiness, and environmental impact. Of all the ocean energy resources, OTEC has the highest prospects for near-term commercialization.

Renewable ocean energy technologies provide no panacea for the growing problem of energy shortage. Each of the new alternatives has its own limitations, each poses problems of implementation, each is capital intensive. Some are intermittent in their production, and some require costly storage systems. All must be phased into an existing social and economic infrastructure geared for conventional technologies.

In this chapter, I discuss the potential of renewable ocean technologies to provide for much of mankind's energy need. A growing number of informed scientists and engineers believe that if we are wise in our exploitation of the ocean resource, we can derive substantial benefits. If we are not wise, we can lose the ocean as a resource and, at the same time, seriously damage our environment. Furthermore, our success in exploiting the ocean hinges on the willingness of nations to exercise social responsibility so as not to misuse the ocean for the sake of expediency. As nations begin to explore the ocean frontier, they must recognize that actions taken today may have consequences that affect the quality of life tomorrow.

2 Ocean Current Energy Conversion (CEC)

Some advocates of current energy conversion (CEC) claim that the Gulf Stream near southern Florida can produce as much as 2 GW of electric power if no more than 8% of the available resource is tapped (Duing, 1974). In a world consuming close to 250 quads of energy per year, the generation of 2 GW, or 0.06 quads per year, is minuscule. The Department of Energy (DOE) estimates the resource to be higher, in the range of tenths of quads per year worldwide (DOE, 1980).

CEC requires huge underwater installations in water depths exceeding 200 m. The turbines must be clustered and electrically interconnected underwater to feed a main underwater cable. Depending on the topography of the continental shelf, the main cable may be tens of kilometers long.

Advocates claim that CEC technology is feasible today. Conversion of the kinetic energy of a fluid into shaft power and then into electricity is well understood. A CEC system resembles both a tidal power turbine at sea and a wind turbine on land. In fact, the turbine efficiency and size of all three systems vary with the third power of the fluid velocity. Thus, because the velocity of the ocean current is on the order of 1 m s^{-1} or 3.6 km h^{-1} compared to 70 km h^{-1} for wind, CEC turbine blades must be long and few and their rotational speed slow (a few hundred revolutions per minute). Figure 1 shows a schematic of a CEC installation rated at 85 MW. Each CEC turbine is partially buoyant and is moored to the ocean floor. Although surface current is stronger than subsurface current, as shown in Figure 2, the turbines are submerged so that they do not interfere with shipping. Anchoring these large machines in a 1-knot or higher current is beyond present mooring technology. Limitations of manufacturing, material, and installation restrict each turbine to not more than a few ($\sim$100) MW.

Ocean current significantly differs from river flow. The distinction between ocean current and river flow arises from the boundary conditions of the channel and from the forces causing the flow in the first place. River flow is caused by a difference in elevation between the source and mouth of the river. The flow is confined by the banks of the river. Any obstacle to the flow, such as a dam across the river, causes the water level upstream to rise because the water has no alternate route. Surface ocean currents are caused by the Coriolis force of the Earth's rotation. Subsurface ocean currents, which are smaller in magnitude, counter the Coriolis current flow from the equator to the poles and replenish the surface water lost through evaporation. In addition, the velocity vector of ocean current is not uniform throughout and, at times, may reverse completely (Duing, 1974). As a result, like wind turbines, CEC turbines must be continuously reoriented to face the prevailing current.

Although no comprehensive environmental assessment study exists, it is believed that the environmental impact of CEC systems is benign. Engineering proposals abound and economic claims by proponents are rosy. However, DOE has not been enthusiastic about CEC development, for two reasons: the limited magnitude of the exploitable resource and the high development cost. In my opinion, CEC commercialization must await further development in a number of supporting marine technologies, such as mooring, underwater maintenance, and underwater electric transmission. These support technologies are under development as part of the thermal gradient program.

3 Wave Energy Conversion (WEC)

The amount of energy available in ocean waves is far greater than that available in ocean currents. DOE estimates the resource in the range of quads per year (DOE, 1980). Initially, it was assumed that only sites off the northern coasts of Maine and Washington could offer the United States an economically exploitable resource. More recently, and with the advent of new designs, the resource estimate has been expanded to include water masses extending 40 km out from the coastline.

Interest in wave energy dates back many centuries. For decades, inventive minds have proposed clever contraptions designed to capture what appeared to be a limitless resource. Yet today, the number of WEC systems in actual service remains small. Most applications are aimed at powering remote navigational buoys. Slow commercial development is not due to lack of enthusiasm for the concept. Well-thought-out development proposals are plentiful, and many

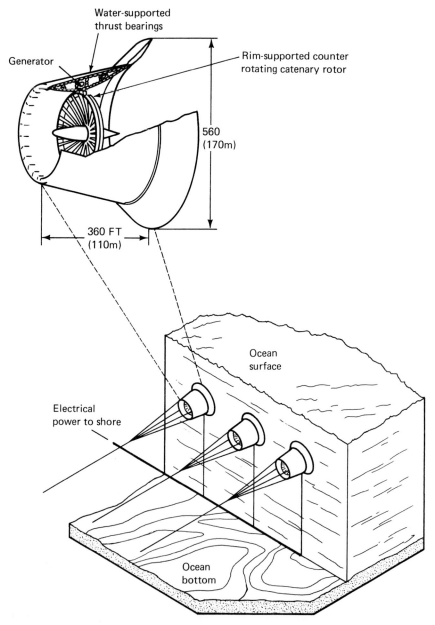

Figure 1. Schematic for a CEC installation rated at 85 MW.

patents have been issued. There is no dearth of ideas or enthusiastic proponents. However, ocean wave energy systems have definite limitations which are commonly overlooked by enthusiasts but which must be addressed by those responsible for research and development budgets.

Like a CEC system, a WEC system requires large equipment that must be moored. Any wave machine, regardless of its operating principle, requires relative motion between two of its elements. Usually, one element is physically constrained so that its motion is minimized or shaped, and its motion relative to

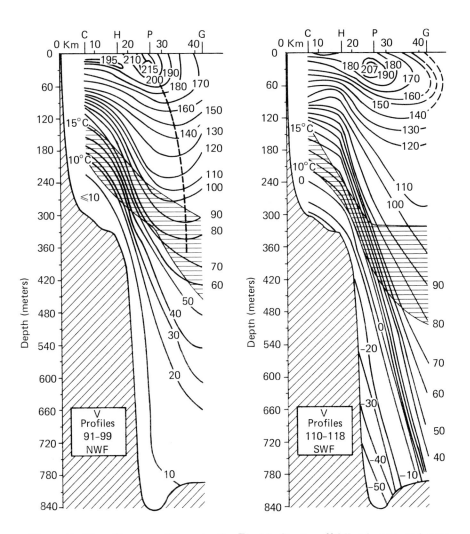

Figure 2. Mean-flow conditions in the Florida Straits off Miami, averaged over one diurnal tidal cycle. Northwest flow profile indicates northward currents during June 14 to 15, 1971. Southwest flow profile indicates southward flowing waters (shaded) on June 17 to 18. Values on the isolines are in centimeters per second (from W. Duing, RSMAS, University of Miami).

the "moving" element maximized. Massiveness of components implies slow dynamic response, low-frequency power generation, and high initial cost. Figures 3, 4, and 5 illustrate some candidate concepts.

Wave Energy Conversion Techniques

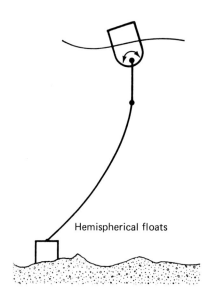
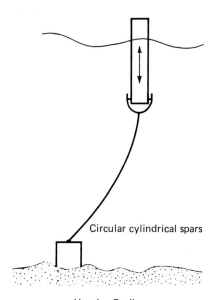

Hemispherical floats	Circular cylindrical spars
Pitching or Rolling Bodies	Heaving Bodies
Basic Principle	Basic Principle
The wave induced rotational motions (about the mooring swivel axis) is converted into electrical energy—device has a resonant roll or pitch.	A body in pure heave resonates with the wave producing amplified body motions.
Extraction Efficiency	Extraction Efficiency
(A) Up to 50%, depending on the system damping and subsystem.	(A) Up to 50%, depending on the system damping and subsystem.
(B) Radiation wave focusing may increase the efficiency	(B) Radiation wave focusing may increase the efficiency

Figure 3. Wave energy conversion techniques. (a) For pitching or rolling bodies. The basic principle is that the wave-induced rotational motions (about the mooring swivel axis) are converted into electrical energy. The device has a resonant pitch or roll. The extraction efficiency is up to 50% depending on the system damping or subsystem; radiation wave focusing may increase the efficiency. (b) For heaving bodies. The basic principle is that a body in pure heave resonates with the wave, producing amplified body motions. The extraction efficiency is up to 50% depending on the system damping and subsystem; radiation wave focusing may increase the efficiency.

Many laboratory models of wave machines have been tested and their efficiencies determined. Both England and Japan have actively pursued WEC. A full-scale sea trial of a "pneumatic" wave-energy system under the auspices of the International Energy Agency has been underway since 1978 in the Sea of Japan (McCormick and Masuda, 1980). The system consists of two 125-kw turbines, one built by England and the other built by Japan. A United States-designed turbine was also to be tested but because of schedule slippage, it was not installed on the test platform, the *Kaimei*.

More recently, passive wave-focusing devices have been proposed (McCormick, 1979). The objective of these systems is to concentrate the wave energy from a large crest length on a relatively small region. The resulting increased intensity implies smaller conversion equipment. The tradeoff is between the added cost of and space occupied by the passive focusing device and the reduced cost of the active components of a WEC system.

Wave Energy Conversion Techniques

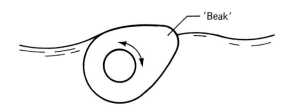

Heaving & pitching
Floats forced to rotate

Combination devices

Basic Principle

Body rotates due to dynamic pressure on "beak",
and also due to beak's buoyancy.

Extraction Efficiency

Greater than 90% under certain wave conditions.

Figure 4. Wave energy conversion technique for heaving and pitching floats forced to rotate. For these combination devices the basic principle is that the body rotates due to dynamic pressure on the "beak" and also due to the beak's buoyancy. The extraction efficiency is greater than 90% under certain wave conditions.

Wave Energy Conversion Techniques

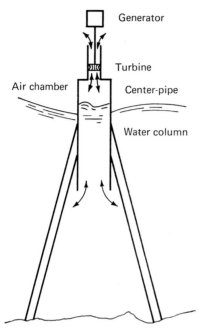

Cavity resonator (pneumatic type)

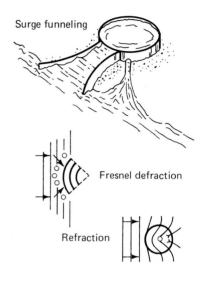

Wave focusing

Basic Principle	Basic Principle
The wave causes a resonance of the internal water column which in turn excites the air in the chamber. This air excites the turbo-generator system.	Using any of the focusing techniques concentrates wave energy from a large crest length on a relatively small region with little wave energy loss.
Extraction Efficiency	Extraction Efficiency
(A) Up to 50% if tuned to wave. (B) Antenna effect may occur increasing the efficiency.	This depends on the mechanical to electrical energy conversion system. Most involve either inductance or turbogenerators.

Figure 5. Wave energy conversion techniques. (a) Cavity resonator (pneumatic type). The basic principle is that the wave causes a resonance of the internal water column which in turn excites the air in the chamber. This air excites the turbogenerator system. The extraction efficiency is up to 50% if tuned to the wave; an antenna effect may occur, increasing the efficiency. (b) Wave focusing. The basic principle is that using any of the focusing techniques concentrates wave energy from a large crest length on a relatively small region with little wave energy loss. The extraction efficiency depends on the mechanical to electrical energy conversion system. Most involve either inductance or turbogenerators.

WEC has certain limitations which are inherent in the resource itself. Ocean waves have broad spectra. This feature has two consequences: to increase conversion efficiency, the wave machine must be tuned so that it extracts energy from a narrow frequency band; but because the spectra are inherently broad, the power available in a narrow band may not be significant. This observation partially explains the fact that although highly efficient devices have worked well in the laboratory, they have not found their way into a working commercial system. Also, wave amplitude can, at times, substantially exceed the design conditions. Thus, components must be designed to withstand hurricanes. Consequently, mooring requirements can become prohibitive. Another limitation is the intermittent nature of the waves. There is no correlation between instantaneous resource availability and load demand patterns. The random nature of the resource makes it difficult to classify WEC installations as either a baseload or a peak power system. This implies that WEC systems can, at best, serve as an auxiliary energy source. To be effective, a WEC system must operate in conjunction with energy storage provisions. Other problems arise from the necessity to restrict the size of each generating unit, to produce low-frequency ac or dc power, and to provide underwater electric transmission.

While there is, to date, no known authoritative study which addresses the environmental issues arising from a WEC installation, proponents assume that WEC systems are environmentally acceptable. Similarly, no comprehensive economic evaluation of WEC has been conducted. The technology is still at the research and development stage. Hence, there are not enough operational data on which to base a reliable assessment of performance and cost. However, because the resource is universally available, its magnitude significant, and the technology simple, WEC systems are bound to become commercial within this century, albeit on a limited scale. Growth will depend on the availability of competing energy technologies. A 10-GW worldwide installed capacity within the century appears realistic.

4 Salinity Gradient Energy Conversion (SGEC)

SGEC is yet another ocean-based energy technology with the potential for producing substantial amounts of power. The osmotic pressure difference between fresh and salt water can be utilized to provide the necessary head to operate a hydroelectric plant. The magnitude of the pressure difference between 0.5-molar sea water and fresh water can range as high as 240 m or 25 atmospheres (Wick and Isaacs, 1976). The greater the salinity gradient between the two water masses, the larger the available pressure head. Of course, not all of the theoretically available head can be realized in a practical system. Nevertheless, the concept is intriguing and offers promising prospects.

Natural salinity gradient sites are located where rivers discharge their flow into the ocean. Man-made lakes have been proposed where salinity stratifications within the lake can be maintained with the aid of solar heat, as illustrated in

Figure 6. SGEC systems are suitable for baseload operation. Even with man-made lakes, continuous operation is possible because the lake stores the solar energy in the form of a salinity difference between surface and deeper water layers.

There are two candidate approaches for implementing a working system. Both approaches require a selective membrane at the interface between the two water masses. The osmosis method relies on a selective membrane which allows water but not sodium, chlorine, or other ions to cross. Through osmosis, a pressure head can be created and maintained if the saline and fresh water masses are dammed. The reverse electrodialysis method requires a selective membrane which allows certain ions to cross in one direction while oppositely charged ions flow in the reverse direction, thereby creating an electric potential difference or a dialytic battery as shown in Figure 7.

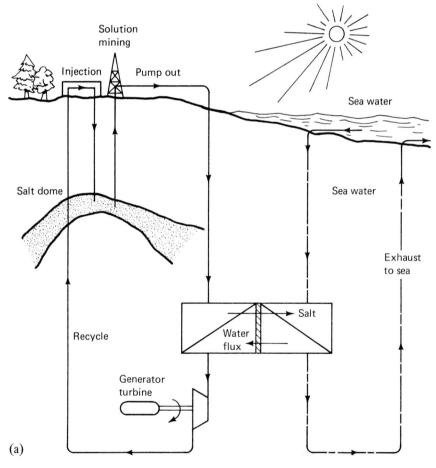

Figure 6. (a and b) SGEC system from an artificial lake with salinity stratification maintained by solar heat.

Estimates of river flow and also of desert land that can be utilized for man-made salinity-stratified lakes indicate promising prospects for SGEC. Proponents claim that the resource is adequate for producing billions of kilowatts. However, the necessary membrane technology is still in its infancy. The membrane area required per kilowatt of output is on the order of thousands of square meters ($\sim$7500 m^2 kW^{-1} at 100% efficiency). To keep such a surface free from biological and chemical fouling poses a major technical challenge. To build the dams required to hold the water several hundred meters above the lake level represents no small undertaking. Nor is it a minor environmental issue to dam a river at its mouth. The electrodialysis approach does produce more realistic membrane figures, but the technology is no more advanced than that of the osmotic approach. Reliable cost estimates cannot be made unless candidate membranes for either approach are developed and thoroughly tested over time to establish per-

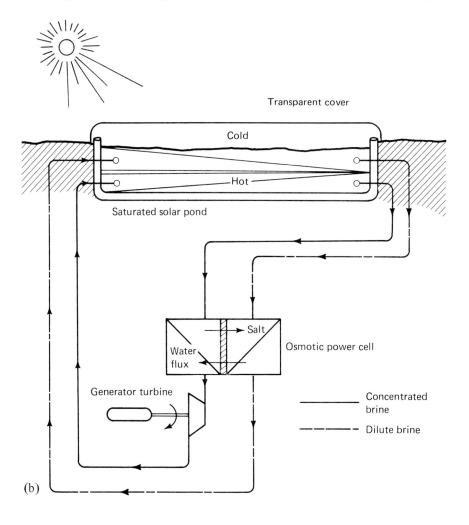

formance and life expectancy. In my view, SGEC remains in the earliest stage of development compared to other ocean technologies. It would be premature to place high hope in its near-term prospects. At the same time, its potential contribution should not be discounted. The magnitude of the resource is sufficient justification for investing in its research and development.

5 Ocean Thermal Energy Conversion (OTEC)

Of all the ocean energy systems, OTEC is probably closest to worldwide commercialization. OTEC relies on the thermal gradient between surface and deep tropical ocean water, as illustrated in Figure 8. It employs a Rankine cycle heat engine with either water (open cycle) or a refrigerant (closed cycle) as the cycle fluid. A schematic of a closed cycle is shown in Figure 9.

In the early 1930s, George Claude conducted a number of experiments on the open cycle. Ever since, the technology has received sporadic support (Marchand, 1980). The Andersons (1966) proposed a number of innovations: use of a refrigerant (propane or a halocarbon) as cycle fluid; a floating platform moored to

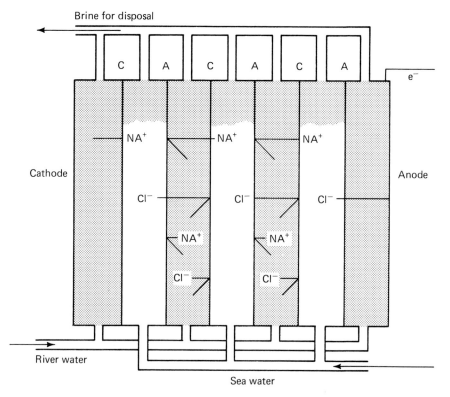

Figure 7. Reverse dialysis method of SGEC. See text for explanation.

the ocean floor; a pipe extending vertically downward to bring the cold water to the condenser. Later, Lavi and Zener (1973) selected ammonia as the cycle fluid because of its superior thermodynamic properties. In the United States, closed-cycle system designs generally use ammonia as the cycle fluid.

What distinguishes OTEC from other ocean (and solar) energy systems is its day-night and year-round operation. Of all the solar options, OTEC alone produces baseload power, relying on the ocean for thermal storage. While the plant operates year round, its output varies seasonally as the available temperature difference between surface and deep water varies. This seasonal variability depends on the distance of the plant site from the equator. Interestingly, for plants located in the Gulf of Mexico and connected to the southeast electric grid, the seasonal variation of plant output closely tracks the seasonal demand for electricity, high during the summer and low during the winter (Cohen, 1979).

Economic analyses indicate that OTEC plants can compete with conventional power plants in practically all tropical islands where imported oil is the main fuel. However, the magnitude of this island market is insignificant (a few gigawatts worldwide). A considerably more significant market is the southwestern United States. This market has a power demand of approximately 100 GW, of which OTEC can readily supply 30%. Other nations possess suitable sites for OTEC plants, but most of these nations are classified as developing, which implies that their electric demand is relatively small.

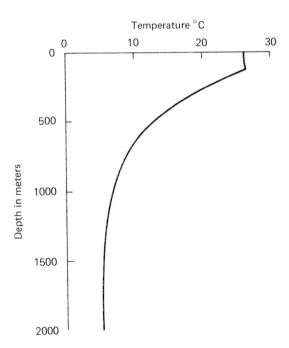

Figure 8. The thermal gradient between surface and deep tropical ocean water.

The worldwide ocean thermal resource is estimated at 30 quads per year or 1 TW (1000 GW). The world's present total energy consumption is approximately 8.2 TW and is expected to increase to 26 to 47 TW by the year 2020 (Haefele, 1979). To exploit this huge ocean thermal resource, Dugger et al. (1975) have proposed to transport OTEC's energy chemically instead of transporting it electrically by cable. Floating OTEC plantships sited near the equator can produce chemicals (principally ammonia), reduce alumina, or charge electrolytic cells. OTEC ammonia can relieve the demand for scarce natural gas which is currently the main feedstock in ammonia production. The ammonia can be barged worldwide for the manufacture of fertilizers, and, once on land, it can be transported by pipeline to major load centers to be "cracked" to hydrogen and nitrogen. The hydrogen feeds a fuel cell for on-site dc power generation and the nitrogen is released to the atmosphere.

6 OTEC Commercialization and Environmental Factors

At first glance, proposals for OTEC commercialization may seem premature. On the other hand, comprehensive analyses indicate that electricity produced from an OTEC ammonia/fuel cell system may compete with electricity generated from oil-fired plants pressed into service to meet peak and intermediate loads for cities like New York, Boston, and Los Angeles. The higher cost of electricity from an ammonia fuel-cell system is offset by its minimal environmental impact and its low-loss transmission cost. As oil prices continue to rise and as oil supplies grow more precarious, as issues of nuclear waste and safety continue to grow, and as the burning of coal becomes more environmentally objectionable,

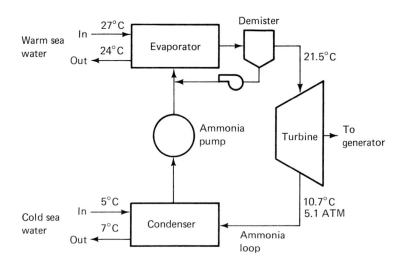

Figure 9. Schematic of a closed-cycle Rankine heat engine applied to an OTEC system.

proposals to develop OTEC cease to raise as many eyebrows as they did only a few years ago.

Although OTEC technology is relatively straightforward, requiring no major breakthroughs, and although a working system at sea has been tested (albeit on a small scale), OTEC commercialization will require major engineering development and substantial capital investment (Lavi, 1980). Technical problems arise for two reasons. First, the small temperature difference translates into low system efficiency and in turn, into components enormous in size compared to those of more conventional power systems. Second, OTEC systems must operate in an ocean environment and most candidate OTEC sites are prone to hurricanes. Clearly, the ocean is not the most hospitable environment in which to locate platforms larger than modern aircraft carriers. The cold water pipe for a 100-MW plant is about 20 m in diameter and almost 1 km long. Mooring a platform for a 400-MW OTEC plant is probably beyond current technical capability. Cabling the electricity to shore from plants located 100 km offshore will require the development of long underwater high-voltage dc cables that do not exist today. It should be recognized that the reason these technologies are not at hand today is because the need for them did not exist earlier. Actually, much of the platform and cold water pipe construction and deployment technology can be adapted from North Sea oil exploration technology. Nevertheless, much work remains.

Like most solar energy systems, OTEC is capital intensive. If OTEC installations were to grow to the point of producing the equivalent of 30 quads/year, the total investment (excluding accumulated depreciation) would be at least $3,500 billion in 1980 dollars. The estimate is based on the assumption that facilities dedicated to the manufacture of OTEC components are already in place and fully operational. The required capital to build these facilities is estimated in hundreds of billions of dollars. The figures do seem staggering, but OTEC plants are expected to live 30 years or more and they do not consume fuel for power generation. The cost should be compared to the discounted present value of energy produced at the rate of 30 quads year^{-1} for 30 years. Whatever the energy source, the cost would be as staggering as OTEC's.

In the 1970s, critics of OTEC were preoccupied with problems of heat transfer and biofouling control of heat transfer surfaces exposed to sea water. As it turned out, the large heat exchangers (15 m in diameter) required by an OTEC power module rated at 50 MW can be produced. Biofouling control with mechanical devices appears adequate. Improvements in heat transfer technology have been realized to the point where a reliable performance prediction of full-scale components is possible. The progress to date has encouraged a once-skeptical administration to initiate the 1-MW heat exchanger tests off Hawaii, propose the design of multiple 10-MW pilot plants, and offer legislation for financing commercial OTEC projects. However, OTEC commercialization and development encompass issues that range beyond engineering problems, performance prediction, capital availability, and product cost.

Advocates of OTEC technology claim that OTEC is environmentally benign.

But their claim cannot be readily substantiated or refuted. An OTEC plant processes enormous volumes of sea water. Each kilowatt of output requires approximately 4 liters s^{-1} of warm water and an equal amount of cold water. A 400-MW plant ingests water at the rate of 1600 m s^{-1}. The processing of these enormous volumes of water may harm both large and small marine organisms. Also, there is concern about changing the surface temperature because of condenser water discharge. To minimize adverse environmental impact, it has been proposed to discharge the evaporator and condenser outflow below the thermocline. But there is no way to avoid some mixing of the volumes of warm and cold water flowing out of the plant. In addition, there is no way to avoid local lowering of ocean surface temperature by about 2°C. Granted that the water flow from a cluster of OTEC plants constitutes but a small fraction of the ocean water mass, the accumulated adverse long-term weather effects, both locally and downstream, are not known. If all OTEC plants were confined to equatorial waters, the weather effects would probably be negligible. But, if many plants were located farther north (or south) of the equator, site insolation may not be strong enough to restore ambient ocean thermal conditions. Weather modification cannot be ruled out. If plants are located near the Gulf Stream, Northern Europe may attribute changes in its climate to OTEC. Hence, OTEC installations may lead to international squabbles. Also, OTEC installations may interfere with shipping, posing particular hazards to submarines. Some critics express concern about the need to protect these remote, multimillion-dollar installations (vulnerable to terrorists and saboteurs), thereby raising national defense issues.

The issue of massive carbon dioxide release into the atmosphere has been raised. The deep cold water is supersaturated with carbon dioxide; as the cold water rises, the hydrostatic pressure reduces and in turn, may release carbon dioxide into the atmosphere. It is believed that this problem is not serious in the case of a closed-cycle system because the cold water is discharged below the thermocline where the water is not saturated with carbon dioxide. Open-cycle systems using spray condensation can result in the release of as much as 30% of the carbon dioxide released by a fossil-fuel plant producing equal power. We thus may conclude that the carbon dioxide release is not an issue and should not hamper OTEC's commercial development. OTEC does not introduce carbon dioxide into the atmosphere in substantial amounts; rather, any significant release of the gas into the atmosphere should be attributed to the burning of carbon-based fuels.

Another environmental issue centers on the possible use of chlorine to control biofouling. Even a few parts per million of chlorine may prove environmentally objectionable. There is also the concern of cycle fluid discharge into the ocean. Clearly, an open-cycle system which uses sea water for its cycle fluid poses no environmental problem. Release of propane is objectionable; release of halocarbons (freons) may be totally unacceptable because of their possible effect on the ozone layer of the upper atmosphere. However, the discharge of ammonia is not of great concern because ammonia dissolves in water and does not pose a long-term environmental threat. Its damage is reversible.

Even if we take the technical feasibility of ocean energy systems for granted, and even if we assume that their environmental impact is benign, the road to commercial implementation is bound to be difficult. I have alluded to some international problems that may surface as OTEC proceeds to full-scale commercialization. Interference with shipping, adverse environmental impact downstream of clustered OTEC plants, weather modification, and security of installations are all issues of concern. Their satisfactory resolution will require time. Jurisdictional disputes between nations and between local and national governments and among government agencies with conflicting charters can hamper development. In my view, the most serious institutional barriers are political and economic.

7 Ocean Energy Systems: Institutional Factors

Development of new energy technologies is both costly and risky. Proof-of-concept and pilot-plant experiments require multimillion-dollar investments. The heat exchanger development program for OTEC has already consumed funds in excess of $60 million. The forthcoming pilot-plant demonstrations are estimated to require upwards of $200 million each. To private industry, these sums are not in themselves prohibitive. However, private investors are reluctant to commit their capital to the development of products that are heavily regulated by government unless there is a potential for a payoff commensurate with the perceived risk. Furthermore, since government plays a significant role in providing financial incentives and subsidies to competing energy options, giving preferential funding to some alternatives while others go begging for funds, industry, aware of the political game of energy economics, attaches a high risk to investment in new energy technologies. Because the expected return on investment does not seem commensurate with the level of risk, energy research and development generally falls into the domain of government bureaucracies. I question whether government agencies can successfully bring a new energy technology into the marketplace. The very notion of a market in our (professed) capitalist system implies active and profitable involvement of the private sector.

I have illustrated that ocean energy technologies tend to require substantial investment in plants and equipment in order to produce cost-effective new components. But capital is scarce. Industries that may be expected to undertake multibillion-dollar ventures have already staked their financial resources in more established energy technologies: oil, coal gasification, and nuclear power. Rightly or wrongly, these industries have committed substantial resources to what they perceive as worthwhile investment opportunities. Once committed, resources cannot be easily uncommitted or diverted. Witness the increased American dependence on oil and the persistence of the nuclear lobby in spite of repeated technical and economic setbacks. Earlier commitments must run their course before new ones can be made. These are factors which I believe are governing and will continue to govern the commercial development of ocean energy systems. The alternative is for government to embark, without regard to economic considerations, on the research, development, and implementation of whatever energy technology it deems beneficial to the nation's well-being. While America and other industrial nations do concede the existence of an energy crisis, so far they do not seem to consider the crisis acute enough to threaten their society.

8 Conclusion

I have mentioned some major criticisms of large-scale OTEC deployment. These criticisms should not be taken lightly. But as we begin to face up to the dilemmas posed by energy shortage, by our dependence on oil imported from unstable foreign sources, by objections voiced against fossil fuel and nuclear alternatives, OTEC—as well as other ocean energy options—scores high marks. After examining the issues that may impede commercial development of the ocean technologies discussed in this paper, I have concluded that the main problems are institutional and environmental. With time, institutional issues can probably be resolved. However, environmental problems may be less tractable. Although I have focused on damages that may result from large-scale deployment of OTEC plants, an equally serious damage, not attributable to OTEC or any of the other technologies, may influence their success or failure. It is the damage to the ocean ecology brought about by careless use of the ocean resource. There is the danger that nations may so abuse the oceans that a discussion of ocean energy prospects becomes but an academic exercise. The disposal of solid waste into the ocean may prove far more harmful to marine life than all the OTEC plants that can be built and operated. An oil spill upstream of an OTEC power plant can be detrimental to OTEC's expensive heat exchangers. An oil film deposited on the heat transfer surface can cause a prolonged shutdown. If we hope to exploit the ocean as an energy resource, we must devise ways to protect that resource. If we are wise in our exploitation of the ocean, we can alleviate the energy shortage facing industrial nations in the decades ahead. If we abuse the ocean either by polluting it with refuse or by carelessly harvesting its resources, we may someday discover that we have killed the only goose left to man that could lay a golden egg.

References

Anderson, J. H., and James H. Anderson. 1966. Thermal power from sea water. Mech. Eng. 88, 41–66.
Charlier, R. H. 1977. Harnessing the energies of the ocean. In: Energy from the Sea (B. L. Gordon, Ed.). The Book & Tackle Shop, Watch Hill, R.I., pp. 115–161.
Cohen, R. 1979. An Overview of the U.S. OTEC development program. IEEE J. Ocean Eng. 16, 3–29.

DOE (Department of Energy). 1980. Ocean Energy Systems Multiyear Plan. Ocean Energy Division, DOE, Washington, D.C.

Dugger, G. L., H. L. Olsen, W. B. Shippen, E. J. Francis, and W. H. Avery. 1975. Tropical ocean thermal power plant producing ammonia and other products. In: Third Annual OTEC Workshop, Houston, May 1975 (G. L. Dugger, Ed.). Applied Physics Lab of Johns Hopkins University, Laurel, Md., pp. 106-114.

Duing, W. 1974. Measurement of ocean current profile in the Gulf Stream off Miami. RSMS, University of Miami, 1972.

Haefele, W. 1979. Global perspectives and options for long-range energy strategies. Energy 4(5), 745-760.

Lavi, A. 1980. Ocean thermal energy conversion: A general introduction. Energy 5(6), 469-480.

Lavi, A., and C. Zener. 1973. Plumbing the ocean depths. IEEE Spectrum 10(10), 22-27.

Marchand, P. 1980. The French ocean energy program. 7th OTEC Conference, DOE Publication No. 800633-2, Vol. 2, Paper No. 2.7. Washington, D.C.

McCormick, M. 1979. Waves, salinity gradients and ocean currents—alternate energy sources. 6th OTEC Conference "Ocean Thermal energy for the 80's," DOE Publication No. 790631/1, Vol. 1, Paper No. 2B-1. Washington, D.C.

McCormick, M., and M. Masuda. 1980. Review of the wave energy conversion project in the sea of Japan. 7th OTEC Conference, DOE Publication No. 800633-2, Vol. 2. Washington, D.C.

Trimble, L. 1979. OTEC goes to sea (a review of mini-OTEC). 6th OTEC Conference, DOE Publication No. 790631/1, Vol. 1, Paper No. 3A-2. Washington, D.C.

Wick, G. L., and J. D. Isaacs. 1976. Utilization of the energy from salinity gradient. In: Wave and Salinity Gradient Energy Conversion Work Shop, University of Delaware, May 1976. ERDA Report No. C00-2946-1, pp. A-1-34.

Aquaculture: Potential Development

Hillel Gordin

1 Introduction

Toward the end of the twentieth century, the human population on earth is still increasing at an average rate of 1.8% per annum (Global 2000 Report, 1980). The world population is expected to reach 6 billion just before the turn of the century. The rate of human population increase is not evenly distributed in time or space. We assume in developed countries this rate will be 0.6% per annum during the next 50 years. In the less developed countries, human population is expected to increase by 2.1% per year between 1980 and 2000. This rate is expected to decrease to 1.5% per year during the period between 2000 and 2030. Summing these rates yields the frightening figure of 9.3 billion people on this planet in the year 2000 (Table 1). Unchecked human population growth is the major vector dominating life processes on earth, including man's. The negative repercussions of this vector on the environment are numerous: water, land, and atmospheric pollution; destruction of natural habitats, resulting in the extinction of species; uncontrolled hunting of wild animal populations (Africa); and overfishing.

The exploitation of nonrenewable resources by the exploding human population will bring many of these resources to their limits in the next 50 to 100 years. Fossil fuels, fertilizers (especially phosphates), and different metals are on the verge of exhaustion as a result of their accelerated exploitation by man (Global 2000 Report, 1980).

Any available resources on this planet will have to be shared by a human population twice as large as the present one. Therefore, unless new sources are

Table 1. Projected world human population growth, 1975 to 2030, in millions*

Year	Less developed countries	Developed countries	Total
1975	2959	1131	4090
1980	3283	1165	4448
1990	4041	1237	5278
2000	4975	1313	6288
2015	6220	1437	7657
2030	7776	1572	9348

*Projections based on the following assumptions: (1) In developed countries, the human population growth is 0.6% per year during the period 1980 to 2030. (2) In less developed countries, the human population growth is 2.1% per year from 1980 to 2000 and 1.5% per year from 2000 to 2030.

found or developed, humanity is bound to approach scarcity in many of its essential supplies, including food.

Already production and distribution of food do not satisfy large masses in the less developed countries. According to the Global 2000 Report to the President (1980), 30% of humanity is ill-fed. In many of these countries, calorie consumption is only 94% of the required amount, according to the United Nations' Food and Agriculture Organization (FAO) standards which allow for good health and normal development of children (Global 2000 Report, 1980). With respect to animal protein, the situation is the same or worse in most of the less developed countries, especially in Africa and Latin America.

The increase of food production on a global scale during the last two or three decades is primarily due to improved and advanced technologies. These technologies are based mainly on intensification of farming, with high energy expenditure, which means exploitation of fossil fuels through mechanization, fertilizers, pesticides, herbicides, and drugs; advanced genetic selection techniques; improved farmed species; and their yields. As fossil oil becomes prohibitively expensive and its global reservoirs diminish, major efforts must be made to find new types of economically feasible energy. New arable lands are limited. The Global 2000 Report (1980) predicts only a 4% addition to the land used for farming over the level of the 1970s. By the turn of the century, however, arable land will undergo severe deterioration around the world through improper management of the soil, salination, pollution, and confiscation for urbanization and industrial usages. Thus, even without major climatic catastrophes such as floods or long-range droughts, satisfying human food needs through agriculture on a global scale will be very problematic in the future (Fig. 1).

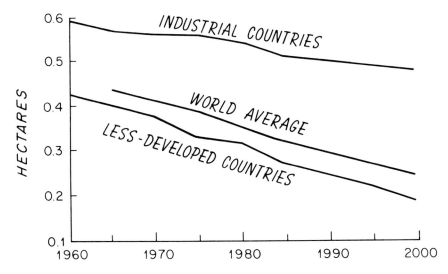

Figure 1. Arable land per capita on a global basis.

2 Fisheries' Role in World Food Production

Much of the world animal protein comes from fisheries. World fishery landings showed an impressive increase in the period between 1960 and the early 1970s, rising from 40 million metric tons per year to 68 million metric tons (FAO, 1979; U.S. National Marine Fisheries Service, 1980). This was the result of increasing effort, modernizing fishing fleets, and using advanced technologies to locate fish schools. Since then the world catch fluctuates around 70 million metric tons (Fig. 2).

Whether landings of wild fish and other marine organisms will increase in the future and to what extent is debatable. Landings as high as 400 million metric tons per annum (Chapman, 1969), through 160 million metric tons (Ricker, 1969; Nägel, 1979) to the lowest figure of about 100 million metric tons (Global 2000 Report, 1980) have been predicted. The last decade's fishing record does not support an optimistic projection (Fig. 2). A few marine organisms have fishing potential, such as the krill in the Antarctic Ocean and the midwater cephalopods, but have not yet been exploited. However, they are not expected, if ever fished on a large scale, to have a great effect on the total world fisheries landings. No matter how immense some fish populations are, they are of a finite size. They can be fished only at an intensity which will allow them to maintain their stocks. Otherwise, overfishing, superimposed on species' intrinsic cycles and fluctuating conditions in nature, may bring these species to the verge of extinction. Management of fisheries has to take into account not only fishing efforts, oil prices, and labor costs, but also the intricate biological processes in

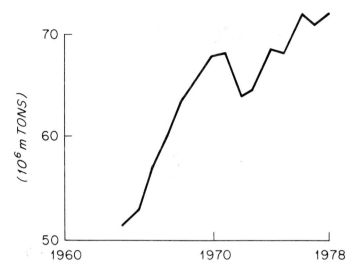

Figure 2. World fisheries catch from 1965 through 1978.

the oceans and their response to physical phenomena. Failing to do so, fisheries around the world will experience an increasing rate of failures.

Many of the commercially fished marine and freshwater species reproduce in nearshore waters, estuaries, and river runoffs. Some spend one or another of their life cycle stages there. In the last few decades, due to water pollution caused by human populations living along many stretches of the world's seashores, increasing numbers of fish species either shy away from shore or fail to reproduce. Fish populations dwindle there. Fish become chemically polluted or bacteria- and virus-infested, unfit for human consumption. Many areas in Europe, the United States, and Japan serve as good examples of this phenomenon. Other man-made, large-scale changes in nature's structure have an adverse influence on fisheries. The erection of the Aswan Dam reduced the flow of the Nile water into the eastern Mediterranean, thus depriving it of much-needed nutrients. The result at present is a noticeable reduction in fishery landings in the area.

To recapitulate the subject of fisheries potential in the next 50 years, it can be said with some certainty that the contribution of fisheries to man's food basket will hardly maintain its present level.

3 Aquaculture Attempts

In order to avert widespread starvation in the years to come, a worldwide, well-planned and executed campaign should be launched. Control over human population growth must be the most emphasized. An increase in food production is another front, in which aquaculture must play an important role. Along with the ever-improving conventional agriculture and fisheries, aquaculture should be recognized as one of the main contributors of food in the future, and so treated. Aquaculture is a very old method of animal husbandry, which for centuries did not develop much in comparison to agriculture.

Most of aquaculture production at present comes from fresh water, some from brackish coastal waters, and some from mariculture. Not surprising is the fact that over 80% of the global aquaculture production comes from South and Southeast Asia, including the People's Republic of China, since this is where aquaculture started (Villaluz, 1953; Pillay, 1976; Lovell et al., 1978). Mainland China and Taiwan produced 43% of the global aquaculture production in 1975, most coming from fresh water; Japan produced 15.5%, most coming from mariculture (Table 2). One can learn from Table 2 that 65.2% of aquaculture production was finfish, 17.2% molluscs, 17.4% seaweeds, and only 0.3% was crustaceans. Aquaculture production from the total fisheries landings was around 8.9%.

This paper addresses itself mainly to the potential development of mariculture and not to freshwater aquaculture. The latter is not expected to increase dramatically over the next 50 years on a global scale, mainly due to shortage of fresh water and land, and to water pollution resulting from cities and industries. The increase in production will come through intensification of farming methods by fertilization and manuring of the ponds directly, or by integrating fish ponds with poultry, cattle, and pigs. More and more fish ponds will serve partially as operational water reservoirs for nearby agriculture. In Israel, due to the extreme shortage of fresh water, fish are grown in reservoirs where winter rainfall is collected, while during the summer the water is used for the irrigation of different crops (mainly cotton). Toward the end of the summer, about 1.0 to 1.5 m of water is left in the reservoirs until the fish are harvested (mainly common carp and *Tilapia* hybrids) (Sarig, 1978, 1979; Tal and Ziv, 1978).

Intermediate water—brackish water of terrestrial origin—has some potential as a medium for aquaculture. Many agricultural lands around the world have been maltreated for centuries. Salty aquifers were connected with the irrigation water table, allowing the salt to be transported to the surface. Since arable land is going to be very precious, underground drainage systems would have to be installed in order to reclaim or save the topsoil. Such water can be used for aquaculture. In Kibbutz Yotvata, which is situated in the southern part of the Arava Valley in Israel, such a system was installed under a 30-hectare date plantation, increasing its annual yield by 50% and allowing enough water [10 g TDS (total dissolved solids) per liter] for a small experimental fish pond system. The large deltas of the Euphrates, Tigris, Indus, and Ganges rivers, which once supported agricultural cultures, have degenerated. The above proposed solution would reclaim agricultural lands and provide significant quantities of brackish water for aquaculture.

Mariculture, thus far, has produced only 35 to 40% of the total aquaculture production (Table 2). Most of it is in the form of shellfish and macroalgae. Mariculture is still in its infant stage but has an immense food-producing potential.

Table 2. Aquaculture production—1975, in thousands of tons*

Region	Finfish	Molluscs	Seaweeds	Crustaceans	Total	%
Asia	1202	460.3	754.8	14.76	2431.8	39.8
China	2200		300		2500	41
Europe	422	399.5			821.5	13.5
North America	23.4	134.1			157.5	2.6
Latin America	26	47.1		0.9	74	1.2
Africa	107.1	0.25			107.35	1.7
Australia		10.05			10.05	0.2
%	65.2	17.2	17.3	0.3		100
Total	3980.5	1051.3	1054.8	15.66	6102.2	

*Source: Pillay (1976).

Sea water, coastal marshes and waste lands, lagoons, and atolls are plentiful, especially in the tropical and subtropical regions where climatic conditions are very favorable for mariculture. Many marine and brackish water organisms lend themselves to the possibility of cultivation. Thus far, however, mariculture has not advanced to a major role in producing, through controlled farming, large quantities of food organisms.

Many farming techniques are used in mariculture. Many of them were developed in Japan, which is the world leader in mariculture production. Some were developed in Europe (mussel and oyster culture), some in the United States and Canada (salmonid sea ranching), and some in South and East Asia (coastal tidal fish ponds—milk fish and mullet cultivation). A number of reviews were published in the last decade describing the state of the art (Ryther and Bardach, 1968; Bardach et al., 1972; Hansen, 1974; Nash, 1974; Ryther, 1975; Pillay, 1976; Webber and Riodan, 1976; Hepher et al., 1978; Lovell et al., 1978; Matsuda, 1978, 1979). Studying these reports, one must draw inferences from mistakes in the past and present and emphasize potential avenues, hardly touched, for future development of aquaculture.

4 Problems in Aquaculture Expansion

In general, management of the basic problems of mariculture, by scientific institutions, governments, international agencies, and industries, was characterized in the last decades by a lack of patience. There was a desire for a quick return on research and development invested efforts. Many failed to understand that one cannot extrapolate from small-scale experiments to commercial large-scale operations. The path lies through gradually increasing scales.

The approach to basic problems such as reproduction, larval rearing, nutrition, diseases, and ecology of culture systems was to achieve fast successes; in many instances this proved counterproductive. Somehow, the understanding that one dealt with a new field, with new problems which required a very intensive, expensive, and long-term research range was lacking; it is still lacking in many parts of the world. This has to change if mariculture is to be developed to its potential scale in the shortest time.

In many cases, socioeconomic considerations in establishing aquaculture enterprises have been neglected in the past. This was due to a lack of understanding of the complex interactions between aquaculture and the social and economic infrastructure at any given place (Matsuda, 1978; McGoodwin, 1979). Unfortunately, such lack of understanding has an unfavorable impact on the development of aquaculture; in the immediate range there is failure, and in the long range these failures discourage governments and agencies from going into aquaculture in a massive way.

Adopting the right farming techniques, scale of operation, and suitable species for any given population should be based on the tradition and culture of the people and their level of technology and working habits. The development should be accompanied by long-range extension and training programs on the one hand, and by attractive financing and funding on the other. Consideration must be given ahead of time to the needed infrastructure for processing, transporting, and marketing the products, whenever the production is not only for self-consumption.

The successful development of mariculture depends on the careful choice of the candidate organisms and their culture systems. The organisms' biological characteristics and their interactions with the ecological structure and dynamics of the farming techniques must be understood at a satisfactory level. Whenever possible, the organism to be selected for culture should be low in the trophic pyramid. Such organisms would ensure maximum yields drawn directly from the primary productivity of the ponds. Milk fish, mullets, brine shrimp, and oysters are good examples of this principle. Unfortunately, it cannot always be implemented. If a given population refuses to consume certain species (such as mullets in the United States and Europe, and milk fish in Japan) there is no sense in culturing them there. If the price a species gets on the market is below production cost, then its mariculture system cannot be economically viable (rabbit fish in Israel). In most industrial countries, mariculture is taking the direction of culturing predatory finfish which receive high prices in the markets, but these species require high-protein diets and high water quality. Such culture systems tend to be intensive, with high densities, and therefore require aeration, high-rate water flow, and heavy feeding.

5 Some Possible Solutions

Mariculture in less developed countries should be promoted without delay, based on expansion of existing systems, such as growing milk fish in marsh lands located along the shores of many Southeast Asian, Latin American, and African countries. Even farming very extensively just by stocking the ponds with wild fry will yield a few hundred kg ha^{-1} yr^{-1} (Pillay, 1976). Estimates of the available potential area vary, but all agree that a 10-fold increase can be achieved without difficulty (Ryther, 1975). At 500 kg ha^{-1} yr^{-1}, 30 million ha will yield 15 million metric tons of fish. This would triple the world total aquaculture production (Table 2). Such development requires national government as well as international funding aid. However, it would not necessitate any foreign expertise in Southeast Asia. The technology is there, mastered by the local people. Aid, planning, and management for processing and marketing the crops should be given to the farmers on a regional government level.

At the same time, intensification of the culture techniques should be started by integrating fish farming into the agricultural system. Recycling domesticated animal wastes and other organic materials into the ponds will increase their primary productivity and thereby their fish yields (Schroeder, 1978). The technology is simple, known, and practiced in many parts of the world (Lin, 1954; Bardach et al., 1972). Implementation of this semi-intensive, low energy, low-cost technique should increase pond production to 3 or 4 metric tons $ha^{-1} yr^{-1}$. Extrapolating this yield to the 30 million potential hectares mentioned above will render some 100 to 120 $\times$ 10^6 metric tons per annum. This point is emphasized here and has been discussed in the past (Ryther, 1975; Pillay, 1976; Nägel, 1979) because it is felt to have the highest priority in developing mariculture on a global scale. Ryther (1975) goes even further, claiming that 40 $\times$ 10^6 hectares can be converted into such productive systems. The investment that should go into developing it according to Nägel (1979) is at least 2.5 $\times$ 10^9 United States dollars annually, which is only one-tenth of the investment given to development of agriculture in less developed countries.

In order to enhance and secure future expansion of mariculture, some scientific and engineering obstacles should be removed. Controlled reproduction and larval rearing of the present and future cultured organisms is considered to be the most difficult obstacle to remove. Many of the marine organisms cultured today are still collected in nature, such as *Chanus chanus* (milk fish), *Mugil cephalus* (grey mullet), and other species of this genus. Gravid females from the *Penaeus* genus are collected in the wild and forced to spawn in captivity. It would be futile, in the long range, to rely on collection of fry or gravid females in the wild. Although, at present, nature supplies the extensive systems of aquaculture in Southeast Asia, the Middle East, and in other parts of the world with fry and fingerlings, it is not a reliable and sufficient source for the expanding industry. Pollution of spawning and nursing grounds, overfishing, and indigenous or intrinsic fluctuation of population sizes from year to year are some of the vectors which make the supply of recruitments from nature unreliable. Controlled reproduction and successful larval rearing will eventually permit the provision of stocking recruits throughout the year. At least along the tropical belt of the world, it would be of great advantage to the farming strategies. The above will allow a thorough genetic selection to start for the better fit, more efficient, and disease-resistant types for any given farming technique. Having mastered the reproduction and larval rearing processes of cultured species, mariculture would be like other husbandry industries in which the farmers are ensured of stocking material of high quality whenever needed. This would cause the farms to operate more efficiently, and products might reach market throughout the year.

When considering intensive mariculture systems, advances should be made in maximizing the efficiency of the resources' utilization. There is some justification in the claim, directed at the development of third and fourth trophic level mariculture, that this is really a food reduction operation rather than a food production one (Ryther, 1975). However, this criticism applies whenever culturing organisms for human consumption are dealt with and whenever these organisms are not from nature's second trophic level, such as poultry, nonherbivorous fish, and shrimp. This criticism is valid, and future research and development should devote a major effort to reducing this inefficiency. It is assumed here that carnivorous organisms will be cultured in the future, and in a major way, because this is what the wealthy markets want. There are a few routes to approaching this problem. On the organism level, fish and shrimp foodstuff should be improved so that food conversion inefficiencies will be minimal. and so that protein in the food will be incorporated as protein and not used as an energy source. On the system level, efforts should be made to maximize utilization of resources through recycling, increase efficiency of the system, minimize energy expenditure, and reduce installation construction costs and other capital investments.

6 Future Development

As to future development, besides the subjects covered above, there are a few more directions into which mariculture is bound to proceed.

Sea ranching: Salmonid species have been used successfully in sea ranching since the turn of the twentieth century. This seems to be a very sound practice. Returns of hatchery-reared fry of 2% are common. A ratio of 1:7 investment to marketable returns in terms of dollars is claimed for coho salmon *(Oncorhynchus kisutch)*, and 1:3.5 in the same terms for chinook salmon *(O. tshawytscha)* in North America (Pillay, 1976). On Hokkaido Island in Japan, 2.5% returns of salmon have been recorded in the 1970s (Matsuda, 1979). Such practices could and should be followed by a thorough study of the interaction between the released animals and the hosting ecosystem. The understanding of such interaction would put sea ranching on a rational basis without disturbing the ecosystem and would ensure successful production of the salmonid species. Sea ranching started in the Northern Hemisphere, and the salmonids are mostly absent from the Southern Hemisphere. Although some trials to introduce them there in the past failed, the prospect is promising and should be tried again in a well-organized experiment. If successful, high-quality edible fish would be added to the overall fish landings in the Southern Hemisphere.

Juveniles of other fish species, and postlarvae of shrimps and blue crabs are being released into Japanese waters annually in order to support and enhance the natural recruitment of these species. There, it is claimed to be successful (Matsuda, 1979), and in the future this technique could be widely spread; controlled genetic selection could be incorporated into it.

Open sea mariculture—artificial upwelling: A thorough theoretical study of the possibilities of culturing marine organisms using artificial upwelling was published in the mid 1970s (Hansen, 1974). The study discusses the principles of open sea mariculture, the approach that should be taken to bring about its realization, and its main and marginal benefits. Once installed, the system is run on low energy and uses upwelled nutrient-rich water and solar radiation for primary productivity (mainly growing phytoplankton), on which other organisms are cultured in sequence. This idea has received further consideration with the OTEC (ocean thermal energy conversion) program in which huge quantities of deep, cold water (nutrient rich) are planned to be pumped to the surface for energy production. Such water, after its temperature change has been used, can be used for mariculture. However, though theoretically possible, the technology of growing a food chain on the open sea is not yet in existence. On the other hand, artificial upwelling for land-based mariculture has been experimented with on St. Croix, the U.S. Virgin Islands, yielding encouraging results (Othmar and Roels, 1973; Roels et al., 1975; Sunderlin et al., 1975; Rodde et al., 1976; Langton et al., 1977). Such technology can be developed in the future wherever land borders deep seas (atolls, volcanic islands) primarily along the equatorial belt, where solar radiation is plentiful.

Power plants and mariculture: Any intensive mariculture farm needs a large quantity of fresh sea water in order to supply oxygen and to remove excreta and accumulated substances. Pumping this water is an energy-expensive operation. However, this expenditure could be curtailed if mariculture farms could be built around power plants, conventional or nuclear, which pump large quantities of water daily to cool down their turbines or reactors and then discharge the water back to the sea. Should such water be used in mariculture operations, the power plants' cooling systems must be designed accordingly. (For example, facilities must be installed to prevent massive kills of farmed organisms with antifouling chemicals used to clean intake pipes.) If water is pumped, wherever possible, from under the thermocline, fewer fouling organisms would be introduced, less water would be pumped, and the water would be richer in nutrients. The temperature of the outflow water would always be higher than the ambient and would enhance growth rates of most of the marine cultured organisms. Since power is a basic commodity in western civilization, and its consumption is ever growing, more power-producing plants will be constructed the world over, mainly along the oceans' land and air interface. The cooling water discharged from power plants, charged with heat waste, should be used for mariculture around the world.

Desert mariculture—integrated culture system: It is not surprising that in Israel the development of mariculture is directed toward the deserts of the country. More than 50% of Israel is desert. Fresh water is used to the maximum (96% of the renewable freshwater resources of the country is in use). Most of the desert is nonarable due to poor soil quality. Means of livelihood are scarce and are mainly based on winter agriculture (no rain to be considered), date palm plantations, light industry, and to a small extent tourism. The combination of high temperatures, abundant solar radiation and sea water should be turned into a powerful system for culturing marine organisms. Here is a possibility for another, badly needed, means of livelihood for desert dwellers. Here, sea water may replace fresh water and nonarable land will become productive. If the IOLR (Israel Oceanographic and Limnological Research) Mariculture Laboratory's research is economically successful, then it has a good chance to serve as a model for many countries with deserts along their shores (North Africa, East and West Africa, Middle East countries, Australia, or western South America).

The evaporation rate in desert climates is extremely high (around 1 cm day^{-1}); therefore, seawater mariculture ponds have to be continuously flushed at a rate which will keep salinity in ponds at a physiological level. However, this

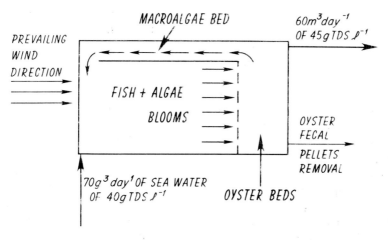

Figure 3. Schematic presentation of fish/algae/oyster/macroalgae integrated sea water pond. Numbers refer to 1000-m^2 pond (TDS equals total dissolved solids).

flushing consumes energy, which is proportional to the surface area of the pond. Thus, the economic rationale dictates a highly intensive culture system: as high a biomass as the system will allow. Animals, fish or shrimps stocked at high density must be artificially fed. Only 10% of the food, at best, is incorporated into the fish/shrimp. The rest is returned to the water as feces, urine, and CO_2.

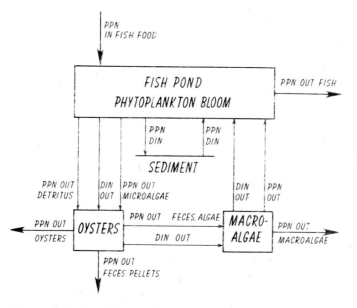

Figure 4. Nitrogen flow in integrated pond system (PPN equals particulate nitrogen; DIN equals dissolved inorganic nitrogen).

Degradation of the first two will result in an abundance of nutrients which, coupled with intensive incident radiation, will allow phytoplankton to bloom. By circulating the water from the fish pond through an oyster bed, the phytoplankton will be removed and will allow the oysters to grow at commandable rates. The water is then returned to the fish section through a macroalgae trough which serves as a nutrient trap and permits the system to have another crop (Fig. 3). The proposed integrated system would maximize crop per every cubic meter of sea water which goes through it. The system would utilize nutrients given as fish food, but not incorporated into the fish, at two or more levels (oysters and macroalgae) (Fig. 4). The best results of the different components of the described system, not yet integrated, were about 9000 kg ha^{-1} yr^{-1}, 13 tons of oyster per hectare (of fish pond) per year. Primary productivity as measured by ^{14}C incorporation (light and dark bottles) was 33 tons of algal dry weight per hectare per year. There are reasons to assume that yields of fish and oysters can be increased several fold (Gordin et al., 1980).

Israel is planning a hydroelectric plant which will drop sea water from the Mediterranean 400 m down to the Dead Sea (Fig. 5). There are plans to use some of this water for a few thousand hectares of intensive mariculture.

To conclude, one can say at present that mariculture has great potential in producing food for human consumption in the future. The problems of food and starvation on our planet cannot be solved by mariculture alone, but it can help. If research and development in mariculture can be widely supported, it might live up to our expectations by the turn of the century and exceed them in 50 years.

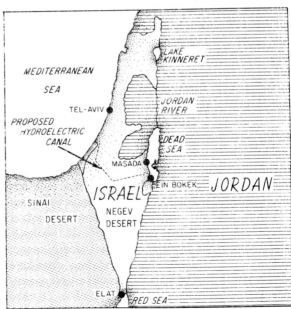

Figure 5. Path of proposed hydroelectric canal.

References

Bardach, J. E., J. H. Ryther, and W. O. McLarney. 1972. Aquaculture: The Farming and Husbandry of Freshwater and Marine Organisms. Wiley-Interscience, New York.

Chapman, W. M. 1969. Some problems and prospects for the harvest of living marine resources to the year 2000. J. Mar. Biol. Assoc. India, 11, 1-39.

FAO (Food and Agriculture Organization). 1979. Yearbook of Fishery Statistics, 1978. Report No. 46, Rome. Unipub, New York.

Global 2000 Report to the President–Entering the 21st Century. 1980. The summary report, Vol. 1. 47 pp. The technical report, Vol. 2. U.S. Government Printing Office, Washington, D.C.

Gordin, H., F. Motzkin, W. L. Hughes-Games, and C. Porter. 1980. Seawater mariculture pond–an integrated system. German-Israeli Aquaculture Symposium, March, 1980. Proceedings of the European Mariculture Society, In press.

Hansen, J. A. Ed. 1974. Open sea mariculture–Perspectives, Problems and Prospects. Dowden, Hutchinson & Ross, Inc., Stroudsburg, Penn., 410 pp.

Hepher, B., E. Sandbank, and G. Shelef. 1978. Alternative protein sources for warm-water fish diets. In: Proc. World Symp. on Finfish Nutrition and Fishfeed Technology, Hamburg, June 20-23, 1978, Vol. I (J. Halver and K. Tiens, Eds.). Heenemann, Berlin.

Langton, R. W., J. E. Winter, and O. A. Roels. 1977. The effect of ration size on the growth and growth efficiency of the bivalve mollusc *Tapes japonica*. Aquaculture 12, 283-292.

Lin, S. Y. 1954. Chinese system of pond stocking. In: Proc. Indo-Pacific Fisheries Council, pp. 65-71.

Lovell, R. T., R. O. Smitherman, and E. W. Shell. 1978. Progress and prospects of fish farming. In: New Protein Foods, Vol. 3. Academic Press, New York, pp. 261-292.

Matsuda, Y. 1978. The growth of aquaculture in developing countries: potentials, patterns and pitfalls. Fisheries 3, 2-6.

Matsuda, Y. 1979. Factors limiting the development of aquaculture: a Japanese experience. Woods Hole Oceanogr. Inst. Tech. Rep. WHOI 79-47.

McGoodwin, J. R. 1979. Aquaculture development in rural atomistic societies. Woods Hole Oceanogr. Inst. Tech. Rep. WHOI 79-53.

Nägel, L. 1979. Aquaculture in the Third World. Animal Res. Devel. 9, 77-115.

Nash, C. E. 1974. Crop selection issues. In: Open Sea Mariculture (J. A. Hansen, Ed.). Dowden, Hutchinson & Ross, Inc., Stroudsburg, Penn.

Othmar, D. F., and O. A. Roels. 1973. Power, freshwater and food from cold, deep sea water. Science 182, 121-125.

Pillay, T. V. R. 1976. The state of aquaculture 1976. In: Advances in Aquaculture (T. V. R. Pillay and W. A. Dill, Eds.). FAO Tech. Conf. on Aquaculture, Kyoto, 1976. Unipub, New York.

Ricker, W. E. 1969. Food from the sea. In: Resources and Man. National Academy of Science-National Research Council. W. H. Freeman & Co., San Francisco.

Rodde, K. M., J. B. Sunderlin, and O. A. Roels. 1976. Experimental cultivation of *Tapes japonica* (Deshayes) (Bivalvia: Veneridae) in an artificial upwelling culture system. Aquaculture 9, 203-215.

Roels, O. A., K. C. Haines, and J. B. Sunderlin. 1975. The potential yield of artificial upwelling mariculture. In: Proc. 10th European Symposium on Marine Biology. Vol. 1 (G. Persoone and E. Jaspers, Eds.). Universa Press, Wetteren, pp. 385-394.

Ryther, J. H. 1975. Mariculture: how much protein and for whom? Oceanus 18, 10–22.
Ryther, J. H., and J. E. Bardach. 1968. The status and potential of aquaculture. Vols. I and II. NITS PB 177-767 and PB 177-768. American Institute of Biological Science, Washington, D.C.
Sarig, S. 1978. Fisheries and fish culture in Israel in 1977. Bamidgeh 30, 91–103.
Sarig, S. 1979. Fisheries and fish culture in Israel in 1978. Bamidgeh 31, 83–95.
Schroeder, G. L. 1978. Autotrophic and heterotrophic production of microorganisms in intensely manured fish pond and related fish yields. Aquaculture 14, 303–326.
Sunderlin, J. B., P. T. Baab, and E. M. Patry. 1975. Growth of clam and oyster larvae on different algal diets in a tropical artificial upwelling mariculture system. In: Proc. World Mariculture Society, 6th Annual Workshop (J. W. Avault, Ed.). Louisiana State Univ., Div. Continuing Education, Baton Rouge, pp. 215–228.
Tal, S., and I. Ziv. 1978. Culture of exotic species in Israel. Bamidgeh 30, 3–11.
U.S. National Marine Fisheries Service. 1980. Fisheries statistics of the United States, 1979, April 1980. NOAA-S/T80-148.
Villaluz, D. K. 1953. Fish Farming in the Philippines. Bookman, Manila.
Webber, H. H., and P. F. Riodan. 1976. Problems of large-scale vertically-integrated aquaculture. In: Advances in Aquaculture (T. V. R. Pillay and W. A. Dill, Eds.). FAO Tech. Conf. on Aquaculture, Kyoto, 1976. Unipub FN-80, New York.

Technology and Communications: New Devices and Concepts for Ocean Measurements

D. James Baker, Jr.

1 Introduction

The focus of this paper is the development of technology and communications for ocean measurements as we look to the future. The technology is our means of observing the ocean; observations are central to understanding how and why the ocean works. Peter Kapitsa, in his collected works (1980) nicely made the point by saying: "Theory is a good thing, but a good experiment lasts forever."

For a history of the development and use of instruments in oceanography, the reader is referred to three excellent articles that were prepared for the history symposium (Sears and Merriman, 1980). The first is by Hendershott (1980) on the role of instruments in the development of physical oceanography; the second is by Herman and Platt (1980) on the coevolution of concepts and instrumentation for the study of mesoscale spatial distribution of plankton; and the third is by Spiess (1980) on the origin and development of deep-ocean instrumentation. In addition, the recent article which I prepared on ocean instruments and experiment design reviews the present state of instrumentation in physical oceanography (Baker, 1981). With these articles so recently in the literature, yet another review of the state of the art in instrumentation seems superfluous.

We face trying to develop the necessary technology for observations of the ocean in an economy where the growth of basic science has markedly slowed in recent years. Inflation, particularly in fuel and service costs, has driven up the costs of doing science at a time when state and federal governments have more and more demands placed on them. The following story could be titled "The director and the ONR team," or "The chairman and the dean":

One day a philosopher was walking in the woods and unexpectedly came face-to-face with the figure of God himself. The philosopher was awed only a moment, then came directly to the point. "You are the Lord, I presume." "Yes," said God, "I am." "Well, then, my Lord, I wonder if you would be good enough to answer for me a few simple questions that have been troubling me for a long time." "Certainly, my son." "Is it true, Almighty, that what is for us a million years here on earth is for you nothing but the merest moment?" "Yes, my son, quite true." "And is it also true," the philosopher went on, "that a million dollars here on earth is for you nothing but a paltry penny?" "Also quite true." The philosopher paused only a moment. "Then, I wonder," he said, showing some anxiousness, "if it would be possible for you, if it is not too much trouble, to give me a penny?" "Why certainly, my son," said God. "I'll be back in just a moment." (E. F. Schumacher, as quoted in Sale, 1980).

But money is not our only problem, although many innovative ways for getting more are being developed continuously by oceanographic institutions. Most important to us are new ideas, and people who are committed to carrying out the development and use of these ideas in the ocean. In oceanography we may be shorter on people than we are on ideas, and this is one of the limiting factors in the advance of the field.

My discussion of technology and communications for oceanography is presented in the context of automatic data collection. This kind of technology addresses specifically the studies of the climate, the general circulation, and ocean dynamics addressed earlier in the symposium. We are not yet advanced enough to help solve the sampling problems in biology, as so beautifully illustrated by G. R. Harbison in his discussion of biological communities in this volume, but we can note that new acoustical techniques are showing promise for monitoring biological communities *in situ* (Greenblatt et al., 1982). The future will certainly see automatic data collection become increasing important to the biological problems of the ocean as the need for long time series increases.

In several of the papers presented earlier in the symposium, we have seen a curious blend of the old and the new: historical distribution of properties presented side by side with new direct measurements. Often the agreement is not apparent. Part of the reason for the disagreement is that the direct measurements tend to be of relatively short duration and concentrated over small areas. The answer is to obtain longer direct measurements over larger areas. This is not easy; it requires much technical development; and it will not necessarily be cheaper than the way we do things now. And yet, new measurements are absolutely necessary if we are to advance our understanding of the ocean.

Three areas can be singled out as important in the development of automatic stations: new and improved sensors; the new microelectronics that permit increased data processing and storage *in situ* at lower power levels; and new techniques of data transmission and communication.

Because of the rich literature on sensors, I will not discuss that subject here but refer the reader to the articles noted above. I will discuss the other two areas and then conclude with a discussion of cost strategies and general points for the future.

2 Velcro, the Digital Orientation

At the beginning, thinking about how to utilize the microelectronics revolution (Noyce, 1977), one must orient oneself to a digital context. Digital technology is cheap and getting cheaper, but it has not had its full impact on the field yet because scientists often are not ready to think in these terms. An example of this reorientation in thinking is given by the invention of "velcro," a fastener.

In 1941, George de Mestral, a Swiss inventor and engineer, had spent the day hunting in the Jura Mountains with his Irish pointer. After he returned home, he noticed the annoying burdock burrs that clung both to his clothing and his dog. Many of us have done this, but de Mestral went a little further. He took his microscope and saw under it that each burr has hundreds of tiny hooks that grab onto loops of thread or animal hair. The commercial woven fastener that de Mestral developed is called "velcro" (from velour, velvet, and crochet, hook) (Kent, 1959a, b).

It is instructive to compare the velcro fastener to an ordinary snap fastener. The snap fastener, or a button, or any device like that, requires that an accurate and relatively complicated device be constructed. In addition, it must be reliable. The snap fastener is made of metal and has a little spring to hold the projecting knob. Why not use two little hooks of nylon thread instead? The reason is that as often as not it will not hold. One would need to use many such hooks, so that if a few did not hold it would not matter. The idea of many connectors is the essence of velcro. A typical sample has two parts: one with about 100 hooks per square centimeter, and one with about 10 times that number of fine nylon fibers that form a thick mat. As the advertising says, the parts go together easily, hold tight, yet open and close with a gentle touch. And it is better than a button or snap fastener because you can adjust the exact closing spot giving your sleeve or other clothing a perfect fit. Moreover, it is easier to connect and to sew on; these are unexpected benefits.

The point is that we can get the job of fastening done with either a few relatively complicated and expensive connectors, or with many simple ones. And the latter gives us some unexpected benefits. The concept is like the difference between a digital and an analog computer. The digital computer does its work by summing the results of millions of individual simple operations: switches that are either on or off. By focusing on doing many cheap operations, the designers have created a whole new way of handling information.

We can also think of the analogy as suggesting the importance of many relatively cheap measurements that cover a broad area as opposed to expensive, detailed measurements at a point. Each has its good points and each is used for looking at different things. We do tend, however, to think in the more detailed terms. Yet the broad measurements, even if some of them are relatively inaccurate, do provide patterns or maps that can give new insights into problems not available from point measurements. Moreover, the data contribute collectively to the accuracy of the maps, as the objective analysis technique shows (Gandin, 1965).

3 Microelectronics in Oceanography

We are led naturally to the use of microelectronics from the previous discussion. In the past, conservative instrument design has suggested that measurement devices must be very accurate, and that it is expensive to carry out *in situ* calculations on signals and data storage.

But we can challenge these notions now. Accuracy required depends on experimental design and tradeoffs between spatial and temporal coverage; the systems analyst asks how little you need to know. In terms of expense, the cost of electronics for any given purpose is dropping faster than, for example, the number of graduate students going into physical oceanography.

In fact, for integrated circuits, the doubling of usage because of the growth of the industry occurs about every $1\frac{1}{2}$ years. This means a cost reduction of about 30% during this time. Annual use of solid state electronics has doubled 13 times since 1960 when the first integrated circuit was introduced by Fairchild (Noyce, 1977). The implications for capacity and costs are enormous: in 1971, Intel introduced the 1000-byte (usually eight or nine bits, represented by 0 or 1) random access memory. The 64,000-byte unit will be available later this year. The cost of storing a single bit has gone from 1.3¢ to 0.026¢, a factor of 50 in the last 5 years. If the cost of a Rolls Royce had evolved in comparable fashion since 1971, it would be about $10 today.

This reduction in costs means that now in most devices the integrated circuit cost is only about 10% of the total: in a $10,000-minicomputer, there are about $1,000 worth of integrated circuits, and in a $300 television set about $30 worth.

Microelectronics had led naturally to the development of computer central processing units on a single chip—the microprocessor. With its arithmetic and logic units, program memory, random access memory, and interfaces for input and output signals, the microprocessor can improve sensing, process data, and store data for later retrieval.

A good example of sensor improvement has been suggested by V. Suomi of the University of Wisconsin: a barometer from a beer can. A crude aneroid with microprocessors to compensate for the nonlinearities and temperature dependence, which could be programmed into the system, could be made into a very precise instrument indeed.

But the microprocessor allows us to do more than just store data; whole programs can be written, stored, and used to process data *in situ*. Thus, the processor, if properly programmed, can do many of the actions that formerly required hardware. In other words, software can replace hardware. This is an important point of instrument design, and one that will become increasingly important in the future. You can see that with the new microelectronics the distinction between sensing and data processing begins to be blurred—data processing helps to improve the sensor.

A good example of this has been suggested by W. Munk of Scripps. He has proposed a tsunami recorder that would be programmed to look only at the long, fast waves typical of that phenomena, and then to send out a warning. The simple pressure device with the microprocessor programming is a much simpler and more reliable system than one that uses mechanical valves and filters.

In data storage, microelectronics has had a major effect, and we can expect this effect to continue. Tape recorders now customarily have tape cassettes providing 10^6 to 10^7 bits, and tape reels allowing storage of more than 10^8 bits. The low power associated with microelectronics allows the cassette recorders to record 10^7 bits with only 4 watt-hours of power required for the entire record. The new tape recorder is clearly a big step up from Ekman's rotating compass and ball bearings dropping into compartments. But it still has one disadvantage— moving parts. Moving parts are notorious for being the source of reliability problems.

Microelectronics may have the answer here, too. A whole variety of memory devices are now available and being developed; among these the magnetic bubble devices are perhaps the most promising. The devices exploit the mobility of microscopic domains of magnetic polarization in thin films of orthoferrite or garnet. The domains are moved in the plane of the film by applying weaker magnetic fields generated by arrays of electrodes. The data are stored and accessed serially as in a tape recorder (in other words, no random access), and are retained without power. The price of these devices is reasonable: in October of 1980, a one megabit memory (on a 10-cm by 10-cm board) would cost about $1,000, and the cost by August of 1982 is estimated to be down to $300. You need to buy 5000 to get these prices, but the advantage of the solid state device is clear. While the tape recorder costs slowly edge upwards because of non-electronics-related costs, the microelectronic storage costs are going down. The bubble devices use too much power to be useful at the present for long term *in situ* storage, but the direction is clear.

Looking toward the future in microelectronics, the experts feel that the physical limits are imposed by electric breakdown and heat production. In order to apply signals, the applied voltages must be greater than the random thermal energy of electrons. As the devices get smaller and smaller, the breakdown of the dielectric is reached. Moreover, as more and more devices are added to the circuit boards, more and more heat is generated. Both of these problems can be addressed with lower temperature operation, a solution that is feasible for computers but not so obvious for oceanographic instruments *in situ*. The costs of producing microelectronics has worked against custom-designed devices that would find only a few applications, but we can expect to see more custom design in the future as markets expand.

It is important to note that this decreased cost of microelectronics does not reduce our total cost of doing business. It actually means that we are able to do more with the resources we have. The expanding capability of instruments gives us new pictures of the ocean, and these in turn suggest new experiments and new kinds of instruments, all of which require new funds.

4 Satellites and Communication

Satellites are important to oceanography because of their potential for global synoptic views of the ocean and for providing measurements of some parameters that would be difficult from ships or buoys. J. Whitehead has summarized some of these types of measurements earlier in this volume, so I will not go into that particular aspect of the subject here. However, I was able to find a good example of this global-scale view which is immediate and revealing: a text of a recorded conversation between Soviet cosmonauts aboard SALYUT-6 and flight control center in Moscow (July 4, 1979):

> (Cosmonaut): "At 10:44, Moscow time, East of Africa, about 250–300 km there was an agitation in the water, up to 100 km long. It was hard to be sure of the width, maybe one km, maybe two, but the upsurge of the water was clearly visible."
> (Grechko, at flight control center): "How did it lie? On a meridian, or along the Equator?"
> (Cosmonaut): "Along the Equator."
> (Grechko): "Maybe it was the Equator you saw!"
> (Cosmonaut [laughing]): "No! The sea was quite calm, the water was calm, and two waves could be seen colliding against the background of this calm water. A strip like this is quite astonishing. It's the first time we've seen anything like it."
> (Second cosmonaut): "It was an amazing sight."
> (Grechko): "What happened? Did the wave subside, or did you leave the area?"
> (Cosmonaut): "We left."
> (Grechko): "A pity you were going so fast! It would have been a good thing if you could have slowed down!"
> (Cosmonaut): "So it would."
> (Grechko): "But, you know, one thing you did have time to do. Acting on your instructions, three trawlers caught 38–45 tons of fish each day for three days in succession."
> (Cosmonaut): "Where was that? What area?"
> (Grechko): "In the Atlantic, just where you had said. It's nice to know these things, isn't it?"
> (Cosmonaut): "Yes, it is."

But I want to focus here on satellites and communication, as separate from the subject of remote sensing. For oceanographic purposes we want to know what is going on inside the ocean just as much as we need to know surface parameters, and in many cases, the surface parameters need to be supplemented by *in situ* measurements, either for calibration or for completeness. For example, measurements of wind stress by satellite scatterometer require *in situ* stress measurements for assessment of accuracy (NASA, 1980a), whereas altimeter measurements of surface ocean topography require density measurements in the water column for estimation of the total geostrophic velocity (NASA, 1980b).

Moreover, the biologist, chemist, or geologist will need on the whole measurements which cannot be made by satellite; their use will involve primarily data communication. (The question of chemical measurements by satellite has been discussed in some detail by Anderson, 1980.)

One important aspect of data communication is platform location and data collection. The ARGOS system currently in use is a cooperative project between the Centre National d'Etudes Spatiales (CNES) of France, and NASA (National Aeronautics and Space Administration) and NOAA (National Oceanographic and Atmospheric Administration). The onboard equipment package is in orbit on board the TIROS-N and NOAA-6 satellites. Other packages are planned to be placed on board the six operational satellites designated NOAA-B through NOAA-G which will be launched at such times as to ensure that two operational satellites are available at all times through 1986 at least.

The orbits of the two satellites are circular, sun synchronous, and polar. At any given moment, each satellite sees all the platforms located within a circle of 5000-km diameter; as it orbits, the ground track of this circle corresponds to a swath 5000 km in width encompassing the Earth.

The system works: during the First Global GARP (Global Atmospheric Research Program) Experiment (FGGE) year of 1979, participating countries supplied some 368 drifting buoys measuring barometric pressure and sea-surface temperature. The buoys were deployed from 41 ships by 15 nations, mainly in the Southern Hemisphere. At the same time, 300 balloons were released near the equator and were tracked at an altitude of 14 km.

The overall capability of the ARGOS system is an accuracy of location of about 1 km, and a capacity of 16,000 platforms for data collection only or 4000 platforms for location and data collection, assuming that both types are reasonably evenly distributed over the Earth's surface. Over 80% of the results have been available to users at the Toulouse processing center within 6 hours after the onboard recording of the corresponding message. As noted above, the system will continue at least to 1986, but costs to the user will rise. Data processing had been provided free of charge to the users during the 1979 FGGE year as part of the French contribution to that experiment; since the end of 1979, all countries must contribute both for the continued processing of FGGE drifting buoys that remain and for new operations.

The scientific and operational need for such satellite transmission of oceanic and atmospheric data is great. Wyrtki (1980) has summarized the requirements for monitoring the oceanic and atmospheric environment by means of buoys, both moored and drifting.

Of particular importance for the future is an ocean data management and handling system that can handle the large amounts of data that are now and will be coming through these satellites and monitoring systems. The recent report from the Office of Technology Assessment of the U.S. Congress (OTA, 1981) finds that federal programs have not given adequate attention to the handling and distribution of oceanographic data. The report notes that large amounts of data are stored unused because their nonstandardized formats are incompatible with user needs and because they are too difficult for many to retrieve from the archives. At present, the NOAA archives seem to be unable to handle the digital data stream from existing platforms in near-real time. New programs like the National Oceanic Satellite System (NOSS) that generate new data streams will only exacerbate the problem, and in fact, a major new data management system is being planned as an adjunct to the NOSS program.

The report of the National Advisory Committee on Oceans and Atmospheres (NACOA Task Group on Ocean Operations and Services, 1981) has recommended that a fast-response, technologically advanced ocean environmental data archival and dissemination system be established by the end of the decade. Elements of such a system include both an integrated data center system that pulls together the separate ocean environmental data centers, and full use of new technology of data base management including mass storage, microcircuitry, rapid communications, and flexible displays. The emphasis that these two major studies place on the data management problem shows the growing importance of the subject.

A second aspect of satellite communication is person-to-person or telecommunications. Here we see the link between computers, microelectronics and satellites. The growing interconnection between computers and telecommunications in information management has been recognized nowhere better than in France, where their term *télématique* summarizes a new technology reality.

The term was coined in a recent report to the French Government by Nora and Minc (1980). Their thesis is that the new innovations in electronics and links between computers and communication systems could transform society in the way that railroads and electricity did in the nineteenth century.

Nora and Minc argue that just as roads, railways, and electricity are stages along the way from the family unit to local, national, and multinational organizations, the link of telecommunications and computers provides an additional network blending pictures, sounds, and memories with the capability of transforming our culture. For individual countries, this increasing interconnection between computer and telecommunications has important implications for economic balance, power relationships, and social structure.

The French in particular are worried about IBM. They see this major private company moving from a manufacturer of machines to a telecommunications administrator. As it does that, they see it encroaching upon communications, which they see as a traditional sphere of government power.

In the absence of a suitable policy, they see alliances developing between France and IBM instead of between France and the U.S. Government. In their view this requires immediate and strong action by France to develop a strategy for telecommunications and the use of computers (*télématique*) so that they are not left out in the cold. In addition, they do not want to see American technological development affecting the social structure of France any more than necessary.

The rapid changes in electronic components, as discussed above, have had implications across the board in the computer field. The distinctions between large and small computers, between access terminals and processing centers, and even between hardware and software are beginning to disappear. The Apple computer company makes an important point in their literature: the first electric motors, invented in the late nineteenth century, were large and expensive and used to power entire shops with pulleys and belts. Only with the advent of the fractional horsepower motor could power be brought to where it was needed. The new microelectronics and associated microcomputers like the Apple allow distribution of computer power in a similar way.

This distribution of power to the people is a modern example of the problem noted by Helland-Hansen, who, when asked which of his many contributions to oceanography he himself considered to be most important, replied: "the demonstration that work at sea can be undertaken from a small vessel which can be operated inexpensively" (Sverdrup, 1956). Before the use of the 70-foot ketch *Armauer Hansen,* work at sea had been conducted from large ships placed at disposal for special expeditions. Helland-Hansen's initiative is still reflected in the character of research vessels, and we can see this theme reflected in computer usage. Clearly the trend is toward smaller computers, linked as required to larger ones for specific purposes. Administrators would do well to note this change away from group use of large computers to individual use of mini- and microcomputers.

The electronic mail systems now coming into use, the "living atlas" concept proposed by H. Stommel many years ago, and the access to all literature now available to in medicine and law but not yet in oceanography, are all good examples of how computerization for communication of information can be effective in scientific work.

The associated new developments in communications are important for us. For example, the Woods Hole/MIT Joint Program is considering the establishment of a microwave link between Woods Hole and Boston for transmitting class lectures. The technology is there for an interactive system; all it requires is the funds. One could take this a step further—why not broadcast the best courses from the appropriate institution to all those educational and research institutions that are interested? Special seminars could be done the same way. The impact on groups that feel isolated or understaffed could be enormous. And this does not have to be done necessarily in real time: lectures, seminars, etc., could be videotaped and traded back and forth among institutions. This exchange of videotapes of special seminar series could be an appropriate activity for a scientific society to maintain.

In terms of meetings, new communications technology could be an enormous help. The currently available technology could reduce the cost of travel and the associated wear and tear by at least a factor of two; all we have to do is use it.

AT&T has a "Picturephone" system which consists of special studios where customers can see one another and talk over a closed circuit television link. The system uses standard TV monitors to show the parties talking. Voice-actuated cameras zoom in automatically on the person doing the talking among the many who may be in the studio. Current costs are from $150 to $400 per hour. How does that compare in cost with a typical meeting in Washington, D.C.? Twelve people meeting for a day and a half, three from the west, three from the midwest, and three from the east, will have a travel cost of about $6,000; lost time due to travel (not counting the meeting) is about $1,500 for a total of about $7,000. Effective meeting time is about 10 hours. At a Picturephone rate of $150 per hour, the total cost for three centers is $4,500, almost half. The breakeven point is at about eight people. Even when we add the cost of traveling to the Picturephone center, and the fact that a certain amount of travel is required for individual discussion which would still have to go on, the cost by Picturephone is significantly less.

The teleconference is happening in industry more and more. Holiday Inns has established a subsidiary called Hi-Net that uses the Home Box Office cable television system for private meetings during daytime hours among customers gathered at various Holiday Inns nationwide.

Not only is the cost less right now, but it should be abundantly clear to everyone now that the costs of individual travel are going up, being primarily fuel and service related, while the cost of communications is going down (or at least going up more slowly) because the latter is related to the costs of data handling. The current Federal Communications Commission (FCC)-approved rates for Picturephone expired in June 1981. At that time AT&T had to decide whether to offer a widespread Picture meeting service, abandoning the current "experimental" service, and if so, what the service would cost. Government agencies and oceanographic institutions should take notice. There are some good opportunities here for reducing our cost of doing business.

5 Cost Strategies

Reduction of costs brings me to my final point. The new technology will be available, but will we be ready to use it and can we afford it? There are four points that can be emphasized here: the long leadtime for development of instrumentation; the hidden costs of new ways of collecting data; rising platform costs; and new ways of jointly funding expensive field programs.

In his paper on eddies and ocean circulation in this volume, Rossby notes the long leadtime necessary for developing instrumentation, typically 10 years from idea to reliable operation. Spiess makes similar points in his paper in the companion volume to this, and he goes on to point out the difficulty in getting funding for such development. This difficulty is a problem that must be solved if we are to have a vigorous instrumental development that is commensurate with the needs. The need for long leadtimes must be built into the funding structure.

The hidden costs of measurements lie in the data archival and dissemination systems that I discussed in the previous section. This part of the total system is necessary and must also be considered part of the cost. The important point is that the data systems must be built in at the beginning. This will probably require new institutional arrangements, and joint agreements between oceanographic and atmospheric institutions with participation from government laboratories. Since the current data stream is already too hard to handle, it is not too early to start so that the data from the operational and experimental satellites already up and planned will be readily available to investigators.

What about ship costs, which are rising faster each year than the funding available? The major culprits are salaries and fuel. Institutions have been trying hard to make the fleet cost effective, and the University National Oceanographic Laboratory System (UNOLS) is working with NSF and the U.S. Navy to reduce costs and still provide a useful facility to the community. New strategies are being discussed: we may have to go to regional centers, and more charter of private vessels. We may see more charter of institution vessels by government and private industry.

In terms of new ships, two directions for economy can be noted. One is to take a more careful look at the virtues of design using existing hulls, as opposed to design from the keel up. One sacrifices some freedom in design but gains savings in costs by use of standard designs. Are the new ships significantly better than some of the old hand-me-downs? In the past, oceanographers have argued strongly for design of their own ships, and the agencies have complied. I believe that not everyone would agree that the result has been entirely happy. In any case, over the next 20 years the federal fleet of about 80 research and survey ships will require replacement, at a cost of about $1.5 billion in 1980 dollars. Some new policies need to be established. Both the OTA and NACOA reports mentioned earlier address the question of ships, their projected uses, and funding requirements.

One obvious saving in vessel cost is fuel. In 1982 the estimated cost of fuel will be about 3.6 times the cost of fuel in 1978, representing 26% of the cost of operation instead of 13%. Can we go back to sail? The Japanese tanker *Shinaitoku Maru* is already in operation using computer-controlled sails to assist its engines (Renner-Smith, 1980). This vessel is the third stage of a program conducted by Nippon Kokkan, a shipbuilding firm, and the Japan Marine Machinery Development Association.

Renner-Smith notes that the Japanese vessel has both sails and a longer, finer hull than a conventional freighter; both factors contribute to fuel savings. Each sail is mounted to a thin steel frame that widens to form an airfoil at each edge. The mast pivots, rotating the sails so that they are always at the best angle to the wind, using an onboard microprocessor that constantly adjusts mast position, engine speed, and propeller pitch in order to save fuel. The sails and engineer work together to keep the ship sailing at a constant 12 knots. Sails have also been used on offshore oil rigs to add some extra speed when the rig is towed to a new location. A number of other commercial sailing vessels are being planned.

The U.S. Maritime Administration has awarded the Wind Ship Development Company a contract for an analysis of the potential for sail on 63 shipping routes. Preliminary indications are that a proper combination of screw propulsion and sail power can propel ships more economically and cost effectively than is possible under either power or sail alone. The Ocean Sciences Board of the National Research Council has also commissioned a study into the use of sail on research vessels.

A fourth area for cost reduction is joint funding between government and private industry. A bridge between the universities and industry is required. A succinct statement of the needs was presented by Fusfeld (1980), who pointed out that with the growth of government support, industrial research has become stronger internally. At the same time, the universities have leaned more and more in the direction of government, away from industry. With the slowing of federal support, a maturing sense of mutual benefits and interdependence has emerged. The universities and industry are now building toward long-term relations that are mutually beneficial.

Fusfeld notes that future growth looks promising, with a special NSF program for stimulating joint research proposals. Oceanographers have been in the lead here, thanks to Press (1981), who in his capacity as the President's Science Advisor, urged the community to look at joint funding with industry of the Ocean Margin Drilling Program, a deep-sea drilling program to extend the work begun by the *Glomar Challenger*. To date there has been a significant commitment for the first phase of the program from the petroleum companies, with a total of 10 companies already agreeing to provide funds. Industry will fund half of this program. We expect to see more of this in the future.

6 Summary

I have addressed here some of the issues that are important in thinking about the future in the context of needs for automatic data collection and information transfer. Microelectronics and computers help us collect the data; new communications systems transmit data and allow scientists to talk to one another about it; and the whole activity must be cost effective.

In the context of this symposium, what can be said about the future and our planning for it? Most planning is done by linear extrapolation; yet as the future comes upon us, we see things change by fits and starts. Quantum jumps in understanding, in capability, and in needs are the rule rather than the exception. Changes often seem to be more exponential than linear.

The practical problems for the near future, mostly related to population growth, have been laid out in a number of recent documents: *The Limits to*

Growth (Meadows et al., 1972) looked at the long-term issues of global population growth, agriculture, resource use, industry, and pollution, and discussed their interaction and possible scenarios for the future. *The Global 2000 Report to the President* (Council on Environmental Quality and Department of State, 1980) reinforces the fact that the world faces decades of difficulties in maintaining its physical environment. *Human Scale* (Sale, 1980) examines ways to shape a more efficient society in the face of population and institutional growth. We need strong steps to improve social and economic conditions, reduce population growth, manage our resources, and protect our environment.

Scientists and engineers will be required to help in these efforts, and in fact will be central to many of them. We see increasingly a push for oceanography to do only that science related to practical problems. But the practical side is notoriously short sighted, and must be, because specific problems must be solved. Since the future is in fact not to be predicted linearly, basic research, without specific direction from practical problems, is necessary. In 50 years our society may be different—for example, less urban—and we need the science base to address the different problems. By responding to the fundamental scientific needs we achieve this basis for the long-term solution to practical problems.

References

Andersen, N. R. 1980. The potential application of remote sensing in chemical oceanographic research. Office of Naval Research, ONR West, 80-1. University of California, San Diego, 82 pp.

Baker, D. J., Jr. 1981. Ocean instruments and experimental design. In: Evolution of Physical Oceanography (B. A. Warren and C. Wunsch, Eds.). MIT Press, Cambridge, Mass., 623 pp.

Council on Environmental Quality and Department of State. 1980. The Global 2000 Report to the President. U.S. Government Printing Office, Washington, D.C., 766 pp.

Fusfeld, H. I. 1980. The bridge between university and industry. Science 209, 221.

Gandin, L. S. 1965. Objective Analysis of Meteorological Fields. Israel Program for Scientific Translation, Jerusalem, 242 pp.

Greenblatt, P. R., E. Shulenberger, and J. H. Wormuth. 1982. Small-scale distributions of zooplankton biomass. Deep-Sea Res. Submitted.

Hendershott, M. C. 1980. The role of instruments in the development of physical oceanography. In: Oceanography: The Past (M. Sears and D. Merriman, Eds.). Springer-Verlag, New York, pp. 195-203.

Herman, A., and T. Platt. 1980. Meso-scale spatial distribution of plankton: Co-evolution of concepts and instrumentation. In: Oceanography: The Past (M. Sears and D. Merriman, Eds.). Springer-Verlag, New York, pp. 204-225.

Kapitsa, P. L. 1980. Experiment, theory, practice: Articles and addresses of P. L. Kapitsa. Boston Studies in the Philosophy of Science, Vol. 46 (Reidel), 429 pp.

Kent, G. 1959a. Velcro: newest magic fastener. Reader's Digest, March 1959.

Kent, G. 1959b. Velcro. Die Weltwoche, February 6, 1959.

Meadows, D. H., D. L. Meadows, J. Randers, and W. H. Behrens, III. 1972. The Limits to Growth. Pan Books, Ltd., London, 205 pp.

NACOA Task Group on Ocean Operations and Services. 1981. Ocean services for the nation: National goals and objectives for services to ocean operations in the 1980's. National Advisory Committee on Oceans and Atmosphere, Washington, D.C.

NASA. 1980a. Guidelines for the air-sea interaction special study: An element of the NASA Climate Research Program. Publication number 80-8, Jet Propulsion Laboratory, California Institute of Technology, Pasadena, 56 pp.

NASA. 1980b. NASA Oceanic Processes Program. Status Report Fiscal Year 1980. NASA Technical Memorandum 80233. 157 pp.

Nora, S., and A. Minc. 1980. The Computerization of Society. MIT Press, Cambridge, Mass., 186 pp.

Noyce, R. N. 1977. Microelectronics. Scientific American 237, 62–69.

OTA (Office of Technology Assessment). 1981. Technology and Oceanography: An assessment of Federal technologies for oceanography research and monitoring. OTA, U.S. Congress, Washington, D.C.

Press, F. 1981. Science and technology in the White House, 1977 to 1980: Part I. Science 211, 139–145.

Renner-Smith, S. 1980. Computerized high-tech sailing ships. Popular Science 217(6), 78–81.

Sale, L. 1980. Human Scale. Coward, McCann, and Geoghegan, New York, 588 pp.

Sears, M., and D. Merriman (Eds.). 1980. Oceanography: The Past. Springer-Verlag, New York, 812 pp.

Spiess, F. N. 1980. Some origins and perspectives in deep-ocean instrumentation development. In: Oceanography: The Past (M. Sears and D. Merriman, Eds.). Springer-Verlag, New York, pp. 226–239.

Sverdrup, H. U. 1956. Transport of heat by the currents of the North Atlantic and North Pacific Oceans. Festskrift til Prof. Bjorn Helland-Hansen. A. S. John Griegs Boktrykkeri, Bergen, Norway.

Wyrtki, K. 1980. Scientific and operational requirements for monitoring the ocean-atmosphere environment by means of buoys. NOAA Data Buoy Office, U.S. Dept. Commerce, Washington, D.C., 43 pp.

Institutional and Educational Challenges

John H. Steele

1 Introduction

The title of this section is "Institutional and Educational Challenges." Immediately we are faced with definitions of these terms: what institutions? Do we include not only academic institutions, but government laboratories and federal agencies? Changes in the interactions between organizations will become as important as the developments within institutions. What education? Do we mean only our graduate students or do we include other scientists, other disciplines, and even the general public? Should education be framed for a particular purpose or as an end in itself?

These questions about interactions arose when Peter Brewer and I discussed the structure of this meeting. We considered basing it on the various disciplines with, say, separate days for physics, chemistry, biology, and engineering. We thought some sessions might be devoted to the "basics" and others to the "applications"; we might divide the problems by geography or latitude.

The Symposium eventually used the concept of scale to separate different topics. In this way we try to unite the disciplines, but this separation produces its own artificial divisions. If one feature identifies oceanography, it is the interaction between events with very different dimensions in space and time.

2 The Structure of Institutions

The problems in structuring this meeting epitomize the difficulties in organizing and institutionalizing oceanography. Any particular pattern has definite advantages but, in turn, creates unnecessary and sometimes unfortunate divisions.

At present we use the classical disciplines as the main method of managing the research: here in Woods Hole (and, I believe, in most other institutions) we have departments of physics, chemistry, and biology. There are two reasons for such divisions: in the old saying, "birds of a feather flock together;" also, each discipline displays a different pattern in its development. We have seen this in the talks during this meeting. Geophysics has the paradigm of plate tectonics giving a basis for an almost monolithic approach to organization. Physical oceanography displays best the division into scales. Biology is in a state rather like geophysics before plate tectonics. We hear pleas for diversity of individual biological studies and statements of need for large programs. If we accept the ideas of Thomas Kuhn, we might hope for a new paradigm in biological oceanography. Conversely, should we expect some fragmentation in the monolithic structure of marine geophysics? There are good reasons why the patterns for each discipline should be different and separate.

In several of the agencies, corresponding divisions into disciplines generate close ties between individual scientists and program managers. This system encourages flexibility and innovation in the direction of basic research within a discipline. It has proved to be a sound framework for the expansion of ideas into large programs such as **POLYMODE, GEOSECS**, and Deep-Sea Drilling Project. It is especially significant that the management of these programs, and others being planned at present, is undertaken by the scientists working directly in the projects. Thus we can meeet many future needs with multi-institutional studies in physics, chemistry, or geophysics.

This can be contrasted with what may be called the European system, which tends to discourage collaboration between different institutions because of the block funding to each particular laboratory. The laboratory then has more freedom to define its own directions but less incentive to work with other groups. However, this prototypical European system can make it easier to cross other boundaries within an institution—the disciplinary boundaries and the "pure vs. applied" boundary. As an example, there are difficulties in international cooperation in research on Antarctic marine ecosystems, where the questions involve both ecological problems and resource management—the survival of endangered populations. Some of the difficulties in collaboration appear to arise more from the differences between European and United States institutional structure than from divergence in scientific aims. In Europe the same institution and even the same individual may be responsible for organizing and implementing these different aspects, whereas here different bodies are given responsibility for the different aims.

3 New Concepts and Patterns

This example is not intended to set one institutional pattern against another but to indicate the way in which such patterns in the interactions of scientist, institution, and funding source influence the development of the science. The great achievements of the last decade are strong evidence in support of our present system in the United States. The new problems for the future, of which climatic change is the outstanding but not the only example, suggest that the system must be extended to include other methods of integration. As a rough guide, the intrinsic problems of oceanography tend to exist within a single discipline and be supported by a single agency, whereas those posed extrinsically cross the disciplinary boundaries and involve several agencies not only as sources of funds but also as participants in the work. It is not clear that we have devised completely adequate means of structuring such interactions.

In this context, we must consider the limitations on oceanographic institutions not only as contributors to important social questions, but also as sources of knowledge for its own sake. We have accepted, on the one hand, that we will always need input of new ideas and techniques from the classical subjects. On the other hand, we are now conscious of the interrelations of the terrestrial, aquatic, and atmospheric environments. Our understanding of the transports of materials and energy within natural systems depends critically on knowledge of the transfers between all three compartments. The carbon and nitrogen cycles are prime examples of intrinsic as well as social problems. Thus Earth Science (used in its broadest sense to include air and water, and also chemistry and biology) must become the unit within which we shall need to look at the practical problems that shall face us in the future. Is such a grouping too incoherent or too amorphous to be manageable at the level of research or funding institutions?

Such a question brings us back from practical problems to consideration of the conceptual unity of our areas of study. The common feature of the Earth Sciences, which we tried to capture in the structure of this meeting, is the immense range in space and time scales of the events that we study. I believe it is this feature which separates our work from other branches of science, providing its unique quality and creating its special problems in our attempts to define the links between different scales. In particular, we have come to realize that some of the superficial uncertainties, appearing often as a lack of knowledge, prefigure deeper questions of indeterminacy. The best known examples are in weather prediction, but the same concepts are being studied for aquatic and ecological systems. The intellectual and conceptual development of this picture of an inherently indeterminate reality may be more significant, to society as well as to ourselves (and to our funding), than the application of our present knowledge to our immediate environment problems. I hope we can ensure a flexibility and variety in organization so that we can match the structure of future institutions to a science which is still growing, expanding its vision, and certainly changing rapidly in response to its own internal stimuli as well as to external pressures.

4 Conclusions

Many immediate problems face our institutions. We are concerned about the future of our research fleet in relation to rapidly increasing costs and this concern reflects the technological constraints on our understanding of the deep oceans. Improvements in technology will always be a short- and long-term need requiring planning across the whole oceanographic community. We worry about an adequate supply of graduate students in particular disciplines, and maintaining the quality of our community must be a paramount consideration. Because of the increasing number of practical problems involving or affecting the oceans, we emphasize the useful components of results, in industrial development and in regulation.

These are essential concerns but it is also necessary to develop a coherent view of oceanographic research which may have different patterns from the more classical disciplines. Also, we need to see these patterns in the context of other studies of the total environment. This view of the world can be and should be as rigorous and exacting as a training in mathematics (or Greek). It should be considered as valuable in itself, and not merely a basis for practical use in industry or research. Thus we must be involved in the education, not only of our future colleagues within oceanography, but of other scientists and of the larger society, in terms of the essential nature and unity of the Earth Sciences.

Index

Acoustic remote sensing 122
acoustic scattering layer 59
acoustic telemetry 57, 58
acoustics 110, 111, 112, 121, 122, 123, 212
Adams, J.W.R. 99
Aguilar, A. 283, 300
air-sea exchange 316, 318
Allen, C.M. 78, 84
Aller, R.C. 178, 187, 188
altimeters 262
Alvin 46, 130, 133, 220
Andersen, K.P. 196, 203
Andersen, N.R. 369, 375
Anderson, E.R. 74, 76, 85
Anderson, J.H. 344
Andrews, J.B. 86
animal behavior 59, 61, 66
Aniti, D. 254
Antarctic studies 50, 141, 243, 268, 270, 315, 317, 318
Apel, J.R. 280
aquaculture 347, 350ff, 357
Araskog, R. 165, 191
Arctic studies 184, 215, 268
ARGOS 265, 369
Armands, G. 228
Armi, L. 180, 187
aromatic compounds 9, 13
Arons, A.B. 311, 326
Arp, A.J. 135

Arthur, M.A. 228
Atlantic studies 90, 180, 233
Aston, S.R. 176, 177, 187
Atmospheric carbon dioxide 309, 315
Austin, R.W. 279
Avery, W.H. 345
Ayala-Castañares, W. 296, 300
Azoy, A.M. 211, 216

Baab, D.T. 361
Bacastow, R. 308, 311, 315, 324, 325
Bainbridge, A. 135
Baker, E.T. 279
Baker, D.J., Jr. 363, 375
Ball, D. 279
Ballard, R.D. 229
Ballard, T.D. 135
Bardach, J.E. 353, 354, 360, 361
Barnes, A.T. 135
Barth, T.F.W. 219, 228
bathypelagic animals 133
Baxter, L. III 125
Bay of Fundy 73
barrier islands 91
bathythermograph (BMT) 112
Behrens, W.H. III 375
Behringer, D.W. 252
Bennett, C.L. 187
Benninger, C.K. 188
benthic communities 17
Berner, R.A. 228

Beukens, R.P. 187
Beverton, R.J.H. 203
Bien, G.S. 191
Bigelow, H.B. 70, 72, 73, 78, 84
Biggs, D.C. 33
biochemistry 11, 13
biogeochemistry 305, 307
biological oceanography 20, 23, 57, 127, 128, 130, 131, 133, 349
biomes 308
biosynthesis 8
Birdsall, R.G. 112, 115, 126
Bischoff, J.L. 220, 228, 229
Bjorkstrom, A. 321, 324, 325
Blaxter, J.H.S. 196, 203
Blumberg, A.F. 74, 76, 77, 84
Bolin, B. 228, 308, 311, 320, 321, 323, 325
Boone, R.D. 325
Boothyroyd, D.D. 96, 99
Bougault, H. 229
Bourgeois, J. 96, 99
Bourgeon, P. 188
Bowden, K.F. 85
Bowen, V.T. 165, 166, 169, 171, 172, 174, 175, 176, 179, 185, 188, 189, 190, 191
Bowers, D. 77, 86
Bowman, T.E. 27, 33
Bowman, R.E. 196, 203
Bradley, A.M. 254
Bradshaw, A. 112
Brazier, O. 68
Brescher, R. 190, 325
Brewer, P.G. 226, 228, 323, 325
Briscoe, M.G. 36, 53, 77, 85
Broecker, W.S. 173, 179, 184, 188, 191, 311, 317, 319, 321, 325
Brooks, N.H. 85
Brown, B.E. 204
Brown, C.S. 191
Brown, Neil 36, 112
Bruland, K.W. 175, 184, 188
Brink, J.L. 190
Brusca, R.C. 14
Bryan, K. 315, 325
Bryden, H.L. 161
Buehler, B.G. 113, 125
Bumpus, D.F. 72, 80, 85

Burchfield, R.W. 209, 216
Burke, J.C. 188
Bunker, A. 236, 241, 247

Cailliet, G. 135
calcium 221
Caldwell, D.R. 37, 44, 53, 238, 253
Cambon, P. 229
carbon 225, 305, 307, 308, 311, 317, 320, 321, 323
carbon dioxide 308–10, 315, 342
carbon-14 172
Carey, F.G. 67, 67, 68
Carmack, E.C. 45, 53
Carranza, A. 229
Carson, R.M. 84
Cartey, A.J. 14
Casey, J.G. 68
Cavanaugh, C.M. 135
cesium-137 165, 169
Chau, Y.K. 14
Chaney, R.E. 253
Chapman, W.M. 349, 360
Charlier, R.H. 327, 344
Charmes, G. 211, 216
Cheh, A. 15
Chemical oceanography 219, 220, 222
Cheney, R.E. 279
Chiang, R. 14
Childress, J.J. 135
Choukroune, P. 229
Christensen 111
Chudyk, W. 15
circulation 70, 71, 78, 79, 116, 137, 172, 202, 231, 245, 247, 249, 321
Clark, D.K. 279
Clark, J.G. 113, 121, 125
Clarke, G.L. 68
Clarke, W.B. 169, 178, 183, 186, 188, 189, 191, 228
Claude, George 338
Claypool, G.E. 225, 228
Clover, M.R. 187
Clutterbuck, P.W. 3, 14
coastal studies, 69, 87, 88, 89, 93, 102, 103, 175, 288, 289, 290
coastal zone color scanner (CZCS) 101, 104, 283
Cochran, J.K. 176, 187, 188

Cohen, E.B. 71, 72, 85
Cohen, R. 339, 344
color, plankton 273
Cook, J.C. 15
conductivity probes 36, 46, 112
continental shelf 69, 70, 95
convection 44
convergence zone 111
copepods 25
Corcos, G.M. 54, 37
Cordoba, D. 229
Coriolis force 79, 238, 329
Corliss, J.B. 135
Cox, C.S. 40, 54
Cox, C. 112
Cox number 42, 52
Craig, H. 161, 173, 188, 221, 228, 326
Crane, K. 135
Crawford, R.L. 9, 14
Cresswell, G.R. 55, 237, 252
Crutzen, P. 310, 326
current studies 248, 328

Dagley, S. 9, 14
Damian, M. 190
d'Ancona, U. 195, 203
Dashen, R. 125
Davis, R.A. 90, 96, 99
Davis, R.E. 236, 238, 252, 253
Degens, E.T. 325
density 45, 50, 234
Demaison, G.J. 224, 228
deMestral, G. 365
Denman, K.L. 43, 53
Department of Energy, 328, 329, 345
deSzoeke, R. 253
Deutches Hydrographisches Institut 176, 188
Dickson, F.W. 228, 220
Dietrich, G. 72, 85
diffusion studies 46, 49, 50, 73, 237
Dillon, T.M. 37, 44, 53, 238, 253
distribution
  of carbon 321
  of food 348
  of population 348
  of salinity 232, 242
  of temperature 232
Dizikes, L.J. 14

Dorsey, H.G. 190, 325
Doubek, D. 14
Dreisigacker, E. 169, 188, 191
Drever, J.I. 220, 228
Druffel, E.M. 191
Dugger, G.L. 345, 340
Duing, W. 329, 345
Duursma, E.K. 173, 188
Duvigneaud, P. 325
Dyer, R. 185, 188
Dymond, J. 135
dynamic topography 250, 263

E G and G 83, 85
Eagle, R.J. 190
economics 283, 286, 363
eddies 48, 50, 71, 76, 77, 82, 115, 137–39, 141, 158, 160, 233, 236, 247
Edmond, J.M. 135
Edwards, K. 14
Edwards, R.L. 196, 203
Eimhjellen, K. 135
Ekman layers 235, 236, 238, 318
Elderfield, H. 182, 188
Ellenthorpe, A.W. 113
Elmore, D. 187
Emery, K.O. 88, 89, 99
El-Sayed, S.Z. 279
Engh, R. 188
environmental policy 101
Erez, J. 220, 229
Eriksen, C.C. 37, 53
Erickson, E. 228
erosion 93
Essig, R.J. 204
estuaries 74
Ethington, R.L. 90, 96, 99
Eulerian flow 80, 233
euphotic zone 175
Evans, D.L. 48, 54
Experimental Data and Information Service (EDIS) 258
Ewing, M. 109, 113, 114, 118, 125, 325
Ewing, G.C. 257, 279
Eyring, 111

fallout 164
Farges, L. 188
Farmanfarmian, A. 135

Farrington, J.W. 189
Fedorov, K.N. 47, 49, 51
Felbeck, H. 135
Fenical, WO 14
Ferber, G.J. 168, 191
Fevrier, M. 229
finestructure 39, 42, 48, 50, 112, 113
Finn, D.P. 71, 85
First Global GARP Experiment (FGGE) 369
Fischer, H.G. 82, 85
Fischer, K-H. 191
fish, pelagic 57, 58
Fisheries 71, 193, 194, 195, 199, 200, 349
Fisheries Research Board of Canada 203
Flagg, CN 86
Flatté, S.M. 113, 125
Fleming, R. 111
FLIP 32, 122
Fofonoff, N. 112
Folsom, T.R. 185, 188, 191
Fomina, C.S. 229
Food and Agriculture Organization (FAO) 195, 360
Forster, G.R. 85
Forstner, U. 14
Foster, T.D. 45, 53
Fox, P.J. 229
Frances, E.J. 345
Francheteau, J. 220, 229
Frankenberg, D. 293
Franssen, H.T. 300
Fraser, G. 167, 190
Freeland, H.J. 237, 253
Frick, T. 14
fronts 48, 73
Frye, D. 86
Fu, L.L. 241, 253
Fuglister, F.C. 143, 145, 153, 155, 160
Fusfeld, H.I. 374, 375

Gabrielson, G. 68
Galapagos, hot springs 220
Gamble, E. 189
Gandin, L.S. 366, 375
Gaposhkin, E.M. 281
Gardiner, S.L. 135
Gargett, A.E. 42, 49, 53

Garrels, R.M. 220, 229
Garrett, C.J.R. 36, 39, 40, 44, 53, 73, 74, 77, 85, 173, 180, 189
Gatien, M.G. 155, 160
Gatrousis, C. 190
general circulation models 315, 317
geochemistry 219, 233
geoid 262
geophysics 236
Georges Bank 70, 79
Georgi, D.T. 53
GEOS 258, 273
geostrophic shear 250
GEOSECS 164, 174, 175, 181, 226, 315, 322
Gerard, R. 325
Gibson, C.H. 42, 53
Gill, A. 141, 160
Gill, A.E. 43, 53
Gleason, F.K. 14
Global Atmospheric Research Program (GARP) 318, 369
Global 2000 Report to the President 347, 348, 349, 360, 375
Goddard, Robert 257
Gold, F. 283, 300
Goldberg, E.D. 186, 189
Gordin, H. 359, 360
Gordon, H.R. 279, 273
Gordon, A.L. 49, 53, 318, 325
Gornitz, V.O. 88, 99
Gorshkov, S.G. 205, 216
Gould, W.J. 253
Gove, H.E. 187
Green, K. 135
gradients
  ocean 37
  density 40
  temperature 46
  salinity 46
gravity 273, 274
GRAVSAT 263
Greenberg, D.A. 85
Greenblatt, 364, 375
Gregg, M.C. 36, 44, 49, 53, 77, 85, 112
Griffin, J.J. 189
Griffiths, R.W. 46, 47, 53
Griffiths, D.K. 73, 85
Gros, R. 188

Index

Gross, M.G.   173, 188
Gruner, H.E.   27, 33
Guerrero, R.   14
Guerrero, J.   229
Gugelmann, A.   190
Gulf of Maine   73
Gulf Stream
  as energy source   328
  eddies   50, 158, 160, 247, 248, 265
  exchange   149
  images   159, 275
  measurements   264
  "Gulf Stream '60"   153
Gulland, J.A.   195, 203
Gullicksen, S.   325
Gustafson, P.F.   189
Gustafson, L.B.   46, 55

Haeckel, E.   18, 33
Haefele, W.   340, 345
Hager, L.P.   3, 4, 8, 14
Haines, K.C.   360
Hajash, A.   220, 229
halogenating agents   4–6, 8, 9
Halstead, M.   209, 216
Hamilton, G.   113, 118, 125
Hamner, W.M.   19, 33
Hampicke, U.   309, 325
Hansen, J.   99
Hansen, J.A.   353, 356, 360
Hansen, P.G.   126
Harbison, G.R.   20, 23, 24, 25, 33, 364
Hardy, A.C.   18, 33
Hardy, E.P.   165, 189
Harper, J.L.   196, 203
Hartline, B.K.   194, 203
Harvey, G.   189
hatcheries,   201
Haury, L.R.   44, 53
Hawkins, J.   229
Haymon, R.   229
Hayes, M.O.   96, 99
Hedges, J.I.   135
Heezen, B.C.   325
Hekinian, R.   220, 229
helium-3   169
Helland   257
Hempel, G.   196, 203
Hendershott, M.C.   363, 375

Hensen, V.   18, 33
Hepher, B.   353, 360
Hepworth, A.   182, 188
Herman, A.   363, 375
Hersey, J.B.   111, 125
Hessler, R.   135, 229
Hetherington, J.A.   174, 175, 176, 189
Hewson, W.D.   4, 14
Hicks, S.D.   88, 99
Hill, G.W.   96, 99
Hobbie, J.E.   325
Hogg, N.G.   243, 253
Holmen, K.   325
Holland, H.D.   220, 224, 229
Holland, W.R.   189, 236, 247, 248, 249, 254
Hollenberg, D.F.   14
Holligan, P.M.   85
Holling, C.S.   199, 203
Holser, W.T.   228, 229
Holt, S.J.   195, 203
Honjo, S.   220, 229
Hopkins, J.   135
Hoppenheit, M.   190
horizontal flow   48, 71, 78, 80, 233, 237, 242
Houghton, R.A.   325
Hovegaard, W.   210, 216
Hovis, W.M.   273, 279
Howard, J.D.   96, 100
Huang, N.   280
Hughes, D.G.   86
Hughes-Games, W.L.   360
Hunter, J.R.   72, 73, 83, 86, 96, 99
Huppert, H.E.   46, 48, 53, 54
Hutchinson, T.C.   15
Huthnance, J.M.   79, 82, 85
hydrogen   10
hydrography   209, 241
hydrothermal vents   131, 132
hyperiid amphipods   20, 22, 25

ice   269, 270
imagery   276
Imberger, J.   54, 85
infrared radiometer   266
Inman, Douglas   96
in situ sampling   18, 19
instability   36, 37, 39, 40, 46

institutional concepts 377
Intergovernmental Oceanographic Commission (IOC) 297, 300
internal waves 39, 40, 48
International Atomic Energy Agency (IAEA) 177, 189, 174
International Decade of Ocean Exploration (IDOE) 293
intrusions 48, 83
Isaacs, J.D. 335, 345
Iselin, C. O'D. 247, 253
isopycnals 40
Izac, R.R. 6, 14

James, I.D. 74, 77, 85
Jannasch, H.W. 135
Jeffries, D.F. 189
Jenkins, W.J. 169, 180, 183, 188, 189, 243, 253
Johnson, D. 99
Johnson, Martin 111
Johnson, R.D. 14
Johnston, D. 283, 300
Johrde, M. 292
Joint Air-Sea Interaction (JASIN) 44, 268
Joint Oceanographic Institutions, Inc. (JOI) 295
Jokela, T.A. 190
Jones, E.C. 99
Jones, M.L. 135
Jones, R. 196, 203
Joseph, A.B. 165, 166, 189
Joyce, T.M. 49, 54, 77, 86
Juszko, B.-A. 85
Juteau, T. 229

Kanwisher, J. W. 67, 68
Kapitsa, P. 363, 375
Kaplan, I.R. 228
Kastner, M. 229
Katsouros, M. 291
Kautsky, H. 167, 189, 190
Keeling, C.D. 308, 311, 321, 324, 325
Keffer, T. 241, 253
Kelley, M.G. 68
Kelley, J.R. 85
Kelvin-Helmholtz 37, 39, 44
Kempe, S. 325
Kent, G. 365, 375

Kent, R.E. 73, 74, 76, 85
Kewalo Basin Laboratory 58
Kigoshi, K. 189
Kilius, L. 187
Kim, K. 160
kinetic energy 36, 43, 76, 247
King, C. A. M. 90, 100
Kirwan, A.D. 237, 253
Knauss, J. 300
Koblinsky, C.J. 242, 253
Koh, R.C.Y. 85
Koide, M. 176, 189
Komar, P. D. 96, 100
Kossinna, E. 135
Kraus, E. B. 42, 54
Krejcarek, G. E. 14
Krey, P. W. 189
Kromer, B. 191
Kronengold, M. 113, 125
Kundu, P. K. 238, 253
Kuo, H.H. 185, 189
Kuroshio current 275

Laajoki, K. 229
Labeyrie, L.D. 174, 189
Labrador Sea Water 180
Lacis, A. 99
Lagrangian flow 80, 173, 185, 233
Lambert, R. B. 48, 54
LANDSAT 257
Langmuir cells 42
Langton, R.W. 357, 360
Larkin, P.A. 201, 202, 203
Larson, R. 229
Lasaga, A.C. 229
Lasker, R. 196, 203
Launder, B.E. 42, 54
Laval, P. 20, 33
Laver, M.B. 135
Lavi, A. 339, 345
Lawson, K.D. 67
Leatherman, S.P. 180
Lebedeff, S. 99
Lee, P. 99
Leetmaa A. 236, 241, 247, 253
Legeckis, R. 269, 279
Leggs, E. 279
Lehn, H. 191
Leibovich, S. 43, 54

Leventhal, J.S. 229
Li, Y.-H. 325
Likens, G.E. 308, 322, 326
Limeburner, R. 279
Lin, S.Y. 354, 360
Linden, P.F. 46, 53, 54
Lipscomb, J.D. 14
List, E.J. 85
Liston, J. 297, 300
Litherland, A.E. 187
Livingston, H.D. 167, 172, 174, 176, 181, 183, 185, 188, 189, 190
Local Dynamics Experiment (LDE) 146, 149
Loder, J.W. 78, 79, 80, 81, 82, 84, 85
Lonsdale, P. 135
Loosli, H.H. 177, 190
Lovell, R.T. 351, 353, 360
Lovelock, J.E. 14
Lovseth, K. 325
Lulu 32
Lupton, J.E. 147, 161
Luyendyk, B. 229
Luykx, F. 167, 190
Luyten, J.R. 161

Macdougall, J.D. 229
Mackenzie, F.T. 220, 229
Madin, L.P. 20, 25, 33
Maggs, R.J. 14
Magnell, B. 47, 54, 79, 86
magnesium 220, 221
Mahan, A.T. 205, 216
Manabe, S. 308, 325
Marchand, P. 338, 345
Marder, 211, 216
mariculture 351, 356
marine biota 3
Marsh, J.G. 279
Marshall, E. 88, 100
Martin, J.M. 189
Martini, L.A. 253
Mason, C. 14
Masuda, M. 333, 345
Mathieu, G.G. 325
Matsuda, Y. 353, 356, 360
McClure, W.O. 14
McCormick, M. 333, 345
McDougall, T.J. 45, 46, 53, 54

McDowell, S.E. 139, 161
McEwan, A.D. 39, 40, 54
McGoodwin, J.R. 353, 360
McLarney, W.O 360
McMurtie, F. 216
McNally, G.B. 237, 238, 253
McWilliams, J. 160
MccGwire, M. 216
Meadows, D.H. 375
Measurement 363, 364, 366, 373
Melillo, J.M. 325
Mellor, G.L. 77, 84
meltwater 48
Menzel, D.W. 199, 203
Merriman, D. 363, 376
mesoscale variability 114
metabolic processes 3, 9
Metzger, K. 126
microstructure 35, 36, 40, 42, 50
Mid-Ocean Dynamic Experiment (MODE) 115, 121, 237
Miles, E. 283, 300
military use 205
Millard, R.C. 55
Miller, S. 229
Minc, A. 370, 376
mineral resources 327
Mitchell, N.T. 167, 189, 190
Mixed Layer Experiment (MILE) 44
mixing 36, 42, 43, 70, 71
Miyake, Y. 165, 190, 238
Miyaki, M. 43, 53
Modeling 39, 40, 43, 78, 185
 carbon 219, 308, 311, 322, 323
 circulation 233, 236, 249, 315
 fisheries 196, 199
molecular processes 3, 36, 44, 48
Molinari, R. 237, 253
Mooers, C.N.K. 54
Moore, B. 309, 310, 325
Moore, G.T. 224, 228
Mottl, M.J. 220, 229
Motzkin, F. 360
Mueller, J.L. 279
Mukhopadhyay, S.C. 14
Munck, E. 14
Munk, W.H. 39, 40, 53, 74, 76, 85, 113, 118, 125, 190, 253, 321, 325, 367
Munnich, K.O. 173, 190

Murphy, J.R.   122, 125
Murray, C.N.   176, 190

Nägel, L.   349, 355, 360
Nakamura, E.L.   68
Nash, C.E.   353, 360
National Advisory Commission on Oceans and Atmosphere (NACOA)   370, 376
National Aerunautics and Space Agency (NASA)   376
National Marine Fisheries Service (NMFS)   349, 361
National Oceanic Satellite System (NOSS)   370
naval oceanography   205, 206, 210
Neal, W.J.   100
Needham, H.J.   229
Needler, G.T.   180, 190
Niiler, P. P.   42, 54, 160, 236, 241, 242, 253, 254
Nora, S.   370, 376
Normark, W.   229
Norris, N.C.G.   86
North Atlantic   235, 237, 240, 241, 243, 247, 311
North Pacific   237, 250
Northrop, J.   126
Noshkin, V.E.   169, 174, 175, 176, 178, 185, 188, 190, 191
Noyce, R.N.   365, 366, 376
Nozaki, Y.   175, 190
NIMBUS   265, 268, 270, 273
nitrogen   223
nutrient flow   195
Nydal, R.   325

O'Brien, M.P.   100
ocean circulation   115, 165, 262, 266
ocean dynamics   111, 164
ocean energy systems   327, 343
Ocean Policy Committee   290, 291, 298, 300
Ocean Thermal Energy Conversion (OTEC)   338ff, 356
ocean transport   143, 163, 184, 262
ocean weather   115
oceanic biology   130
Office of Technology Assessment (OTA)   376

Oeschger, H.   173, 177, 190
oil and gas   71, 101, 102, 327
Olla, V.L.   68
Olsen, H.L.   345
Olson, P.   14
Okubo, A.   82, 85
Orcutt, J.   229
Orr, M.H.   53, 125
Osborn, T.R.   40, 54, 112
Ostlund, H.G.   169, 173, 178, 179, 188, 190, 191, 311, 317, 325, 326
Othmar, D.F.   357, 360
Owens, W.B.   148, 254
Oxford, A.E.   14
oxygen   61, 153, 223, 224, 225
oxygenases   9

Pacific studies   221, 233, 311
Palmer, J.E.   204
Pannetier, R.   168, 191
Paolucci, S.   43, 54
Parkes, O.   211, 216
Parker, R.R.   55
Parker, P.L.   189
patchiness   20 ff, 196
Patry, E.M.   361
Patti, F.   188
Paulik, G.J.   194, 203
Pawson, D.L.   135
Pearson, R.   12, 14
Pedlosky, J.   253
pelagic organisms   25, 133
Peng, T.-H.   173, 188, 316, 317, 321, 325
Pentreath, R.J.   189
Peterson, B.J.   325
Peterson, W.H.   191, 317, 326
phosphorus   223, 224, 225
phosphates, marine   227
physical oceanography   194
Picot, P.   229
Pilkey, O.H. Sr.   100
Pilkey, O.H. Jr.   89, 100
Pillay, T.V.R.   351, 355, 356, 360
Pingree, R.D.   73, 77, 78, 85
Pinkel, R.   122, 125
Pirlot, J.M.   20, 33
plankton   17, 19, 20, 23, 30, 31, 273
Platt, T.   363, 375
polar oceanography   268

Pollard, R.T. 43, 54
politics and research 284, 291, 298, 378
pollution 3, 13, 103, 186, 226, 310ff, 342, 350
POLYMODE 115, 138, 146, 237
Pope, J. 196, 203
Porter, C. 360
potential energy 44
Pratt, H.L. 68
Press, F. 376, 374
Price, J. 147, 161
Price, J.F. 43, 54
Price, J.R. 254
Price, M.H. 135
Pritchard, J.A. 279
Pritchard, D.W. 73, 74, 76, 85
production 195, 199, 200
    of food 348
    of hydrogen 10
    of oxygen 223
public policy 101, 104ff
Pugh, P.R. 85
Purser, K.H. 187
pycnocline 44

radionuclides 163, 164, 172, 173, 179, 186, 232
Raistrick, H. 14
Raitt, 111
Randers, J. 375
Rangin, C. 229
Rand-Meir, T. 14
Rau, G. 135
Reed, W.J. 200, 204
Reid, J.L. 143, 161
Reineck, H.E. 96, 100
remote sensing 22, 39, 214, 255, 280, 364
Renner-Smith, S. 373, 376
Revelle factor 317
Revelle, R. 111, 125, 228, 229
Reynolds, P. 236, 253
Reynolds stress 79, 137, 242
Rhines, P. B. 54, 137, 161, 169, 180, 189, 236, 242, 243, 247, 249, 253
Richardson, P.L. 237, 245, 253, 254
Richardson number 37, 39, 74, 77
Richmond, H. 205, 206, 207, 216
Ricker, W.E. 195, 204, 349

Ridley, W.P. 13, 14
Riley, G.A. 69, 72, 85, 311, 326
Rind, D. 99
Rinehart, K.L. Jr. 14
Riodan, P.F. 353, 361
Risebrough, R.W. 189
Riser, S. 143, 161
Riveillere 211, 216
Robertston, W. 189
Robinson, A.R. 137, 161
Robinson, R.M. 39, 40, 54
Robison, B.H. 67
Rodde, K.M. 357, 360
Rodi, W. 77, 86
Roels, O.A. 357, 360
Roether, W. 169, 171, 173, 188, 190, 191
Rooth, C.G. 169, 190, 191, 317, 326
Ropp, T. 211, 216
Ross, D. 298, 300
Rossby, H.T. 123, 126, 138, 139, 141, 143, 147, 160, 161, 253, 254, 372
Rotty, R. 309, 326
Ruddick, B.R. 49, 53, 54, 77, 86
Rude, J. 14
Russell, G. 99
Russell, I.R. 189
Ryther, J.H. 194, 204, 353, 354, 355, 360, 361

Sakai, H. 228
Sale, L. 364, 375, 376
salinity 37, 40
    distribution 139–141, 232, 242, 318
    gradient 45, 48, 49, 335, 336, 337
salt fingers 47
satellites 260, 368, 370
satellite altimetry 122, 145, 198, 238, 273
satellite tracking 238, 245
Sandbank, E. 360
Sanders, H.L. 135
Sanford, T. 112
Sarig, S. 351, 361
scale, concepts 199, 377
scanning multichannel microwave radiometer (SMMR) 268
Scarlet, R.I. 79, 86
Schevill, W. 111, 125
Schlee, S. 135

Schleicher, K. 112
Schley, W.S. 210, 216
Schmitt, R.W. 48, 52, 54
Schmitz, W.J. 236, 243, 247, 248, 250, 253, 254
Schneider, D.L. 190
Schneider, E. 189
Scholz, W.S. 290, 291, 300
Schott, F. 235, 254
Schotterer, U. 190
Schroeder, G.L. 354, 361
Schröder, J. 168, 191
Schuert, E.A. 189, 191
Schulkin, M. 122, 125
Scotti, R.S. 37, 54
SCUBA 17, 20, 23, 31
Sears, M. 363, 376
SEASAT 255, 263, 266, 273, 276
SEASAT-A scatterometer system (SASS) 273
Sea water 220, 305, 317
sediments 90, 96, 175, 176, 182, 220, 223, 227, 323
Segall, H.J. 15
Séguret, M. 229
Seiler, W. 310, 326
Seiwell, H.R. 161
Shanks, W.C. 229
Shaver, G.R. 325
Shaw, P.D. 3, 14, 15
shear 36, 37
Sheldon, R.W. 200, 204
Shelef, G. 360
Shell, E.W. 360
shellfish farming 351
Sherman, F.S. 36, 37, 39, 54
Sherman, J.W. III 280
Shippen, W.B. 345
Shockley, R.C. 126
Shor, G. 114
shoreline research 87, 88, 93, 96
Shulenberger, E. 375
Siebenaller, J.F. 135
Sieganthaler, U. 190
Sillén, L.G. 220, 229
Simpson, J.H. 72, 73, 77, 83, 84, 86
Sims, J.J. 6, 14
Sissenwine, M.P. 196, 204
Skidaway Institute of Oceanography 100

slope water 155
Smith, K.L. Jr. 135
Smith, L. 298, 300
Smitherman, R.O. 360
Sokolovskij, V.D. 216
Somero, G.N. 135
solar heating 76
sound fixing and ranging (SOFAR) 111, 138, 155, 237, 247
species abundance 197
species identification 23, 25
Spiegel, S.L. 79, 86
Spiesberger, J.L. 126
Spiess, F. 220, 229, 363, 376
Spindel, R.C. 115, 126
Spiro, T.G. 15
Sreekumarau, C. 185, 188
Stanners, D.A. 176, 177, 187
Steele, J.H. 196, 199, 200, 203, 204
Steinberg, J.C. 112, 126
Stenhouse, M.C. 191
Stevens, W. 135
Stewart, R. 245, 254
Stokes drift 80
Stommel, H. 235, 247, 250, 252, 254, 311, 326, 371
Stouffer, R.J. 308, 325
Strøm, H. 20, 33
Strong, A.E. 279
strontium-90 165, 169
Stuiver, M. 173, 188, 190, 191, 311, 322, 325, 326
Sturges, W. 48, 54
Sturm, B. 279
submarines 212
Suess, H. 229
Sugarman, D. 135
Sugihara, T.T. 174, 191
Suida, J.F. 4, 14
surface studies 42, 232, 238
 temperature 71, 266
 water 179, 318
 wind 238, 262, 270
Sunderlin, J.B. 357, 360, 361
Suomi, V. 366
Sverdrup, H. 111, 371, 376
Sverdrup balance 235
Swallow, J. 247, 254
Swanberg, N.R. 33

Index   391

Swingle, H.S. 200, 204
Swift, D.J.P. 90, 91, 100
swordfish 59, 61
synthetic aperture radar (SAR) 276
systematics 200

Taft, B. 160
Tal, S. 351, 361
Tamplin, A. 189
Taylor, H.W. 53
Taylor, L. 14
Taylor, S. 135
Teal, J.M. 67
Telegadas, K. 168, 191
temperature 37, 40, 43, 45, 64, 71, 266
Templeton, W.L. 179, 191
Territorial Group. 284
Thayer, J.S. 15
Theiler, R.J. 7, 15
thermohaline circulation 137
Thompson, R.O.R.Y. 43, 54
Thomson, J. 190
three-dimensional flow 240
Thorpe, S.A. 37, 39, 44, 54
tidal currents 79, 80, 81
tidal power 327
TIROS 369
tomography 118, 121
Toole, J.M. 54
Toonkel, L.E. 165, 191
Top, Z. 188, 191
topography 250
tongues 140, 233, 241
Tourtelot, E.B. 229
trace elements 145, 221, 225, 226
transport, ocean 143, 163, 184, 262
Tranter, D.J. 51, 55
tritium 311
Trumble, L. 345
turbulence 42
Turekian, K.K. 188, 190
Turner, J.S. 36, 39, 43, 44, 46, 48, 49, 53, 54, 55, 76, 86
tuna 57

United Nations 300
United Nations Conference on Law of the Sea (UNCLOS III) 283, 285

United Nations Draft Convention on Law of the Sea 285, 287
undersea research vessels 31
University National Oceanographic Laboratory Systems (UNOLS) 31, 373
upwelling 194, 196
Ursin, E. 196, 203

Valdes, J.R. 254
van Andel, T.H. 135
Van Blaricom, G.R. 14
van Leer, J.C. 54
Vannucci, M. 297, 300
variable ocean structure 121
variability 196, 233
Veizer, J. 225, 229
velocity of waves 42
vents 222
Veronis, G. 185, 189, 315, 326
vertical circulation 232, 233, 237, 238, 266
   exchange 73, 321
   mixing 71, 72, 73, 74, 77, 80, 81, 242, 319
   transport 47, 49, 241
Villaluz, D.K. 351, 361
Vine, J.D. 229
Vine, A. 111, 125
volcanism 225
Volchok, H.L. 188, 189, 191
Volkov, II. 229
von Herzen, R.P. 135
Voorhis, A.D. 49, 55, 126, 161
Vorosmarty, C.J. 325

Wade, R.J. 14
Waterbury, J.B. 135
Watson, K.M. 125
Watson, T.L. 100
waves 37, 44, 329, 332, 333
weather sources 247
Webb, D. 115, 126, 161
Webb, D.C. 55, 254
Webb, D.J. 272, 279
Webber, H.H. 353, 361
Webster, 112
Weihs, D. 68
Weiss, R.F. 318, 326
Weiss, W.M. 169, 191

White, R.H.   14, 15
Whittaker, R.L.   308, 322, 326
Wick, G.L.   335, 345
Wilgus, C.K.   229
Williams, A.J.   47, 55
Williams, D.   135
Williams, G.O.   49, 55
Williams, P.M.   176, 191
Wilson, W.H.   279
wind   42, 43, 237, 247, 262, 270
Winn, W.E.   68
Winter, J.E.   360
Wirsen, C.O.   135
Wishner, K.F.   135
Wittmann, G.T.W.   14
Wong, P.T.S.   14
Wong, P.M.   185, 188
Wong, K.M.   190, 191
Wood, J.M.   8, 9, 13, 14
Woodhead, D.S.   189
Woods, J.D.   37, 49, 55, 112
Woodwell, G.M.   325
Wooster, W.S.   292, 298, 301
Worcester, P.F.   121, 126

Wormuth, J.H.   375
Worthington, L.V.   254, 247, 317, 326
Worzel, J.L.   109, 113, 114, 118, 125
Wright, D.G.   85
Wrigley, R.C.   279
Wunsch, C.   39, 55, 115, 118, 125, 160, 235, 253, 254, 281
Wüst, G.   141, 161, 241, 254
Wyrtki, K.   250, 254, 369, 376

Xavier, A.   14

Yentsch, C.S.   279
Young, W.R.   242, 253
Young, J.Z.   27, 33

Zachariasen, F.   113, 125
Zak, I.   228
Zander, I.   165, 191
Zener, C.   339, 345
Zenk, W.   54
Zimmerman, J.T.F.   80, 82, 86
Ziv, I.   351, 361
zooplankton   17